PHENOMENOLOGY/ONTOPOIESIS RETRIEVING GEO-COSMIC
HORIZONS OF ANTIQUITY

ANALECTA HUSSERLIANA

THE YEARBOOK OF PHENOMENOLOGICAL RESEARCH

VOLUME CX

Founder and Editor-in-Chief:

ANNA-TERESA TYMIENIECKA

*The World Institute for Advanced Phenomenological Research and Learning
Hanover, New Hampshire, USA*

*For further volumes:
http://www.springer.com/series/5621*

PHENOMENOLOGY/ONTOPOIESIS RETRIEVING GEO-COSMIC HORIZONS OF ANTIQUITY

LOGOS AND LIFE

Volume CX / Part II

Edited by

ANNA-TERESA TYMIENIECKA

*The World Institute for Advanced Phenomenological Research and Learning,
Hanover, New Hampshire, USA*

Published under the auspices of
The World Institute for Advanced Phenomenological Research and Learning
A-T. Tymieniecka, President

Editor
Prof. Anna-Teresa Tymieniecka
The World Institute for Advanced
    Phenomenological Research
    and Learning
Ivy Pointe Way 1
03755 NH Hanover
USA
wphenomenology@aol.com

Printed in 2 volumes
ISBN 978-94-007-1690-2   e-ISBN 978-94-007-1691-9
DOI 10.1007/978-94-007-1691-9
Springer Dordrecht Heidelberg London New York

Library of Congress Control Number: 2011934269

© Springer Science+Business Media B.V. 2011
No part of this work may be reproduced, stored in a retrieval system, or transmitted in any form or by any means, electronic, mechanical, photocopying, microfilming, recording or otherwise, without written permission from the Publisher, with the exception of any material supplied specifically for the purpose of being entered and executed on a computer system, for exclusive use by the purchaser of the work.

Printed on acid-free paper

Springer is part of Springer Science+Business Media (www.springer.com)

TABLE OF CONTENTS

Acknowledgements xi

INAUGURAL LECTURE

ANNA-TERESA TYMIENIECKA/Inspirations of Heraclitus from Ephesus Fulfilled in Our New Enlightenment 3

VOLUME 1

SECTION I PHENOMENOLOGY OF LIFE IN THE CRITIQUE OF REASON

KONRAD ROKSTAD/Was Plato A Platonist? 15

DANIELA VERDUCCI/The Life of Being Refound with the Phenomenology of Life of Anna-Teresa Tymieniecka 23

ELLA BUCENIECE/Critique of Reason Projects with Reference to Antiquity: I. Kant and the Platonic Ideas, E. Husserl and the Mnemosinean Enticement, A.-T. Tymieniecka and the Dyonisian Logos 39

TANSU ACIK/What was a Classic Until the Beginning of 20th Century? 51

SECTION II LOGOS AND LIFE

SIMEN ANDERSEN ØYEN/The Existentialistic Subject Today 61

ROBERT SWITZER/Re-Turning to the Real: Phenomenological Appropriations of Plato's "Ideas" and the Allegory of the Cave 75

ANDREAS BRENNER/Living Life and Making Life 91

MANJULIKA GHOSH/Man's World and *Logos* as Feeling 103

VELGA VEVERE/The Feast of Life or the Feast of Reason – Kierkegaard Versus Plato 111

WEI ZHANG/Gibt Es Ein Materiales Apriori? 123

SECTION III   LOGOS AND EDUCATION

RIMMA KURENKOVA, EVGENY PLEKHANOV, AND ELENA
 ROGACHEVA/The Idea of Paidea in the Context of Ontopoesis of Life   141

ELENA ROGACHEVA/International Dimention of John Dewey's
 Pedagogy: Lessons for Tomorrow   147

MARA STAFECKA/Thinking Conditioned by Language
 and Tradition   171

J.C. COUCEIRO-BUENO/How to Conduct Life (*Arete* and *Phronesis*)   181

SECTION IV   HUSSERL IN THE CONTEXT OF TRADITION

WITOLD PŁOTKA/The Reason of the Crisis. Husserl's Re-Examination
 of the Concept of Rationality   191

KOUSHIK JOARDAR/*Logos* as Signifier: Husserl in the Context of
 Tradition   207

SUSI FERRARELLO/The Axiology of Ontopoiesis
 and its Rationality   217

TÕNU VIIK/Originating the Western World: A Cultural Phenomenology
 of Historical Consciousness   227

MAGDALENA PŁOTKA/The Recovery of the Self. Plotinus
 on Self-Cognition   241

CEZARY JÓZEF OLBROMSKI/Social Connotations of the Category
 of the «Now» in the Late Writings of Edmund Husserl vs. J. Derrida
 and B. Waldenfels   251

SECTION V   COGNITION, CREATIVITY, EMBODIMENT

LARS MORTEN GRAM/Pound, Propertius and *Logopoeia*   269

KIYMET SELVI/Phenomenology: Creation and Construction of Knowledge   279

AYDAN TURANLI/Perspicuous Representation: A Wittgensteinian
 Interpretation of Martin Heidegger's View of Truth   295

MINA SEHDEV/Origin and Features of Psychical Creations in an
 Ontopoietic Perspective   307

SECTION VI NATURE, WORLD, CONTINUITY

FRANCESCO TOTARO/Nature and Artifice in Manifesting/Producing
the Being 317

MORTEN TØNNESSEN/Semiotics of Being and Uexküllian
Phenomenology 327

SİBEL OKTAR/The Place: Where We See the World as a Limited Whole 341

CATIA GIACONI/Lines for Contemporary Constructivism to Revisit
and Reintegrate the Ancient Sense of Continuity between Men and
Nature 355

VOLUME 2

SECTION VII LOGOS AND THE SELF

ANÍBAL FORNARI/Paul Ricoeur: Critical Consent of *Logos* to Life 373

ERKUT SEZGİN/The Phenomenological Significance and Relevance
of the Reminders Assembled as "Language-Games" 395

CARMEN COZMA/"Sophia" As "Telos" in the "Ontopoietic Perspective" 413

OLGA LOUCHAKOVA-SCHWARTZ/The Self and the World: Vedanta,
Sufism, and the Presocratics in a Phenomenological View 423

THOMAS RYBA/Trinitarian Appropriations of the Transcendentals:
Givenness and Intentionality in Levinas, Marion,
and Tymieniecka 439

SECTION VIII CREATIVITY AND THE ONTOPOIETIC LOGOS

WILLIAM D. MELANEY/Blanchot's Inaugural Poetics: Visibility
and the Infinite Conversation 467

HALIL TURAN/Love of Life, Tragedy and Some Characters in Greek
Mythology 485

ANNA MAŁECKA/Humor in the Perspective of Logos: The Inspirations
of Ancient Greek Philosophy 495

BRIAN GRASSOM/The Ideal and the Real: Bridging the Gap 507

FIRAT KARADAS/Historicity, Narrative and the Construction
of Monstrosity in John Gardner's *Grendel* — 521

ALİ ÖZTÜRK/The Power of Dance/Movement as a Means of Expression — 539

SECTION IX INTERSUBJECTIVITY, FREEDOM, JUSTICE

JOHANNES SERVAN/Making History Our Own – Appropriation and
Transgression of the Intentional History of Human Rights — 551

JAN SZMYD/Vitality and Wobbliness of Universal Moral Values in the
Post-Modern World: Creativity and Regulative Function of the Logos
of Life — 559

MICHAEL GUBSER/A True and Better "I": Husserl's Call for Worldly
Renewal — 579

SAULIUS GENIUSAS/The Question of the Subject: Jan Patočka's
Phenomenological Contribution — 599

SILVIA PIEROSARA/"Human Creativity According to the Being"
and Narrative Ethics: An Actualization of Aristotle's Account
of Imagination — 613

SECTION X SEEKING THE LOGOS IN DIFFERENT CULTURES

KATARZYNA STARK/*Theosis* and Life in Nicolai Berdyaev's Philosophy — 631

TSUNG-I DOW/Harmonious Balance: The Ultimate Phenomenon of Life
Experience, a Confucian Attempt and Approach — 645

JONATHON APPELS/"Dance: Walking and Self-Moving
in Husserl and Merleau-Ponty" — 659

BRUCE ROSS/The Songlines: Dreaming the Ancestors and Sustaining
the World in Aboriginal Art — 665

SECTION XI CONTEMPORARY RETRIEVING OF THE PRINCIPLES OF THE UNIVERSAL ORDER

MAIJA KŪLE/Logos and Life: Understanding of Rhythm — 675

CLARA MANDOLINI/Life Powerful Force Between Virtuality and
Enactment — 685

PATRICIA TRUTTY-COOHILL/Visualizing Tymieniecka's Approach
with String Theory 695

NIKOLAY N. KOZHEVNIKOV/Universal Principles of the World and the
Coordinate System on the Basis of Limit Dynamical Equilibrium 701

ATTILA GRANDPIERRE/The Biological Principle of Natural Sciences
and the Logos of Life of Natural Philosophy: A Comparison
and the Perspectives of Unifying the Science and Philosophy of Life ... 711

Name Index 729

Program from the 60th International Congress of Phenomenology,
Logos and Life: Phenomenology/Ontopoiesis Reviving Antiquity,
Held in Bergen, Norway, 2010 741

ACKNOWLEDGEMENTS

We present this volume to the public with considerable pride. The title: *Phenomenology/Ontopoiesis Retrieving Geo-Cosmic Horizons of Antiquity: Logos and Life* plans to bring to light the long chain of fragmentized issues along which we have retrieved the full Greek intuition of man, earth, cosmos which has been in its entire horizon, forgotten since. The horizon of the cosmos called to be retrieved. It is in the ontopoietic foundation of the logos that "the soul is in the cosmos, and the cosmos is in the soul".

Papers collected here were read at the 60th International Congress at the University of Bergen, *Logos and Life: Phenomenology/Ontopoiesis Reviving Antiquity*, held August 10–13, 2010. We owe sincere thanks to all the authors.

The local organizers, chaired by Professor Konrad Rokstad, representing *The Research Group of Phenomenology and Existential Philosophy* at the Department of Philosophy at the University of Bergen, Ane Faugstad Aaro, Anne Granberg, Egil H. Olsvik, Johannes Servan have offered us a warm hospitality for which we thank them wholeheartedly.

Jeff Hurlburt deserves our usual appreciation for his preparation of the volume.

A-T.T.

SECTION VII
LOGOS AND THE SELF

ANÍBAL FORNARI

PAUL RICOEUR: CRITICAL CONSENT OF *LOGOS* TO LIFE

ABSTRACT

The central aim of the article is to make evident the critical relevance of consent in the framework of the phenomenological dialectics of the voluntary and the involuntary, according to Paul Ricoeur. Consent is the theoretical-practical form of knowledge, which aims at a fundamental alternative facing being, trough the absolutely involuntary, where the decision for existence is at stake. After analytically-existentially recovering the forms of necessity that hang upon human existence, to free them of their monist or dualist anthropological marginalization, the analysis aims at grasping them in their sense and thus bring them to the incarnated and broadened exercise of freedom. The theme is introduced by means of the confrontation between consent and creativity, for the latter has been repeatedly proposed by Ricoeur as central issue of the background of his thought. We start from the devaluation that the current concept of creativity would try to exercise upon the value of consent as expression of freedom. We answer back to the shallowness of that questioning, showing the critical conscience of consent, for freedom to appropriate itself creatively of the figures of the absolutely involuntary. In the very hard core of the absolutely involuntary, reflection bumps into the paradigm and the source of creativity: life itself as the sense of being. Finally, the rejection of the absolutely involuntary by a [form of] freedom that [vindicates] itself as abstract, the hyperbolic acceptance of necessity in orphism and the unilateral acceptance of necessity in stoicism are analysed, to discern what integral consent implies to creativity of life as expression of *lógos*. Summary: 1- The paradox of consent and creativity. 2- Consent and critical reason. 3- Topics of consent. 4- Life as *lógos* and paradigm of creativity. 5- From reversed to creative consent.

THE PARADOX OF CONSENT AND CREATIVITY

On multiple occasions Ricoeur declares that rationality of creativity is the permanent question on his fundamental ontology of human reality, attested, verified and widened along the long road of language configurations (poetical-metaphorical, mythical-symbolical, historical-narrative, literary-narrative and aesthetical-theological), of the methodical forms of consolidated knowledge (conflict of systematic interpretations) and of time concatenation in the living present of initiative and respondent-responsibility in the religious, ethical and political fields. Ricoeur indirectly explores the meta-categorial background of "original creation", through the "multiple modalities of regulated creation", articulated at a "philosophical anthropology" level.[1]

Conversely, consent appears, at first sight, in the whole of Ricoeur's work, to be a concept held up in the first stages of his it, at the end of the treatment of the dialectics between the voluntary and the involuntary. Consent would give the impression of being an anti-creativity, the resignation of the lordship over one's own life, and a deficit of freedom and critical rationality. In front of a supposed olympic leap of creativity towards the new, breaking and order and liberating a generative energy causing a certain anarchic chaos of the existing, and of the preventions of an eventual projective and dominant subject, consent would appear to represent the caution of obeying the established with certain resignation, accepting with a pseudo-sensible balance, an order held to be closed and indefectible, in ourselves and in the world. On the one hand, we would not be able to definitely tell, whether it would be a wiser or a more fatuous order than that which perhaps could make a subject prevail, starting from his idealizations. On the other hand, as is very well known, in the end this indefectible order brutally imposes itself on us.

In this preliminary sketch, which, to the meaningfulness of our issue–critical consent–, opposes the subjacent theme that threads up Ricoeur's thought –human creativity–, it is necessary to warn that the latter is not thinkable or operable, but respective of its *other*, of what is given and regulated, the pre-disposition of which causes incarnated reason to be attentive to the emergence of diverse forms, from the possible to the impossible: "Nous ne connaissons pas plus par voie immédiate la créativité –dice Ricoeur- que au début de mes recherches le *cogito* ne m'apparaissait pas transparaint à lui-même; nous ne connaissons la créativité qu'à travers les règles qu'elle explique, déforme ou subvertit".[2] The expressive creativity and the novelty of each *cogito* that breaks into the world is not the laboriously searched for result of an originality produced by the *tabula rasa* of a gesture of evasive breaking. Rather, that creativity is the endowment that comes up towards a fidelity to the deep of being, which inhabits man as retroaction and horizon. The novelty of the singular *cogito* does not manifest itself in the leap to the empty transparency or self-reflection, but when it embraces with intelligent friendliness the ontological density of its incarnation. The cogito is not an act of pure self-positioning, but one that lives in acceptance and dialogue with its own conditions of establishment. Critical distance is at the same time an act of participation in the reactivation of settlement.

The creativity that leaves its mark is neither ephemeral and reactive spontaneity, nor pure idea or sentiment. "Wanting is not creating" Ricoeur says, towards the end of *The voluntary and the Involuntary*. There is no creativity, if the idea and the wishing do not assume in their flying movement the regulated density of the real involved in it. Even more, because of creativity, a given order of existence takes place again; it reappears enhanced in its irreplaceable uniqueness and its attractiveness, for instance, a genial work of thought or music enhances the human signification of a given order of knowledge or art. Creativity does not emerge from the ideal postulated as radical negation of the real, labelled as inert. In this sense, Ricoeur confesses and asks himself in relation to a piece of work: "*L'Être et le Néant* de Sartre ne suscita en moi q'une admiration lointaine, mais aucune conviction: un disciple de Gabriel Marcel, pouvait-il assigner la dimension d'être à la chose inerte et ne réserver que le néant au sujet vibrant d'affirmations en tous les ordres?"[3] What is inert is not the dimension of being as such, but the declining of our relation with it. The moment of negation – that admirable boldness– is necessary but secondary, in that it emerges from a feeling of positivity that progressively turns into a conviction as it goes through the tests or reality. Favouring that precedent positivity, negation turns away prejudices that shrink the gaze of reason and block the way to recognition of the event that renews a poetic relation with the real. There is no creativity without a self inspired and questioned by something *other* that precedes it and towards which it goes. "Sous la pression du négatif, des expériences en négatif, nous avons a reconquerir une notion d'être qui soit *acte* plutôt que *forme*, affirmation vivante, puissance de exister et de faire exister."[4] The co-implication between creativity, as observing distance between the given to follow the suggestion of the new, that emerges form the inspired heart of the self through the impact of being, and consent, as critical distance regarding the prosaic surface of appearances, to let speak the signs of the new appearing of being for the intensity of the self's life, is evident.[5]

CONSENT AND CRITICAL REASON

Our purpose here is to point out the rationality of consent, its criticality attached to the experience of self.

To begin with, consenting is an act that breaks the dividing line between the theoretical and the practical: "Ce qui dèconcerte c'est que le consentement semble avoir le caractère pratique de la volonté, puisque c'est une espèce d'action, et le caractètere theorique de la connaissance intellectuelle, puisque cette action vient buter à un fait qu'elle ne peut changer, à une nécessité."[6] The deep rationality of consent which turns into a higher sphere, expressive of a consolidated and bearing freedom, is fulfilled in the recognition of the challenge raised by the mostly "other" and most irreducible there is in reality, because it is a decisive indicative dimension of the self's own reality that comes to it, [but] without it. Then: "Le consentement est ce mouvement de la liberté vers la nature pour se joindre a sa nécessité et la convertir en soi-même".[7] Recovering the original friendship in and with being as an act, just as it has been quoted in the preceding paragraph, is in its turn, neither a spontaneous act, nor a calm possession. The "yes" of consent is patience, because it is always re-conquered starting from a "no", starting from an intimate factual consent of what "should be" from an own imaginary measure. Active patience and ontological tenderness towards oneself, the deepness of self-esteem, are not initial data with which we receive ourselves as an endowment, but a critical recovery facing a necessity already implanted in my existence. It comes to me, to make it mine in an inimitable way. The passiveness of consent may perhaps be the supreme basic activity by which an existing entity answers the alterative key of existing. As regards my existence, I have the possibility of receiving and adopting the gift I am from the transcendence of the other, to myself and to the world. But it is as well to surprise oneself, acting one's own singularity, what makes the road to consent an irreplaceable work, nobody can do it for me, but inasmuch as I-am-given, and, in that sense, called upon. "La patience supporte activement ce qu'elle subit; elle agit intèrieurement selon la nécessité qu'elle souffre. (...) Consentir c'est moins constater la nécessité que l'adopter; c'est dire oui a ce qui est déjà déterminé".[8] This patience that consents without anticipating anything, without projecting future, is, notwithstanding, the possibility condition of an authentic projection. It arranges the previously given resources of singularity, because without them there is no creativity, it is not possible to want what is new and effectively appreciate it in its difference, without finding myself involved in what is old and previous.

Knowledge has its connections in the global dialectics of human existence facing Being. Ricoeur's first metaphysical reflections, that accompany and support his phenomenology of the disproportion of the wish-to-be, first shaped between the voluntary and the involuntary and then in the dialectics between finitude and infinitude, inherent to all intentional experiences of the capable-self and to its affective apperception, recover the anthropological issue in the framework of a general ontology of being, as actuality and potentiality, in two decisive articles: *Vraie et fausse angoisse* (1953) and *Négativité et affirmation originaire* (1956).[9] There he tackles the dramatic [aspect] of the narrowness of our finitude, inasmuch as the trigger for truth and happiness throws itself, in wishing, from [our finitude itself], powerful as well as immeasurable. The structural human disproportion thus implies a legitimate sadness and a positive anxiety that express the grandeur of the self, its non-conformity with something less than everything: its impossible ontical self-satisfaction, the folly of a self-synthesizeing interpretation of the disproportion, which is, because of its own nature, without synthesis projectable and producible. Consent has to be aware of this, so as not to distort itself either as a naturalistic and historicist observance of necessity or become shocked at the anarchical power of negativity, of non-conformity.

To deepen into the rationality of consent is to unravel the precedence of positivity that underlies, and causes, the most violent negation which emerges from the inadequacy between transcendent demands of the self and historical fulfilments. It is not a mere gnoseological inadequacy in which the beginning of freedom puts itself at stake in the knowledge of necessity. At stake is [also] the self-Being original relationship, self-to-Infinite, as source of freedom, experimented in the rational-affective character or the original relation. It is contemplation with no distance, in friendliness, or rather, an active adoption of necessity, recognizing it as a gift, and thus, as dawning expression of freedom: "C'est par là que le consentement est toujours plus qu'une connaissance de la necessité: je ne dis pas, comme du dehors,: 'Il faut'- mais, repassant en quelque sorte sur la necessité, je dis: oui, qu'elle soit. 'Fiat'. Je veux ainsi"[10]

The embracing dialectics of activeness and passiveness in which consent operates, possesses an interesting and decisive critical implication of an ethical-ontological type. On the one hand, this dialectics in its turn upholds "une éthique (...) marquée par la dialectique de la maîtrise et du consentement"; so, ethics, rather the moralism of behaviour norms, implies something much more important: human stance in front of the real, in front of the stars, in front of the other, and in front of itself. On the other hand, this embracing dialectics of activeness-passiveness dismisses every anthropological monism or dualism inasmuch as it shows the opening of an alterative gap in the dramatically maintained unity of human disproportion. "Les implications ontologiques de cette dialectique de l'agir et du pâtir ne me sont apparues qu'à la relecture de ma thèse (VI) à l'occasion d'une conférence à la Société francaise de philosophie: 'L'unité du volontaire et de l'involontaire comme idéelimite' (1951). La phénoménologie du volontaire et de l'involontaire me paraissant ainsi offrir une médiation originale entre les positions bien connues du dualisme et du monisme. (...) Un peu plus tard, écrivans *L'homme faillible*, je devais me risquer a parler, dans un langage emprumté a Pascal, d'une ontologie de la disproportion. L'expresion ne figure pas dans *Le volontaire et l'involontaire,* bien qu'elle exprime correctement la tonalité majeure de la sorte d'antropologie philosophique dont relevait l'arbitrage proposé entre monisme et dualisme".[11] I transcribe this decisive autobiographical confession, because it offers the neccesary categorial context to understand how every thing is bound together and what Ricoeur accurately aims at, when he upholds, against every anthropological monism or dualism, that the effectiveness of the voluntary or the involuntary is disproportion and that the unity of the voluntary or the involuntary is a boundary idea.[12]

What does boundary-idea here mean? Every limit is like skin –R. Guardini says–because in principle and vitally it communicates a nourishing *otherness.* Boundary-idea, because it is the demanding and undetermined presence in ourselves of something really other, the opportunity of occurrence and the physiognomy of which, is a-priori unknown to us. But we do know and sense, that, should it exist and come out, it must be somehow real and totally correspondent, in an unimaginable way, with our disproportionate existential demand of truth and happiness. Boundary-idea refers to a *tertium quid* inherent to the bipolar inadequacy of praxis that questions every form of comprehension and of practical relation with the human that could flatten and smother the disproportion. Dualism and monism express the historically recurrent attempt to accomplish that fading away. Be it by breaking the bipolarity of disproportion to direct it by means of a double register of perfection (dualism), or by turning the disproportion unilateral by means of a materialistic or spiritualistic planning of perfection (monism). The naivety of will to power rests upon the pretence of producing the proportion and offers an answer to man without man, measuring the immeasurable: it is magic as imaginative infinitization of the finite. The shrewdness of the will to power rests upon making the infiniteness of disproportion empirical, sharing freedom in the continuous accumulative circulation of the finite: it is the mirage of the accessible finitization of the infinite, keeping its novelty by means of the eternal returning of different phases of the same. In both cases, it is about the obviation of self-acceptance as disproportion, [about] the hushing of the human question or the deviation of its direction, to make it manageable and bearable, or to excite it by exacerbating the wish-without-being, in a theoretical and practical programmed indecision, regarding every possible event corresponding to the original relation, right up to the weariness of itself as true question and wish-to-be. Is it a true answer the one that trivializes, destroys and exceeds the question, or is it already a dead and deadly answer? Is the proposal corresponding to the structural dual (not monist or dualist) disproportion, to cure ourselves of the disproportion or to praise it without measure, praising the original metaphysical relation in the presence of an existing entity that would be sign and figure of the disproportion itself? Is it not the very ideal of self-sufficiency, and of making oneself the enemy of every real disproportion, of critical consent?

There is a decisive aspect –I would call it blondelian– in Ricoeur's work, extolled by critical sense of consent in reference to truth. Consenting does not refer to anything abstract or anonymous, to a mental by-product of man himself.[13] This getting into necessity of reason and freedom makes explode a healthy crisis in the immaturity of the thinking self, raising a fundamental change of attitude. The self is summoned to a conversion: "le moi, plus radicalement, doit renoncer à une prétension secrètement cachée en toute conscience, abandonner son voeu d'auto-position, pour accueillir une spotaneité nourricière et comme une inspiration qui rompt le cercle stérile que le soi forme avec lui-même".[14] To consent is to adhere through reason, freedom and affectivity to something objective I give myself, but which intimately affects me, and the positivity of which I sense and potentiate through a renewed relation operated by consent. "Or le lien qui joint véritablement le vouloir à son corps requiert une autre sorte d'attention que l'attention intellectuelle à des structures. Elle exige que je participe activement à *mon incarnation comme mystère*. Je dois passer de l'objectivité à l'existence".[15] This ontological feeling of participation in being through flesh, prevents objective analyses from degrading into a naturalism with no deepness, just as the notional clarification of existence prevents rational-affective access to the ontological from disappearing into sentimental confusion. In philosophy, knowledge is inseparable from commitment to existence. The deeply felt epochal need of a new thematically widened Enlightenment of reason, open to the specificity of its fields of objectivation and with a multidimensionality in methods of knowledge, means that philosophy is not there to close itself up in a mythology of knowledge for the sake of knowledge, typical of a soul exiled from its incarnation.

From his first works, Ricoeur questions every pretence of giving an answer to existence –which is total demand of significance, exhaustive and inexhaustible– through the network of a discourse or an ultimate argument. To the life and existence of the self in action corresponds only effectively a living, intelligent and persuasive answer, to which it be possible to consent freely and affectively. The calibre of the "mystery of my incarnation" tunes in with the contemporaneity of a living presence that recognizes and embraces me in consent to my humanity. With the disproportional "questioning flesh" that I am, and not with the conclusive system of an apotheotical discourse, does the event of an encounter take place, the correspondence of which becomes verifiable in act. Critical reason resumes its task as critical conscience of an experience in act. Philosophy re-discovers itself as "science of conscience of the total sense of experience", correlative to the self as "self-conscience of the kosmos". Reason is prepared by consent for this enthralling task. Ricoeur questions a certain tempting rationalism in reason, which gobbles down, expresses and substitutes the occurrence of being in a "science of experience of conscience", in what has already been experienced. The hegemony or the already decanted[16] concept, of reasoning and analysis, above the recurrent and renewed observation of the original occurrence of the real in that field of encounter that experience is, – therefore reduced neither to a mere volatile sensation, nor to the prejudice held up on the already known and delimited– becomes living experience, when those second attributes of reason let themselves be measured by the received alterity. The subject of knowledge grows up as subject, when he judges incorporating what he experiences. The primacy of the experiential encounter, is decisive for experiencing the real intensively, involving reason down to its very bottom. Another misleading possibility is growing spiritually older in a perhaps splendid mental-aprioristic substitution of the experience of the real.

TOPICS OF CONSENT

The present paragraph intends to grasp the nucleus in which the interpellation of necessity condenses and to show that consent is the answer of freedom to what is factually inevitable and unmovable in existence itself, but yet is never fixed in its experienced sense. For the better or for the worse, it is in consent or in rejection where this sense is at stake. Ricoeur calls these topics of consent the *absolute involuntary* –character, unconscious, life– not because they invalidate the permanent dialectics of activity as initiative and of passiveness as reception, but because here reception must be total, for initiative to be real, incident. What absolutely affects us in our own flesh, and thus raises the questioning of the self, with a logic structure that aims at a total response, shows that metaphysical issues are not raised and decided form the clouds upwards, but in everyday flat ground: "le caractère, cette figure stable et absolument non choisie de l'existent, la vie, ce cadeau non concerté de la naissance, l'incoscient, cette zone interdite, a jamais inconvertible en conscience actuelle",[17] they defy freedom and reason in an immediate and permanent way, because they are there, in the nearest, but radically approaching them.

This total pressure of the involuntary, however, does not present itself equivocally, menacingly, without any antecedents, because the involuntary is accompanied and analogically nourishes the two precedent moments, where the creativity of will manifests itself more agility, having, in spite of it, not overcome the road to arbitrariness. The regulated sovereignty of freedom already manifests itself in *decision*: it implies the initiative of a project that intends to carry something out, but the project would never emerge, were it not nourished by the involuntary, providing motives arising from necessities experienced because of our bodily and historical-cultural incarnation in being. Motives that turn into values as soon as they are estimated as leading to the project –estimated in its turn, when it is perceived as significative of our will-to-be, of our destination to happiness– and arise knocking at the door of moral conscience, frequently in a conflictive manner. Decision crowns that reason for acting, which, finally, after deliberative scrutiny, shows itself more valuable, conducive, persuasive and concurrent with the projected will-to-be and with goodness. The same regulated sovereignty of freedom pre-shapes itself in the second sphere of the exercising of freedom, through the living presence of the motivating involuntary.

Inasmuch as decision does not come into the world if it does not become *action*, acting requires that I set my body in motion and make a voluntary effort. But if such movement and effort pretended to be a pure and non ruled creativity, all decisions would remain in the air, without acting, without interfering in the course of the world. To take the step from decision to action, it is necessary to have available a great amount of know-hows, pre-shaped in spontaneity, as well as exercising bodilyand psychical capacities, which we must already count on, and which are involuntarily more or less available to be voluntary involved in action. The alternate rhythm between the living impulse or the paralyzing shock of the different forms of emotions facing the new acting that is coming into the world and the calm and secure, or diverted and nonchalant position of accumulated habit, makes evident the decisive incidence of the involuntary in acting.

This presupposes not only the presence of an incarnated self, which transcends through sensitive-reason that which conditions it, but supposes as well the freedom in relation to which there may only be conscience and experience of necessity suffered-experienced in the flesh. In this way, consent is also a work of freedom, challenged to involve itself, in the interstices of its corporeity, in the consistence of being that presents and signifies itself in different ways, also in practical terms. So, what freedom is it about? A freedom that is not (by principle, though it might be by pretension) pure act, olympic movement, arbitrary or suicidal flight, because it is originally warned and sheltered by the suggestive and strict indications of being, which warns the wake conscience about the consistence of existence, through the involuntary. A freedom that is, in each of its moments, activity and passiveness, initiative and receptivity, which is exercised by taking what it is given, what it finds, and what does not do itself: motives and values, energies and powers. And finally, that imposing strength and that necessitating vulnerability of total nature, that reaches me through my body, without me, for me to be myself: the absolute involuntary. So: "L'acte du Cogito n'est pas un acte pur d'auto-position; il vit d'acueill et du dialogue avec ses propres conditions d'enracinement. L'acte du moi est en même temps participation".[18] Participation to what?: "au mystère central de l'existence incarnée; pour être compris et retrouvé cet mystère que je suis exige que je coïncide avec lui, que j'y participe plus que je ne le regarde devant moi à distance d'object".[19] That is why human freedom is *dependent-independence*: admirable and paradoxical disproportion of an existent that is "maître de soi et serviteur de cette nécessité figurée par le caractère, l'inconscient et la vie".[20]

Character, first figure of the absolute involuntary, does not refer to any collective typology, but to the inimitable uniqueness of the incarnated self. This is the indicator of its singular existence, from it arises the feeling of impossibility of each of getting rid of oneself. In character, freedom and destiny are no longer considered as two juxtaposed kingdoms, one beginning here and the other there. Rather, thanks to character, all determinations manifest themselves as the inimitable way o being of freedom itself, as the irreducible perspective upon values which manifest themselves through motives, while motivation particularizes itself, because its conscience is unique, with a style that distinguishes it from all others. Through character richness of human plurality is accomplished, a great mystery, which precisely because of that, is not in the hands of men's power. Ontological mystery is, always, the only real critique of power. In its turn, character is our effective way of being, of co-existing and of co-inciding on the world; it is also matter to be worked upon, so that its virtualities do not become misappropriated.

The unconscious is the second manifestation of the absolute involuntary, there where it borders and penetrates in the actual critical nucleus of the self and shows that its transcendence in self-reflection is accompanied by an undetermined other, by an undefined matter that questions our very questioning capability, inasmuch as it is exercised with the pretension of total transparency and of original innocence of conscience, which easily accuses the limits of the real. On the one hand, the unconscious is the already given and never ending background of conscience, abyssal rearguard of clear and distinct ideas. But it acts as boiling source of inspiration, nourishing the productive imagination that drives artistic creativity. Its confusing and irreducible alterity accompanies the clear conscience, like the dark bottom ground in which we find ourselves as already given existing entities, and, even as consciences, we already come from near here and somewhere else. The fact that that indefinable matter is mainly affective, referred to the origin and persistently rebellious to discovery, suggests that the immemorial memory of our roots in being is an *original*, generative bond, linked through a predecessor alterative series that intrigues us, that inspires our eager search towards the other through others, always bumping into the elastic barrier of the immemorial. This impassable but backwards flexible barrier, which neither lets itself be placed at a distance, nor evaluate as if it were a motive, nor move as if it were a power docile to our projects, but shows indirectly through signs and symptoms, issues a highly instructive methodological criterion regarding the healthy general and philosophical use of reason. When the cognitive subject throws himself, loyal and coherent towards the foundation of what he finds in experience, to get to know and support its significance, the latter never appears as a directly available object. It manifests itself in experience through discontinuous strokes, discreet and even incongruent to a diverted and superficial look. Just as in the case of the unconscious, to pick up its traces, become aware of its correlations, and follow the conclusive indications regarding its significance, a method of signs is required, the seriousness of a hermeneutical-critical method that could support what it is, and could confirm the breach of escape towards the unknown within the already known, towards the invisible within the visible.

On the other hand, the unconscious protects the conscience against self-creating illusion, against the confusion between the capacity of self-determination and the *aseité* (*ens a se*), to change the sense of autonomy regulated by conscience under the pretext of an ontological self-sufficiency, which is a mould for traumas and every possible arbitrariness. Therefore the unconscious is also the sign of an *original* link with a type of equivocal and sinister alterity, in respect of which all psychical conflicts tie themselves together, increase and repeat themselves. As a last resort, the unconscious is the lodging-place of self-rejection. En efecto: "Je peux donc rester seul à dire non quand toute la nature à sa facon dit oui, et m'exiler à l'infini dans le refus. Mais ma lucidité doit être sans borne. Qui refuse ses limites refuse son fondement; qui refuse son fondement refuse l'involuntaire absolu qui double comme un ombre l'involontaire relatif des motifs et des pouvoirs. Qui refuse ses motifs et ses pouvoirs s'annule soi-même comme acte. Le non comme le oui ne peut être que total".[21] The third figure of this total challenge of the absolute involuntary is life. Life is the basic necessity and most fundamental condition of will. But, is it directly visible at all?

LIFE AS *LÓGOS* AND PARADIGM OF CREATIVITY

We start from the devaluation that creativity claimed to exercise on the signification of consent as expression of freedom. Answering to the shallowness of this questioning, we become aware of the critical consistency of consent, for freedom to appropriate itself of the absolute involuntary. Now, in the same hard core of the absolute involuntary, we discover the paradigm and source of creativity: life as the sense of being. Thus, consent presents itself as critical activation of creativity: of life as *lógos*.

LIFE IS THE GRACIOUS LÓGOS OF AN ORIGINAL GENEROSITY

It is possible to sense the difference and novelty of life when placing it in a correlation with two other figures of the absolute involuntary, and thus, to grasp as well its incidence in this whole: "Si le caractère est la nécessité la plus proche de ma volonté, ont peut bien dire que la vie est *La* nécessité de base. Elle alimente les virtualités de l'inconscient et leurs conflits, elle donne au caractère ses directions privilégiées; c'est en elle que tout se résoud en dernière instance. (...) En moi et pour moi, l'union de l'âme et du corps est l'union de la liberté et de la vie. Je suis 'en vie': comme le suggère le langage, il suffit que je sois 'en vie' pour que je vienne 'au monde', -pour que 'j'existe' ".[22] Life is a plan of existence that wraps up conscience and that comes *from outside myself*, to permeate *everything in me*. It is the reconciling and totalizing topic of necessity and freedom: its occurrence is enough, for everything to be virtually there, for each way of being and each figure of the involuntary to grow according to their own perfection. The sense of *being* as act of existence and of making exist discerns itself in life, as totalizing basic necessity. Life in itself is not vitalist, it is logical, but not tautological. It is so rationalistically demanding, that man is capable of giving his life to uphold that value which gives sense to it and supports it. Suicide itself gives testimony that a life without *lógos* is unbearable; *lógos* is the forceful and evasive answer to the misunderstanding, that, in general, men already produce between life and the adequate reason for living. This non-vitalist demand of life has its pre-figurations in vegetal life, like the grain that dies for the sprouts to be born, and in animal life, in the total exposition of the mother to defend her brood. In the meantime, life goes on being, on the one hand, the basic endowment which allows values to be such for me, and on the other, the fluent organicity of a gracious and silent wisdom that sets essential tasks before the intervention of any human reflection, making possible every knowledge, power and effort.²³ Following Husserl's steps, Ricoeur becomes suspicious of an vitalist and historicist idea of a philosophy of life. The theme bursts in bound to demands of reason: the essential primacy of "birth" pertaining its concealment of the romantic fascination exercised by death and the ethical primacy of the wish to live [according to] goodness, for moral normativity to have sense.²⁴

LIFE IS CREATIVE REASON IN ACTION

Its mode of construing being is not accessible to humans: "Dès que l'on compare en effet l'invention et la finalité en biologie à l'invention e la finalité humaine, elle apparaît nécessairement étonante; comparé au cheminement difficile de la construction humaine, l'edification organique nous stupéfie: alors que l'homme fabrique des outils du dehors par addition des parties, la vie édifie ses organes du dedans par croissance orientée. Tout se passe comme si une intelligence que s'ignore, mais une intelligence infiniment plus clairvoyante et infiniment plus puissante ordonnait la matière".²⁵ This exceptional and inexhaustible intelligence inherent to life manifests itself even more in the bodilydefections of human finitude: "elle se donnne comme une puissance de réparation, de compensation, de guérition; le spetacle merveilleux de la cicatrization, du sommeil et de la convalescence confondent ma volonté, ses faibles moyens et sa maigre patience".²⁶

In front of this spectacle it is necessary to exercise an effort upon reason, very violent, were it not for what the dominant mentality already facilitates, to cut the dynamics of the sign interposing a prejudice, which could prevent the sensitive and sensible intelligence from surprising itself by what occurs before its eyes. This thought arises spontaneously: "si la vie faisait ce que'elle fait volontairement, elle serait una volonté sans comune mesure avec la nôtre, et por tout dire: démiurgique. (...) La vie édifie la vie; la volonté ne construit que des choses. Le spetacle de la vie humille toujours la volonté".²⁷ Life is the expression of a rationality the final plan of which we do not completely grasp, but, at least, as an enormous good for ourselves. In the construction of organisms, life shows a geniality that goes over smallest details, where each organ performs perfectly determined functions and complementary of others in keeping with the living whole. This symphony has its rhythm and works with time, which makes the complete truth of the living entity explicit: it proceeds with method, and keeps growing through time in successive stages, in which the irreplaceable result that each one has been assigned in accordance with the living whole, is carried out. This beauty of life is not separated from the radical contingency of nature: life and nature are marvellous because they're signs, only signs of a presence, which, through them, suggests itself as infinite *lógos*.

THE LÓGOS OF LIFE IS SINGULARIZING SELF-AFFECTION AND INDIVISIBLE UNITY

Conscience is not designed to be exiled form life, to oppose it and contradict it, but to become absorbed in it in each concrete whole in which it develops, and thus to understand it, because life is not merely our infrastructure, but mainly the positive figure of our fate: "elle est une certe nécessité d'exister que je ne peux plus m'opposer pour la juger et la maitriser. Je ne peux pas aller jusqu'au bout de cet acte d'exil qu'est la conscience (...) La vie n'est pas seulement la partie basse de moi-même sur laquelle je règne; je suis vivant tout entier, vivant dans ma liberté même. Je dois être 'en' vie por être responsable 'de' ma vie. Cela que je commande me fait exister".[28] This same enveloping power of life that reaches us is what makes of it an affection: it reveals itself to me as felt rather than as known, wit no intentionality towards something, with no room for perspectives, not observable in itself. And it is, notwithstanding the first form of conscience-of-self, the elemental form of apperception of the self, which originally accompanies every conscience of something, every relation with something else: It is the inevitable conscience of the consistency of my particularity as an unredeemable whole. That is why life is apprehended with no perspective, not through aspects or sides, it offers no foreshortenings. Self-affection is the non-perceptive conscience of my body, it is my body sensed as a concrete and unique whole, i.e., conscience of life is not conscience of something, but self-conscience as individual living totality, which does not make itself, which receives itself.

In affective conscience life reveals itself as indivisible: "I exist as one".[29] I am the unit that circulates among the diverse functions of my organism. I have limbs, feelings, ideas, but not lives. Life is not plural. I find myself already existing: conscience has no right to any previous decision upon existence. Because when the latter wakes up, the former is already given, it is being endowed, and conscience cannot anticipate itself to life, to give it to it.

LOGOS OF LIFE IS ENDOWED CONTINGENCY DIRECTLY RE-BOUND

Life is the cipher of a unique and brimming creative action, which comes from *somewhere else* and sets up as basis of acceptance of every aspect of the existing. "Le 'je' est sur fond de vie, (...) l'indéfini d'une vie donné gracieusement. Cette impuissance de la conscience à se donner l'être et à y persévérer est tantôt souferte comme une blesure originelle ou éprouvée comme une joyeuse complicté avec une élan venu d'ailleurs. (...) ce sentiment d'être debordé par ma vie est augmenté par cette assurance que la vie est une dans le monde, qu'elle vient de plus loin que moi et me traverse seulement en me donnant d'exister".[30] As organism, life is the unfolding in myself of an intelligence absolutely wiser than myself, that supports me without resting, just as "growing" is de manifestation of that endowment towards its expressive completeness, just as "birth" is the occurrence of a mysterious election witch singularizes my total contingency. Life as donation to a free conscience, to a self that is the subject of this living body, implies to reference to nothingness.

On one side, it is positively "made of nothing", from the total gratuitousness that only requires from me the recognition and performance of that contingency, as a strive with reality in the conscience of dependency of the mystery that makes me singular. The only negation birth speaks of is the ex-nihilo of existence, of this unique and not another being that I am, having one day come to this world of life: "Je ne me pose pas dans l'existence; je n'ai pas de quoi produire ma présence au monde, mon être là; la conscience n'est pas créatrice: vouloir n'est pas créer".[31] On the other side, negatively, life as punctual donation to a free conscience implies "nothingness" as possibility of negating my life as an endowment, of considering it as a brute an anonymous fact, which therefore neither tells me anything nor refers me to anything, and which I must therefore defeat according to my own measures and calculations. Now, this alternative and binding decision are already traumatically and continuously at stake, from the beginning of the self, with its reason and freedom, in history.

What does this imply for each of the moments of the effective experience of life, characterized by that structural disproportion between its being within limits and its wish-of-being, its wanting-everything? In general it implies that life as an endowment ontologically loses its potential because of un-binding a decision that breaks the bond between donation and donator. Thus the organism tends to disorganize itself and becoming vulnerable, illness attacks it, and it responds up to where it can, and we experiment all that in physical and psychical pain. At a certain point, growth combines with the decline of vital energies, and is substituted by ageing; "birth", which is the celebration of my positive contingency or of the occurrence of my singularity, with the limits that qualify it, begins being interfered by something strange to life, that distorts it, turning it into the *anxiety of contingency*.

Which fortuitous factor towards life is this? The historical resistance of reason against the freedom to embrace the mysterious singularizing dependency inherent to the gift of existing as an incarnated self distorts the happy contingency turning it into a distressing one, which will become resentment towards contingency. For the latter begins to be linked above all, with the radical foreign strangeness of death. There is neither original experience nor apperception of death, neither has the latter any symmetry with birth. There is no "natural death". There is the need for an accident by which one is thrown out of the scenery of life: "La mort reste un accident par rapport au dessin de la vie; la mort n'est pas tout à fait naturelle; il faut toujours un petit choc pour nous pousser dehors...".[32] Then, suddenly, death becomes "le révélateur privilégié de cette angoisse de la contingence; c'est pourquoi l'idée de la mort est dévenue en quelque sorte l'équivalent objectif, l'amorce et l'excitant de cette angoisse éminemment subjective de *ma* contingence".[33]

LÓGOS OF LIFE GIVES LESS RELEVANCE TO THE SIGNIFICANCE AND THE REACH OF DEATH

Conscience of the unavoidable fact of death darkens the original experience of contingency as glorious occurrence of the unrepeatable singularity of the incarnated self. Contingency is intentionally misappropriated towards resentment against finitude and is affectively emptied of the joy of human disproportion that is a sign of its binding. This radical confusion and this profound affective trauma arise from the impact produced by perception of death as what it is: a tragic incongruence with the superior dynamism of life, a scandalous disintegration of a singularized existing entity, meant to last as such. In the apperception of the self and self-affection, there is no trace of death. It is knowledge that has been learned: "L'idée de la mort reste une idée, toute entière apprise du dehors et sans équivalent subjectif inscrit dans le Cogito".[34] Death is so anti-natural, that not only at the beginning does the child have an idea of it, but also, paradoxically, in spite of all our adult knowledge and suffering near it, and of verifying its unavoidable character that has hurt us so many times, we overlook it, we live as if that certainty were not essentially serious. Ricoeur insists: "cette certitude est un 'savoir' et non pas un'experience, le plus certain de tous mes savoirs concernant mon avenir, mais seulement un savoir".[35]

There is no personal experience of death, but there is a live and suffering knowledge of it, through the death or our loved ones. It hurts us, because it is an exceptional negation that takes place necessary and separates us, interrupts our relations, our limits and our powers. But its necessity has a curious internal lessening which Ricoeur adequately points out: "se donne avec une nécessité irrécusable: 'Tu dois mourir'. Et pourtant cette necessité ne peut être déduite d'aucun caractère de notre existence; la contingence me dit seulement que je ne suis pas un être necessaire dont le contraire impliquerait contradiction; elle me permet au plus de conclure que je peut ne plus être un jour, que je 'peux' mourir: car qui a 'du' commencer 'peut' finir, -mais non pas que je 'dois' mourir".[36] This could-not-have-been of death – on which somehow we always count, when we in everyday life put in first place, in spite of certainty and crying, our devotion for going on working and asserting life – is the diminishing of relevance dictated by our ontological memory of something more original: the occurrence of our contingency.

Death produces in us a deep repulsion and fear because of its offending anti-natural character. But we should be aware that before and behind death, in the historical beginning and in each day in history, there is another big rejection in that entails us all. The rejection of the relation to being as act of existing and making exist, experienced as conscience of consistency of the instant. This is implied in the original self-conscience of our contingency, as joyful belonging to Being, and as dependency nourished from source of existence. It could be said that death, with its humiliating historical weight –that which we are not made for– is the last educational resource for a final alert so as not to lose something much bigger and more decisive for which we are made, given the introductory character of this very life, which, in the wish-to-be, expects life. Death, then has the mission of an alert, which, with its forcefulness, gives us the opportunity to tear ourselves from the illusion of self-sufficiency –that historical illusion, humiliating as well, for which we are not made– and which, under its multiple distortions, is the support of all violence of history.

FROM REVERSED TO CREATIVE CONSENT

THE GREAT REJECTION AS REVERSED CONSENT

To Ricoeur, as different from Heidegger, there is no original being-for-death: death does not constitute me, like *Sorge* or *Mit-sein*; neither does it interfere with me in the way other injuries of life do: suffering, ageing and distress of contingency. Death is always an intruder, as foreign to life as impossible to do away with it. Thus, death takes over the most total and menacing image of the necessity and limit that destroys freedom and life. It is there, where resentment against every other form of limitation begins, in short, against the human and creatural condition itself. Behind the great rejection of the absolute involuntary, hides the confusion between finitude and guiltiness. On the other hand, life contains the trigger for perfection and happiness: from the body it sends its message of its primitive passion to coincide with freedom, and the latter the passion to exist and express itself in a body that totally obeys it. There is human disproportion because life is desire to be and possess the infinite. When this desire loses its intentional direction, it turns against the dimension of finitude of the same subject that carries the disproportion. It does it in an indirect way: by despising the limits (which are like skin: they communicate totality and alterity). To reject the limit is to reject reference to the fundamental alterity. Then, the transcendence of the self, its sovereignty in the world through its connection with the Infinite, exchanges this constituent ideal for the utopia of projecting the vane and empty self-sufficiency. To evade the challenge of sense of the disproportion and the tension between the voluntary and the involuntary, the *Cogito* settles down in the abstract, denying that which, in it, does not obey the project of self-sufficiency. "Le trait plus remarquable de ce refus à triple tête (del límite de un carácter, de las tinieblas del inconsciente y de la contingencia de la vida), c'est qu'il ne se donne pas d'abord comme refus, mais se cache dans une affirmation de souveraineté".[37]

But it is only a speculative sovereignty, postulated to defend itself from finitude, because it has previously hidden what makes it human; the structural disproportion implied by the presence of an Other, on the way of the self to oneself. Then, faced with the narrowness of character of the speculative self, it builds up the promethean purpose of being the concentrated totality of possible humans, in the fashion of a powerful collective subject that represents real humanity. In front of the unfathomable unconscious, it raises the pretence of total transparency of an I-Think in general, which has already absorbed the sea of being and signification in the mirror of conscience. In front of the contingency of life and its condition of non-necessary individual existent, it considers itself as total and exhaustive condition of possibility for something to exist and have sense, becoming the operating subject of the "ideal genesis" of everything else, thanks to the alleged action creator of conscience.

The ontological significance of the fundamental senses and feelings, that show the resistance and transcendence of the body, of the other, of life, of duration of all "constitution" enterprise, is epistemologically discredited as pre-critical instance of reason, which has not yet reached the recognition of its absolute transcendentalism. Which is the task of so many fixations on the purity of method to build the speculative tower? The crouching down of the virus of nothingness, of the hidden fear of death, calls for some mental form of conjuration: "c'est par un geste de puisssance que la conscience réfute sa propre angoisse de ne pas être".[38] That is why, "une philosophie de la conscience trionphante tient en germe une philosophie du désespoir. Il suffit que le refus dissimulé dans le voeu d'auto-position se connaisse come refus pour que la vanité et l'échec de ce voeu transforment soudain en désespoir la prétention de cette liberté titanesque...".[39] Nihilism occurs as deceited idealism due to its lack of reality, but it stops its criticism half-way up. It unmasks the inverted significance of the great idealist theses, but remaining in their game. Where realism risked itself for the idea, nihilism states that it is only a strategy of will for power; where it speculated on the freedom of the spirit, it judges that it is an aesthetic game, to set arbitrary values as conditions of conservation and increase of that odd will of will. Men have arrives at the mature time, in which they only energize themselves for nothing, everything is interpretation to support a fleeting role on the great tragicomedy of the world, to which, in their massive loneliness, they lend themselves.

The self-contradictory parable of the great rejection as reversed feeling, does not need refutation, but returning the human to its place: "Le refus marque la plus extrême tension entre le volontaire et l'involontaire, entre la liberté et la nécesité; c'est sur lui que le consentement se reconquiert: il ne le refuterá pas; il le trascendera".[40] What does this imply? (a) To always redo the road to reason trough a fundamental ontology of human reality, that is, "une méditation directe et concrete sur la condition véritable de l'homme..." (VI 438), having learned form our modenity, by contrast, that "toute genèse *idéale* de la conscience est un refus de la condition *concrète* de la conscience".[41] (b) To rehabilitate metaphysics as ecumenical critical-poetical dialogue, down to the clarifying threshold of the issue of Being, as task of philosophical reason linked to the experience of the self as freedom incarnated in history. The poetical root of consent is neither a capitulation in front of necessity nor an arrogant rejection of its character of sign directed to reason and freedom. It is gratitude before total gratuity of the enormous ontological weight of the gift of existence, which, because of the same, is a total proposal that challenges a just as total response. Because: "¿Comment justifier le oui du consentement sans porter un jugement de valeur sur l'ensemble de l'univers, c'est à dire, sans en apprécier l'ultime convenance à la liberté? Consentir n'est pas capituler si malgré les apparences le monde est le théâtre possible de la liberté. (...) Ainsi le consentement aurai sa racine 'poétique' dans l'espérance, comme la décision dans l'amour et l'effort dans le don de la force".[42]

FROM IMPERFECT TO HYPERBOLIC CONSENT

The actuality of human disproportion and the project of cancelling the given by rejecting it, make it evident that the question about sense, reach and articulation between the absolute involuntary that constitutes life and the demands of freedom cannot be omitted. This emerging breach not only problematizes the self-experience of the unity of the bodilyself, but also affects the whole of relations with reality and defines the ultimate position of the self before Being. Ricoeur points out that

"Réfuser la necessité d'en bas, c'est défier la Transcendance. Il faut que je découvre le Tout-Autre qui d'abord me repousse. C'est ici l'option la plus fondamentale de la philosophie: ou Dieu ou moi. Ou bien la philosophie commence par le contraste fondamental du *Cogito* et de l'Être en soi, ou bien elle débute par l'auto-position de la conscience, qui ha pour corollaire le mépris de l'être empirique".[43] It is remarkable, that the contempt for the "empirical being" is to be verified as well in other attitudes that no longer invert but fracture the sense of consent, which always puts the relation between the individual existence and the totality of being at stake, as is the case of these two figures of consent, recurrent in history: stoicism as imperfect consent and orphism as hyperbolic consent. "Le stoïcisme représentera le pôle du détachement et du mépris (negation of the consistence of the empirical being, indifference towards the individual and concentration of the spirit upon itself before the Oneness); l'orphisme la perte de soi dans la nécessité (de-individuation and assimilation of the vital impetus of the great Wholeness) Mais l'un et l'autre néamoins indiquent à leur facon le nexus du consentement et d'une philosphie de la Transcendence (the existential form of the constituent ontological relation from the self to Being defines itself in the modality of consent)".[44] The effective truth of the inevitable link to Being (and with Transcendence) shows itself in the form in which the self tends to be related everyday with its concrete reality.

In a way stoicism is as well a philosophy of autonomy, but not faced up to rebellion in front of necessity, but from the indifference of the soul regarding the dissolutions introduced by the desire affected by the plurality of the desired things. As huge effort of self-control, it directs itself to reduce the life impulse, which acts through the body, which is therefore left aside as a thing among things. The soul must train itself to a high degree to operate a systematic homologation of all the individuality that affects it, to reduce it to indifference. "L'idée de l'insignifiance des choses que passent est à elle seule purificatrice; jointe à celle de l'ordre total, elle dévient pacifiante".[45] For stoicism there are no passions of the soul but only of the body, while the soul is pure impenetrable act. The soul steadfastly construes its specific spherical shape, placing itself in the reverent and dispassionate admiration of divinized nature. The bonds that move the soul and affect it must be reflectively reduced to mere relative opinion, arising from a circumstantial state of mind. The reality it affects is reduced to doxical sentiment; affection is suppressed by suppressing opinion.

Consent is "un art de détachement et du mépris, par lequel l'âme se retire dans sa propre sphéricité, sans cesse compensé par une admiration révérencieuse pour la totalité qui englobe les choses nécessaires et la divinité qui habite cette totalité".[46] Because of its reference to cosmic divinity, the soul is not the centre of being. Because of its indifference regarding the affections arising from its bodilyreceptivity, it is not mere part of the whole. But the soul is in the whole, and cannot therefore judge it. Its ultimate attitude is to remain contemplatively detained before the necessity of the whole. "La nécessité prise dans sa totalité est aimable; elle est raison, elle est dieu. La force du stoïcisme est de transférer au tout le prestige arraché à la partie. Le changement qui rend chaque chose et mon corps insignifiants est surmonté et conservé dans la substance du tout".[47] The imperfect character of consent, as conceived by stoicism, consists in that the admiration of the cosmic whole absorbs the corps-body while the subject-soul keeps itself in its spherical contemplative shape. This dualism breaks up the rational and affective demand of the unity of life. If the figure of consent needs to deny or reduce fundamental human factors to produce a partial technical-spiritual reconcilement, ethical responsibility is systematically justified. The admiration of the divine cosmic whole coincides with the disdain for the concrete-particular that constitutes the net of the relations in the everyday world. A true significance of existence to which we could consent is that which lights everything up without denying anything, carrying admiration down to the interstices of everyday life. "On ne peut pratiquer à la fois le mépris des petites choses et l'admiration du tout. La limite finale de l'estoïcismec'est de rester aux lisières de la poesie de l'admiration"[48] In orphism, consent is, on the contrary, the loss of subjectivity in the frenzy of necessity. The vital course of the world is a solved problem, and man himself is already determined inside it. In this generic optimism of transforming evolvement, which absorbs life and death undramatically, and where the individual is a disdainful quantity, there are no ontological distinctions or gradations. "L'univers est en travail sur la dure loi du 'meurs et déviens'; cet oeuvre majestueuse, où la ruine, la perte, la mort sont toujours surpassés en quelqu'autre être, est offerte a ma contemplation dans les formes minerales, organiques, qu'ignorent le consentement. (...) Non seuelement la vie en moi, mais le tout est un problème résolu".[49] Certain nietzschean atmospheric turbulences of "loyalty to earth" and "innocence of the becoming", in which the subject submerges to be vindicated by the anonymous stream of necessity, emanate a romantic effusion that does not match a discernment based above all in a discreet conscience of the self: "cet monde unique, unique pour moi, ce monde incomparable, est bon de une bonté elle-même sans degré, d'une bonté qu'est le 'Oui' de l'être. (...) Il est parce qu'il dévient; il dévient parce que toute ruine est surmontée. La bonté du monde c'est 'meurs et déviens', c'est la métamporphose".[50]

But the enthusiastic "yes" of admiration does not make us free from the responsibility of consent in consent, inasmuch as this is the "yes" as act of freedom ofa rational individual. This means that consenting is not gaily admitting the disappearing of ourselves as individuals to become amazed in the great cosmic-naturistic metamorphosis. True admiration does not play with the abolition of the admirer. When nature in its beauty and the kosmos in its wonderful immensity are no longer signs of something else, but the very foundation and the unique entity where things and individuals emerge and sink like waves with no substance in the sea o being, it is reasonable to ask oneself how is so much background obviousness possible, in which a universal solution, thought-out by the self, that leaves out and suppresses the self is established. This self is solved inasmuch as it is dissolved in a total problem; solved for me, without me. "L'orphisme est le consentement hiperbolique qui me perds dans la necessité, comme le stoicisme ètait le consentement imperfait qui m'exilait du tout, que pourtant il m'efforcait d'admirer".[51] On the contrary, in the realistic, rational and affective act of consent, the horizon of totality is not re-mystified and closed, but opened in the recognition of the signs and ciphers of a *Lógos* as evident in the text of universe, as exceeding all possible clarification or translation without very large margins. These constellations of signs that show themselves with reserve, because they require that freedom searches for them attentively, are completely correspondent with the reason and freedom of man. They indicate sense, but do not oblige to recognize it: they also require to be wanted and expected. That is, they propose but so not impose themselves, because they refer to a self that can only recognize truth if it loves it, and if, therefore, it is itself entire like an incarnated self that exercises its sentient reason and its affective freedom. For if "l'admiration ou contemplation me descentre et me réplace parmie les chiffres, le consentement me rende à moi-même et me rappelle que nul pêut me délivrer de l'acte du oui. C'est pourquoi l'admiration et le consentement font cercle".[52] They reasonably draw circles in the self, if reality as sign, in and out of myself, refers to a present "you", although directly in-cognoscible, that is interests itself in me attracting me by means of reason, through the necessity and for the freedom by which I live.

CREATIVE CONSENT

Just as we are realizing it, consent puts different alternatives and metaphysical decisions at stake, which in the essential are few and precise, an in respect of which, the evaluating criterion lies in their capacity to give reason of facts. The facts are: the immeasurable and finite, intelligent and mysterious material reality of the kosmos, the exultant and dramatic reality of man, as reason incarnated in finitude, open to the infinite and pursued by the challenge to freedom as search for its destiny, under the sign or happiness. Here we barely indicate the fourth figure of consent, as critical approval inasmuch as it is gift of creation and demand of freedom. Regarding the three moments of th absolute involuntary this means: "Oui a mon caractère, dont je puis changer l'etroitesse en profondeur, acceptant de compenser par l'amitié son invencible partialité. Oui a mon inconscient que demeure la possibilité indefinie de motiver ma liberté. Oui a ma vie, que je n'ai point choisie, mais que c'est la condition de toute choix possible".[53] As regards the suffering, evil and death that interfere as intruding bodies in these aspects of positive donation, the judgment of reason always imposes discernment and liberating appeal. For we cannot simply change into joy the sadness of the finite in a character that not only distinguishes me, but also separates me from the others, from an unconscious that is not a creative background, but also a traumatic weight, from contingency, that is not only the experience of a joyful communial dependency, but also the factual pain of having to disappear of every kind and pleasant company.

Thus, the implication of freedom and Transcendence in consent adopts a completely new figure. If admiration is possible, it is because the world an analogy of the Transcendence that creates it. And it is such, above all, because there is freedom, and therefore, the desire and the demand for total Goodness. With it, as well, the effective possibility that this finite and oscillating freedom let itself be dragged by evil and barbarism. So, the road to consent refers neither to the mythic admiration of nature summarized in the absolute involuntary, nor to the detachment or reduction of freedom that is infinite desire, nor to the use of this same desire, to deny, through rebellion the collection of factors in which it is given. Inuasmuch as it is the desire of a concrete and carnal existent. It is evident that character, unconscious and life, as determined expression of the bodily self we are, is given to us. But the greatest lies in that, the gift consists in the fact that we are a full freedom to accept and work on that gift or reject it and misappropriate it. Behind us there is Somebody, who has totally risked himself with us. Totally risking is creating. It is clear that in the logic of nature corporeity is transmitted, but freedom is not generated. Neither does the free Transcendence that nature creates and makes the singular self occur automatically generates the assent of the adherence of the self to its creative donation. If the rebellious, the stoic and the orphic fail, where is the way out? First of all, in drawing away from the technical image of oneself in self-position, by which one becomes boss and fabricator of oneself, without realizing everything that is contained and given in one-self. If not, the gift of being runs the risk of being objectivised and thought of as a kind of violation of subjectivity, like a compulsion exercised upon an thing.

The *invincible involuntary* makes us live, because it is what opens up from the immense context in which the face of the other appears, that occurs in the sign because it retires in discretion. This allows us to point out two aspects that pertain the critical status on consent, as gesture of reason *with* freedom. On the one hand, the given being that precedes, inevitable and irremovable, every constitution on the part of reason and every initiative of freedom, has the healthy role of educating subjectivity in the objectivity of the real, in the peculiar weight of the significance of things, in the learning of the transcendental and existential method of signs. On the other hand, that resistance of the real trough the necessary in myself (character, unconscious, being-in life) is also a principle of individuation that I do no set. That which makes us unique and irreplaceable in the world, is also something that does not depend on us. It is given to us, however, receiving all its positive innovative effectiveness depends on ourselves. The independence of our singularity on our circumstantial arbitrariness (luckily), the fact that every intention to arbitrate it in opposition to an active fidelity towards the already-given does nothing but degrade it, means that the critical study of consent cannot be ontologically and conceptually dissociated from the reference to experience of the original creation, which sets, favours and arouses the presence of reason and freedom, in the world, in the universe. And that presence is every carnal self of-woman-born. But neither can consent be ontologically and conceptually dissociated from the problem of evil and death. The imposing solitude of man to answer to that distressing request for liberation brings itself round to the paradox of freedom; the timid foreboding of which of the presence of Goodness in history – as wishing of the *gift of Being* that could heal the mortal injuries of freedom – tends to be content with little. To constitute itself in an unsustainable self-liberating epic or to succumb sober or inebriated in mortality. But freedom itself is intelligible in the world without creative Transcendence. Could it be intelligible as claim of liberation without liberating Transcendence? Philosophical poetics is the hermeneutic encounter between a dramatic wisdom of the occurrence of freedom and of the facts that testify it to the present, and a fundamental ontology of human reality and of the historic human condition. Poetics presents an occurrence in being and not an ultimate reasoning in the mind, thus indicating that metaphysical problems have neither metaphysic nor discursive solution, but require at the same time a historical and transcendent answer. That is why, as Ricoeur wonderfully states: "L'admiration est possible parce que le monde est une analogie (sign) de la Transcendence (mystery); l'esperance est necessaire parce que le monde est le tout'autre que la Transcendence. L'admiration, chant du jour, va à la merveille visible (creation), l'espérance transcende dans la nuit (liberation). L'admiration dit: le monde est bon, il est la patrie *possible* de la liberté; je peux consentir (sign). L'espérance dit: le monde n'est pas la patrie *definitive* de la liberté; je consens le plus possible, mais j'espère être delivré du terrible (mysterium iniquitatis) et, à la fin des temps, jouir d'un nouveau corps (truth is the immemorial nostalgia of the wish-to-be as life in the flesh) et d'une nouvelle nature accordés à la liberté (liberation as, with eminency, re-invented creation)".[54]

CONICET - Círculo de Fenomenología y Hermenéutica de Santa Fe, Laprida 5059, piso 11 - CP 3000 Santa Fe, Argentina
e-mail: afornari@santafe-conicet.gob.ar

NOTES

[1] Paul Ricoeur, *Réflexion faite. Autobiographie intellectuelle,* Éditions Esprit, Paris, 1995, expressions shown in quotation marks, can be found on p. 26.

[2] P. Ricoeur, "Auto-compréhension et histoire", in: *Paul Ricoeur: Los Caminos de la Interpretación,* Tomás Calvo Martínez y Remedios Ávila Crespo (Editores), Anthropos, Barcelona, 1991, p. 21. (In this lecture, Ricoeur synthetically shows the diverse folds which the central issue of creativity acquires thoroughout his work.

[3] P. Ricoeur, *Réflexion faite. Autobiographie intellectuelle,* Éditions Esprit, Paris, 1995, p. 23.

[4] Cfr. P. Ricoeur, *Histoire et Vérité,* Seuil, Paris 1964: "Négativité et affirmation originaire", p. 360.

[5] Cfr. P. Ricoeur, *Du texte à l'action. Essais d'herméneutique, II* – Esprit-Seuil, Paris 1986: "La fonction herméneutique de la distantiation", pp. 101–118 and "Herméneutique et critique des idéologies", pp. 333–378.

[6] P. Ricoeur, *Philosophie de la volonté I. Le volontaire et l'involontaire,* Aubier, Paris, 1950, p. 321. (Onwords in the text, VI).

[7] VI 325.

[8] VI 324.

[9] Cfr. P. Ricoeur, *Histoire et Vérité,* o.c., pp. 317–360.

[10] VI 322.

[11] P. Ricoeur, *Réflexion faite,* o.c., p. 24.

[12] Cfr. Ricoeur, P., "L'unité du volontaire et de l'involontaire comme idée-limite", *Bulletin de la Société Francaise de Philosophie,* N° 45 (1951), pp. 1–29.

[13] My first Reading of Ricoeur, in the 60s, was *Histoire et Vérité,* after my discovery of Maurice Blondel's *L'Action* of 1893. The renewed impact that, of both readings, persists in me, is the growing discovery that the road to truth is, on one side, the result of the deepening in self-experience and, on the other, the result of the occurrence of an encounter with a presence that is word alive, opening a new horizon upon the experience of self. Ricoeur's irradiating article contained there, "Vérité et mensonge", o.c., pp. 165–197, shows how truth is the occasion of an especial experiential signification, the real contradictor of which is not error –that, once detected as such guides back to truth–, but (in St. Augustine's sense) lie. Recognizing truth supposes *a fundamental ethical-metaphysical option of recognition and consent to the self-presentation of being, as existential sense* of a living-life in the interstices of experience, as presence of the whole in the fragment experienced. Critical reason is what makes conceptualizing reason always resume the road of experience and occurrence. Irrationalization of reason consists in narrowing the pretence of truth down to its discursive validity, losing the sensibility to truth as testimony and verifiable judgement in effective reality. The collusion between discourse and violence crouches, according to Ricoeur, in the rationalist-speculative synthesis which dries truth down to the already known and to the sole coherence of concept; in the ideological-political synthesis that systematized prejudice puts across

as mediatisation of every relation with the other and with the real, preventing its possibility of public revelation; in the clerical-doctrinal synthesis, as substitution of the experience of faith, hope and charity through theology as the independent variable. This does no omit the importance of doctrine as critical conscience of the experience in act, of reflective speculation or of the ideological-political proposal. It emphatically states that truth is a living presence which does not let itself be crystallized in [precisely] that, the function of which is to depend upon it, and not to substitute it itself and its accessibility.

[14] VI 17.
[15] VI 17-18.
[16] For this debate of Ricoeur's, in the field of history as well as in that of theological faith, between the great speculative temptation of the primacy of concept and analysis, still under the figure of the dialectical synthesis, or, on the contrary, the critical primacy of experience in itself, shielded under the fragile synthesis or open mediation –that Ricoeur proposes– to the actual presentation of alterity in the occurrence, in reference to the most genial and disguised modern rationalist formulation, the hegelian, cf. P. Ricoeur, "Renoncer à Hegel", in *Temps et récit III. Le temps raconté*, Seuil, Paris, 1985, pp. 280-299 and "Le statut de la *Vorstellung* dans la philosophie hégélienne de la religion" (1985), in *Lectures 3. Aux frontières de la philosophie*, Seuil, Paris, 1994, pp. 41-62.
[17] P. Ricoeur, *Réflexion faite*, o.c., p. 23.
[18] VI 21.
[19] VI 22.
[20] P. Ricoeur, *Réflexion faite*, o.c., p. 24.
[21] VI 541.
[22] VI 384-385.
[23] cf.VI 320.
[24] Cf. Ricoeur, P., *La critica e la convinzione*, Jaka Book, 1997, pp. 138-139.
[25] VI 392.
[26] VI, 393.
[27] VI 392-393.
[28] VI 385.
[29] VI 387.
[30] VI 388.
[31] VI 427-428.
[32] VI 429-430.
[33] VI 435.
[34] VI 429.
[35] VI 430.
[36] VI 430.
[37] VI 436.
[38] VI 437.
[39] VI, 438.
[40] VI 438.
[41] VI 437.
[42] VI 439.
[43] VI 449.
[44] VI 441.
[45] VI 442.
[46] VI 442.
[47] VI 442.
[48] VI 445.
[49] VI 445.
[50] VI 447.
[51] VI 448.
[52] VI 448.
[53] VI 450.
[54] VI 451.

ERKUT SEZGİN

THE PHENOMENOLOGICAL SIGNIFICANCE AND RELEVANCE OF THE REMINDERS ASSEMBLED AS "LANGUAGE-GAMES"

If you do not expect the unexpected, you will not find it; for it is hard to be sought out and difficult.[1]

Psychological – trivial – discussions of expectation, association, etc. always leave out what is really remarkable, and you notice that they talk around, without touching on the vital point.[2]

Only surrounded by certain manifestations of life, is there such a thing as an expression of pain. Only surrounded by an even more far-reaching particular manifestation of life, such a thing as the expression of sorrow or affection and so on.[3]

—Wittgenstein

ABSTRACT

The phenomenological significance of Wittgenstein's reminders assembled as "language-games" are presented here with those aspects relevant to interpreting the ancient idea of *Logos* by understanding how concepts with rules operate and picture the logic of thinking, meaning, naming, intending, showing *any*thing significantly. That "thing" may be something like a means of representation as the Use of a naming sign, as well as the Use of something named as "object" which both signify their identity and difference as a "name" or an "object named" with the signifying Use of other signs in the manifest stream of phenomena, from which reminders assembled as "language-games" present cross-strips. What is elucidated as such is also the trans-historical significance of these reminders as they elucidate the ultimate internal connections of the use of pictures of historical-language-games with the manifest signifying stream of phenomena as the ultimate *limits* of saying, showing, meaning, representing anything in language.

Let us follow Wittgenstein on the way of sharing the insight that trans-historically illuminates how the conventional uses of signs as pictures of language used as means and ends of representation operate as internally connected with the manifest signifying phenomena of life.

Although it is commonplace to speak of the astonishment or wonder that started philosophy, it is ironical to see the form of expressions of thinking, reasoning, arguing habits, so structured historically by the rules of historical language-games as not to be able to express a sense in response to the call of its originating movement. That movement tried to articulate a language of awareness as to the manifest of a single substance, as *physis* or nature as *Natura Naturans*, the aspects of which are reminded and assembled as "language-games" by Wittgenstein as the signifying manifest background of every picture representing the identity and difference of an object or event, pointed, shown, named by the use of an ostensive definition which may be the use of a pointing finger or a naming sign used as such. Understanding the Logos of *Physis*, or manifest nature is intimately connected with such a sense of life of nature in manifest which is not subject to be represented by pictures, names and descriptions of language. It manifests as the possibility of speaking by means of using and operating with representations of language. Hence the manifest nature as *Natura Naturans*, *Physis* in manifest is intimately connected with understanding the "logos" of words, understanding how words mean, name, describe, operate with meaning, with the rule, the "logos" of which is kept and shared for the meaningful application of the word to its object.

According to Platonism narrated by the history of western philosophical tradition, the logos of the sign and the signified is maintained in so far as by soul's reminiscence of the original Idea associated by sensations of the signified which manifests as a copy or imitation of its original Form. By the narrative of Logos of Platonism the word is separated from the signifying Use of the word in internal connection with the signifying manifest of phenomena, to a transcendental realm of Ideas and Soul life which paves the way to narratives of soul substance and subjectivity capable of thinking and applying a priori concepts or rules of thinking to empirical experience conceived in the manner of Kant's *synthetic a priori* or Bertrand Russell's atomic proposition the subject and predicate signs of which are supposed to name a particular and universal obtained to be perceived by the analysis of sense data which is subject to the perception and analysis of the thinking soul substance. The essentialism about the logos and the essentialist presuppositions in the construction of such narratives disguised in the manner of scientific analyses and theories are intimately connected, as they are being narratives, pictures of language as such, constructed by means of following and applying rules of language without however understanding the rules of the game in accordance with which pictures picture in language, i.e., how signs operate to name and represent, mean, show, define, give an ostensive definition of anything in language. Such failure of understanding the *Logos* of words opens a way to a series of confusions the hidden unquestioned presuppositions of which are disguised by the appearance of being a logical theory as epistemology and ontology, the logic of which is conditioned by the hardened operational rules and conventions of the language-game, in which one's thinking habits are operationally structured to react with pictures of language in analysing and projecting pictures – whereas the problem in question actually requires an awareness about the phenomena that manifest the forms of expressions and significations operational with learning the rules of projecting, acting and operating with pictures of language, both as *means* as well as *ends* of language.

What is required is then the sense of awareness of manifest phenomena to save "Phenomena" by elucidating it from the prejudices of language; hence elucidating to shine the manifestation of life as the possibility of operating with the signs of language; namely as the possibility of naming, showing, pointing, speaking, writing, representing (picturing) by means of the use of the signs of language. What is brought to awareness as such is the reactive conditioned forms of expressions of our imagined self-consciousness. They are the form of expressions habit structures operationally structured in reaction to pictures which are expressed by gestures of pointing things or events as if they are meant or differentiated, perceived as such prior to the learning and operating with pictures of language.

This indicates that "intentionality" needs to be elucidated as a phenomena expressive of itself in connection with significations that express and connect it with the rules and conventional uses of signs which are used in both ways: as means and ends of representing; the use of pictures as means as well as of pictures held fast as reality as the standard rule of comparing and measuring the truth of representational forms of expressions. In other words, elucidation of intentionality in internal connection with the use of signs and representational language recovers and illuminates manifest phenomena, as Nature in the aspect of *Natura Naturans* of Spinoza, as "that which is conceived through and in itself".[4]

Therefore the form of expressions clarified as internally connected with manifest signifying phenomena allows us to untie the knots of intentionality the modalities of which are conditioned and structured by objectified representations of language and that such an intentionality in its conditioned state cannot make the required articulations of thinking to untie the knots of its own conditioning by elucidating the phenomena in manifest but perpetuate to react with operational habits of thinking structured by language phenomena.

What is in question is the presupposed background in which "dark" and "light", "language" and "world" and all opposites obtained by naming, affirming and denying are internally connected with significations; that is the *interface*, the chasm between conceptual identities and differences particularized by naming and picturing.

Here we need to dig deeper into the reality of the forms of our expressions to open an interface between our sense of reality structured by conceptual representations, i.e., between concepts representing the reality of our wake life and the reality of our dream life, the reality of the latter of which is subordinated and judged by the rules of the language-game of our wake life. Missing the awareness of manifest background phenomena in the signifying web of which our forms of expressions signify to be shared as the rules of operating with signs, we seem to be acting and reacting as if consciously intending and willing subjects of our operations with signs, as if subjects naming and describing objects and events of a surrounding world horizon, in deep oblivion of the fact that such a horizon is a shared representational horizon historically structured. That is a horizon which appears on the other hand as the objective reality itself, as if representing its own nature or essence; the reality of which as if it's subject to be pointed at, shown, touched, seen as this and that, in the clear light of common sense. That is the light of common sense determined by the rules and habits of language, rather than the light of awareness free of such a determination.

Hence what seems to be our wake life as opposed to dream life, may not be so awake unless we dig deeper into the forms of expressions the signifying shared consequences of which express and sustain the differences and similarities of our wake and dream life by the rules of the game in internal connection with manifest form of expressive phenomena manifested by significations.

A recognition of what is essential and inessential in our language if it is to represent (picture), a recognition of which parts of our language are wheels turning idly, amounts to the construction of a phenomenological language. (Wittgenstein, *Philosophical Remarks*, Blackwell, 1975, I, p. 51.)

Our operations with the use of signs manifest in internal connection with the manifest signifying stream of life, phenomena as such (manifest "suchness" as expressed by the Zen Doctrine of no mind, awareness of life with a freedom of distance to the operationally conditioned intentional consciousness and imagination in reaction to the pictures of language.). Hence clarifying essential aspects of manifest phenomena from what is inessential, allows us to take notice of the forms of expressions of intentionality, the operational habits of operating with representations of language as tools or means of representation. Lack of that awareness of the language phenomena – in the manifest of which "language" and "world", or "subject" and "object" and everything represented by conceptual identities and differences operate in language – results in a language of conditioned intentionality the forms of expressions of which express confusions of misunderstanding how pictures of language and intentionality operate and intertwined historically as they continue to condition each other building up a strengthening circle in which intentional habit reactions and constructing pictures operationally feed on each other. I.e. the words "name" and "object" only seemingly identify and represent the naming sign and the signified as "object" without actually recognizing the signifying phenomena in which these words and other signs operate in internal connection with the signifying use of other words as signs in the stream of manifest phenomena. Therefore, all the suppositions and analyses introduced by descriptions of "subjectivity" and "objectivity", or as to the subject predicate analyses of a proposition to explain the representational connection of a proposition with the actual world are in fact expressions of a confusion resulting from using pictures of language in describing the phenomena essential for representation, in describing the facts of picturing, which are not on the same level with the pictured facts already re-presented as "ready to hand", and which are in the space of expectations and associations of imagination and memory habits operationally structured with the use of pictures of language. The manifest phenomena on the other hand is not in the space of expectation and associations of imagination, is not anywhere of logical and temporal space and horizon of imagination and memory habits. It is a manifestation in the signifying web of which our memory and imagination habits in reaction with the signifying consequences of signs are operationally structured with the operational rules and techniques of operating, acting with the use of signs. In other words, the awareness of manifest phenomena in question changes our intentional modality conditioned as empirical subjectivity in reaction to pictures of language particularized as objects and events in physical and temporal space.

Therefore, the elucidation of the manifest phenomena in which intentionality with pictures of language are operationally structured is of primary importance. On the other hand the human intentionality (the forms of expressions of consciousness, consciousness expressed as phenomena manifested by significations in internal connections rather than consciousness and phenomena pictured by using pictures as means of picturing, i.e., in the manner of subject object epistemologies) structured as such with representations of language seems not to be able to think except by operating and projecting pictures of language in interpreting its own pictures, hence building modules by projecting pictures that paves and structures a way of treading by its own projections. The problem seems as if we are condemned to be determined to move by language habits, by the rules and techniques of operating and thinking with signs of language, rather than with a freedom of awareness as long as we fail to be struck by the manifest signifying phenomena.

It is indeed here, the philosophical insight or intuitive awareness comes in as the key to the elucidation problem concerning manifest signifying phenomena where forms of expressions manifest in internal connections with significations simultaneous and spontaneous as life without a subject and object, or manifest significations in internal connections as the limit of saying, showing, giving an ostensive definition, doing anything in language with signs, hence operating with signs, or pictures of language. That is an awareness which enables us not to react as conditioned by any cultural historical system of beliefs based on the rules and pictures of any historical language-game, but as one which enables us to understand the human form of life by tracing the modifications of intentionality shaped in any historical context of language-game by the rules and techniques of the pictures used as means of representation to represent reality – to picture reality as one's surrounding world horizon whether scientifically or culturally.

However, that aspect of language phenomena in manifest, remain covered and left in deep oblivion owing to the intentional structure of our operating with signs, in speaking, showing, naming, describing things and events in physical and temporal space. It is due to that oblivion or effacing of manifest phenomena, life as "suchness" from one's awareness that our world horizon seem to us as if subject to our intentional operations and perceptions as if they are "there", as "ready to hand", so to speak with the terms of Heidegger, objectified as objects with their conceptual identities and differences in space, and likewise events as temporal occurrences as temporal space. Therefore, the reminders assembled as "language-games" of Wittgenstein; or "*Lifeworld*" of Husserl; or *Vivencia* of Ortega y Gasset; or *Dasein* of Heidegger; or *Virtuality of the Durée* of Bergson, (in which memory reactions are structured and temporalized and spatialized as cause and effect, before and after); or the "*chiasm*" of Merleau-Ponty whose "Visible and Invisible" is an attempt to open an interface between the boundary drawn by concepts representing the visible and invisible, which is an interface opened to elucidate the interplay of significations that internally connect what is visible and invisible the significations of which imply and presuppose each other, hence providing a distance of freedom of intelligence to the interference of the prejudices or imaginings of the subjectivity whose intentionality structured by operational habits of reacting to the pictures of concepts; or likewise the interplay of "Cairos" and "Chronos" of Anna-Teresa Tymieniecka are all philosophical elaborations focused to elucidate and express the same sense of life in manifest, as the presupposed *trans-historical* background of historical time-structure of intentionality, the temporalized behaviour of historical consciousness and imagination as such.

Therefore, the phenomenological terms "lebenswelt", "vivencia", "umwelt" etc. always need to be used in the light of phenomenological elucidation. They are such critical terms which need to be saved always from the associations of operational language and the intentionality conditioned along with the historical backload of such historical languages. Heidegger's introduction of *Dasein* is such a term developed to elucidate the phenomena in manifest in the web of which intentional operational thinking habits are webbed and get structured as one's shared historical consciousness. Here the awareness that calls to be heard by means of a movement of thinking that elucidates phenomena in which the logical syntax of language of intentional consciousness is structured is not on the same level of awareness with the intentionality, whose operational thinking habits are determined historically, by the historical dynamics of the language and culture in which one is trained and educated as an actor of the language-game. We need not to name and categorize this higher awareness as transcendental, or as absolute – this would be a wrong gesture that triggers imagination reactions. What we always need is to keep the way open for ourselves and others by elucidating the ways for that awareness to awaken and move to articulate the forms of expressions to express and manifest itself as it is; hence moving out of the determined historical modalities of intentionality to the unconditioned free mode of awareness, as the possibility of which pointed out by Spinoza.

Here philosophy and poetic language are in need of each other as philosophical elucidations need to guide the way out from the misguided memory habits and reactions of imagination by the associations of pictures of historical language-games. Poetry by itself may not be enough to bring out the ultimate light of awareness, whereas the harmony and the stillness that shines in poetry shines in the light of awareness that traces its articulating movements all at once with the manifest form of significations.[5]

Without that awareness, our thinking with tools of language has to remain imprisoned so to speak by representational language, it has to move within the circle of constructing and projecting pictures of language with the operational habits of applying methods and techniques of language, with the intentionality the thinking and imagining habits of which are already determined by the rules and techniques of the historical language-games.

The difference between thinking (philosophizing) for the elucidation of manifest phenomena and thinking phenomena always by means of constructing and projecting pictures and models with the methods and techniques of comparing pictures with pictures is noted by Wittgenstein in his Foreword below. Elucidating the aspects of manifest phenomena as what is essential for representation, for acting and operating with pictures of language, clarifies not only the *a posteriori* basis of rules and techniques of applying logic and logical thinking in historical language-games, but more significantly the phenomenon expressive of the intentionality, the form of expressions of consciousness in reaction to the images and associations of the representations of language which is a reaction that closes and imprisons one's thinking to a modality of thinking conditioned by pictures of language.

That is a modality of subject object thinking the wheel of which is turned by our operational habits of intentionality structured to operate with pictures of language. Therefore it is crucial and significant to understand how we think/operate with pictures of language rather than picturing facts of picturing by means of constructing and projecting pictures in the name of analysis and synthesis, with the methods and tools of analytical thinking habits that produce science and technological culture with the representational historical world horizon peculiar to the modes of thinking that constructs it by reactions and consequences that project and picture it.

That requires clarifying essential aspects of manifest phenomena from what inessential, i.e., from the imagination pictures expressed by the form of expressions of gestures of meaning, pointing, showing, giving an ostensive definition of anything as if it's the basis of naming and representing something in language, as was the implicit presupposition of Russell's idea of naming in his theory of atomic proposition.

Wittgenstein delineates with a clear awareness his authentic difference that characterizes the sense of motivation behind the movement of his thinking always directed to elucidate phenomena as it is manifested, to save the phenomena as manifest from the prejudices of the intentionality the thinking habits of which are determined by operational habits of applying a logic that works with operating with pictures.

This book is written for such men as are in sympathy with its spirit. This spirit is different from the one which informs the vast stream of European and American civilization in which all of us stand. That spirit expresses itself in an onwards movement, in building ever larger and more complicated structures; the other in striving after clarity and perspicuity in no matter what structure. The first tries to grasp the world by way of its periphery – in its variety; the second at its centre – in its essence. And so the first adds one construction to another, moving on and up, as it were, from one stage to the next, while the other remains where it is and what it tries to grasp is always the same.[6]

Elucidation of manifest phenomena of life, with significations internally connected simultaneous and spontaneous is what is saved and grasped as always the same, namely the same manifest surroundings with its essential aspects of significations internally connected for learning and operating with conventional rules and pictures as means and ends of constructing and describing pictures are always reminded as a cross-strip and saved as the *sub specie aeternitatis* presupposed background of showing, pointing, naming, giving an ostensive definition of anything – hence speaking, meaning, saying something in physical and temporal space. In other words, the reminders assembled as language-games, are such cross-strips saved as aspects of manifest of phenomena of life, the simulteniety and spontaneity of which are always presupposed as what sustain our thinking and operating with pictures, the logical space of which both physically and temporally are segmented with the habit structures of intentionality by operating and learning to operate with signs as they signify internally connected with the spontaneous and simultaneous virtual manifest of significations. Therefore, the awareness that elucidates the internal connections of manifest signifying surroundings for the use of any sign to be used as a picture, either as means or ends of picturing reminds and assembles those aspects as what is essential for a picture to operate as such. Hence the reminders assembled as language-games present cross-strips and strike us, if they *ever strike us*[7] at all as *sub specie aeternitatis,* to touch us with the momentum of touching our whole stance of existential modality to bring the standstill of its habitual operational movement, in such ways as expressed by the form of expressions of the Zen masters: "Moving as unmoved", intending, willing, meaning and so on. That is to say, without the backload of the historical habits of imagining subject the whole intentional psychological modality of which as being determined in reaction to pictures and associations of pictures of historical language-games.

They are forms of expressions that unite poetic expression and philosophical insight and wisdom – as the articulations of which serve to open the *interface* between representations, between the particularizations of memory and imagination habits of pointing, showing, demonstrating, giving an ostensive definition of a thing or an event already conceptualized. The interface opens up for us a different horizon of signs in manifest,[8] signifying simultaneously, as uncaused in the sense of spontaneity from which no conceptual difference is yet pictured, or marked off in space and time. This is the horizon the sense of which is flashed out by Heraclitus' remark: "One cannot step into the same river twice." That is a remark which is poetic as a metaphor and philosophical as it expresses an insight of awareness as to the manifest sense of phenomena of life, in the unfolding significations of which our reactions with their signifying consequences become operational with our use of signs as tools (means) of representation as well as what is represented. As we operate with signs to show, name or describe the conceptual differences signified by the use of signs which are particularized or qualified as objects in physical and temporal space, our operating with signs are likewise segmented and ordered causally and temporally by our operating with signs. Hence the form of reality is first projected by the consequences of our entering to the life of language acting and reacting with significations in the heart of simultaneous and spontaneous shining of virtual manifest of life, without being neither a subject nor an object.

When Wittgenstein reminds us how a child learns the use of words by acting and operating with their signifying consequences, what is expected from us is to take notice of signifying phenomena of life in manifest, which requires a different awareness of phenomena from the intentionality of an adult, whose intellectual habits of thinking with pictures of language are already determined by the techniques and rules of the language and culture in which one is trained and educated as an actor. Hence actors' intentionality, consciousness, thinking habits are a hindrance to take notice of the phenomena in manifest which needs to be elucidated as internally connected with the expressive phenomena of one's consciousness, intentions, desires, feelings and so on, the significations and consequences of which are expressed and become operational with particular objects, desired, felt, used, operated with consequences and technologies and methodologies interlacing in the life of language-game as culture and history of culture. In other words, an adult who may be intellectually a master of such technological and intellectual knowledge as an actor of the language-game and historical culture, who may precisely for this reason be hampered from sensing and noticing the manifest of life as it is, in its simultaneity and spontaneity without a subject and object as such, from the manifest single substance of which such modalities of object/event consciousness, historical intentionality as such is manifested so to speak in Spinoza's terms.

Lack of awareness of how language, intentional consciousness and pictures of language operate as internally connected and condition each other reciprocally contains so many misunderstandings and confusions resulting from the missing of such an awareness of manifest phenomena in the signifying web of which our memory and imagination reactions become operational with signs structuring an intentional consciousness with the use of signs and their signifying consequences which interlace and criss-cross with other significations. Hence language, without the awareness as to the operational structuring of intentionality with the use of signs and representations of language becomes a labyrinth of *doxas* created by language habits, by habits of thinking and operating with the techniques and rules of applying the tools of language which serve as both means and ends of picturing anything as "real" as opposed to "unreal" and so on. Therefore, elucidating the internal connection between manifest signifying phenomena and the intentionality of thinking and operating with signs is crucial to understanding everything in the light of phenomenological elucidation of phenomena, which amounts to responding appropriately to the insight which the word Logos expresses. "Understanding the internal connection" means elucidating the internal connections between operating with the intentionality of memory habits connected with external connections of signs with internal connections of signified phenomena always presupposed as the possibility of saying and showing something by means of an external connection defined and maintained, reminded by Berkeley as God's continuing to perceive as what sustains the mutual agreement in the signifying consequences shared and sustained as the rules of the game, as the possibility of saying and showing anything in the language-game.

The subject of the awareness or intelligence is nowhere of space, as it elucidates how we come to speak in terms of space through operating with signs, but also it is everywhere of the space of memory, as it is an awareness in contrast to thinking and imagining habits conditioned and confined to the place/space of memory of one's operational habits, hence to move only operationally with the rules and techniques of pictures of the historical space of language-game in which one is trained and educated. This is not a denial of historical culture nor historical consciousness but the possibility of the freeing of intelligence by its own movement of awareness that is otherwise confined to move only through the habit structures of thinking conditioned and structured operationally by being trained and educated with the historical rules of the language-game. Here lies the only possibility of understanding history and historical condition of man, with a certain distance of freedom from the reactive imagination and consciousness, the intentional habit structures of which are conditioned by being trained and educated operationally with rules and techniques of pictures of historical language-games which weave and condition human thinking with all the backlog of historical imagination and thinking habits. What I am trying to elucidate is thus meant to contribute to sharing and expressing the insight that started philosophy which seems to have fallen away from its originating awareness with the intellectual technological development of constructing pictures and techniques in the form of theories and hypothesis which served as a model for constructing epistemological and ontological pictures while the real question required a clarity of understanding as to how pictures of language pictured concepts with their represented identities and differences. The same confusion and failure of awareness of the manifest of life of phenomena still infects philosophy and understanding science in the form of philosophizing in accordance with methods of science, in pursuit of scientific pictures of cognition, popularized as cognitive sciences and so on, while the real question requires clarity of understanding how pictures of language picture/represent concepts with their identities and differences, which is the clarity as a key to tracing not only science with its internal connections with language and culture, but tracing the reactive behavior of intentional consciousness under the impact of the pictures of language cultural as well as scientific.

The elucidation of interface elucidates what is meant by the concept of "Lifeworld" (Lebenswelt) which phenomenology arrived to face as the manifest signifying background in which intentionality is structured to be expressed with the form of expressions of consciousness of objects. In other words consciousness of space and time unfolds and is sustained by manifest signifying phenomena as intentionality is structured by the forms of expressions the signifying consequences of which are shared to make up the rules of acting with signs as both means of representations as well as what is represented (pictured) as the reality of the surroundings, as our world picture with its horizons. Such that pictures of language are used to picture all our surroundings with a space of acting and operating with them exhausting all the space of thinking and acting with its own rules and techniques of operating with them, leaving no space for thinking and questioning as to its structuring. This shows that our thinking and imagining habits are so much structured and determined by the language of representations which does not allow for our intelligence any space or interval to take notice of the manifest signifying phenomena expressing the behavior of our operational habits of thinking and imagining with its own forms of expressions peculiar to it. The intentionality expressed by habits of thinking and imagining that are structured with the rules and techniques of pictures of language does not allow for our intelligence any space of movement except by operating with the techniques and rules of pictures, which then results by forming and constructing general pictures, in the form of theories, hypotheses and so on. And which then is closed by its own movement operationally determined by thinking habits with rules and picture constructions, to move so to speak in a spiralling way by describing pictures by means of pictures, by projecting picture constructions where the question requires understanding how pictures operate internally connected with the manifest of phenomena of life.

That closed horizon, is the horizon of "physical space" once read and held to be three dimensional due to the missing awareness of the internal connections between our operations with pictures and the world horizon the appearances of which are read and spaced by these operational activities. That is also the physical space once supposed and re-presented as filled by "matter" defined by its primary qualities as opposed to secondary qualities. The analytical habits of thinking went to analyze objects in terms of essential and accidental properties, or in terms of primary and secondary qualities, imagining a "perceiver" "analyzer" "subject" or "consciousness" in reaction to the particularized identities and differences which operate as pictures of language; hence going away from the sense of manifest of phenomena of life, expressed by the poetic metaphorical language of Heraclitus and Parmenides.

Where does our body end and the physical space of matter start? Or the border of body end and the awareness of self subjectivity start? How do we come to speak with these conceptual descriptions that picture our bodies, or our supposed subjective or objective perceptions? Do they exist independently of our coming to learn to speak with these concepts and descriptions? Or do we also come to experience our life horizon precisely due to our coming to learn to operate with such words and concepts?

Hence, the importance and significance of phenomenology as it represents a movement of thinking that is concerned with elucidating structure of intentionality, the behavior of imagination intertwining operationally with the signifying consequences of signs with regard to phenomena in manifest – that is a movement of thinking contrariwise to the subject object ontological and epistemological suppositions centered around a supposed subjectivity, intentionality as such, with a priori or a posteriori rules conceived either in Cartesian terms or in Kantian synthesis. *The phenomenological elucidation of manifest phenomena requires therefore always opening up the interface between concepts*, which allows us the awareness of the continuity of signifying stream of phenomena between the discontinuities and fragmentations of intentional consciousness due to reactions and their operational consequences in the form of habit and belief structures to pictures of language. Which are the beliefs one entertains as a subject who speaks and reports one's dreams, as opposed to one's wake life, having learnt to speak and report with concepts that are used to describe and report them. As our concepts, such as "dream" and "wake life", or "language" and "world", or "subject" and "object"; "mind" and "matter"; "*res cogitans*" and "*res extensa*", etc. operate as pictures the rules of which are based on our reactions and consequences in the language-game in which we are trained. As long as we fail to be struck by the awareness of the signifying stream of phenomena, our life experience are determined by the consequences of our reactions which become operational with rules and pictures of historical conventional language-games. The narratives and beliefs systems associated then dominate and shape human sensibility as an intentionality empirically and historically shared, as discontinuous and fragmented life experiences experienced, *seen as* in the manner of perceiving, meaning, showing, pointing reality. Such truth beliefs, as Wittgenstein does, are provoked to be expressed as expressed by gestures and gesticulations of meaning, showing, pointing at the reality of anything which may be something like a sensation, or something like Moore's hand, while on the other hand they are carried out to their logical conclusion by reminders assembled as language-games, by clarifying that they remain like an idle wheel, turning nothing with itself, signifying nothing, in oblivion and in isolation of the signifying internal connections that one learns to operate with the manifest signifying Use of other signs. That Use internally connected with the manifest signifying surroundings is presupposed as the ungrounded grounds of learning the conventional rules and pictures of all historical-cultural-conventional language-games. Intentionality remains fragmented and conditioned in oblivion of the manifest of signifying surroundings. That conditioned modality of thinking and imagining in reaction to the identities and differences represented by pictures of language doesn't in turn allow for one the freedom of space (of awareness) to trace back and forth the signifying process that structures intentional operational habits in internal connection with pictures of language reciprocally. Hence while pictures of language are projected by means of reactions and their signifying consequences which arrive to constitute one's world horizon, they operate on the other hand by filtering signifying manifest phenomena of life from the horizon of the intentionality structured to operate in reaction and with the use of the pictures of the language-game.

Such analyses in oblivion of the internal connections of signifying surroundings which operate as the possibility of *mean*ing, showing, pointing, intending, willing *any*thing with the use of signs, follow from intentional operational memory habits structured in reaction to pictures of language, more correctly to the particular images isolated from its signifying manifest surroundings, as they are associated and identified albeit mistakenly with the particularized images of a picture; which is, as a concept, represented by its signifying Use with the signifying Use of other signs in manifest. Such analyses and suppositions result from the deep forgetfulness of the analytical habit reactions that operates by describing and constructing external connections between pictures without the awareness of the Use of the picture as a sign in signifying internal connections with the Use of other signs. The analytical habit reaction manifests itself by gestures and gesticulations of pointing and meaning to the associated images of pictures imagined as objective, supposed to be public as opposed to the image imagined as private. In both cases, imagination reaction manifests itself as a reactive imagination of solipsism which results from analytical thinking habits forgetful of the signifying internal connections presupposed as the possibility of pointing, meaning, showing, picturing anything with its identity and difference in language, whether the images of pictures are supposed to be "private" or "public", as the latter and former are polar concepts presupposing and polarizing each other in the logical space of memory reactions that operates in reaction to the particular images and associations of images that resemble pictures, while missing the Use of the picture that is internally connected with the signifying use of other signs in manifest of life. Therefore, Wittgenstein always elucidates the Use of the picture with its internal connections with the use of other signs, by colliding his reminders with analytical habit reactions that tends to identify the picture as if the picture is representing its own identity; hence in oblivion and isolation of its Use which actually represents its identity and difference in internal connection with the signifying Use of other signs in manifest. Hence, he always reminds the Use of the picture as against and in contrast to habit reactions of imagining what the image of the picture resembles which trigger only the associated images of a picture in isolation from its manifest signifying surroundings.

When G.E. Moore demonstrated his hand by his gesture of showing his hand as part of the external world in order to point out that doubting is excluded as nonsensical in such cases, Wittgenstein proceeds to clarify that such a gesture of demonstration remains as a wrong gesture due to his failure of recognizing signifying phenomena in the weave of which we come to speak of our surroundings in terms of our hands and limbs and what they touch and use as "objects or experience" as "sensations of touching, seeing, or feeling" and so on. Here clarity about the grounds presupposed by our operating with signs in speaking and expressing our propositions is connected with clarity about our forms of expressions expressing certainty and uncertainty with its operational consequences needless of ascertaining by any logical demonstration or verification. On the contrary, the possibility of logical demonstration or verification presupposes the certainty expressed and shared operationally by learning to operate with rules of the language-game, as doubting makes sense only where certainty is operational.

Connected with such elucidation of signifying expressive phenomena of operating with signs and rules, such concepts as "private" and "public" are clarified from the backload of confusion that results from the analytical habit reaction expressed by the form of expressions in the form of a gesture of giving an ostensive definition directed to the images that are associated by their resemblances to the pictures of language in isolation from the use of the picture that is operational with the use of other signs internally connected with signifying phenomena in manifest. Such form of expressions express pictures of imagination which share the same confusion due to the missing awareness of how the use of pictures of language with their conceptual differences are expressed and learnt operationally with their differing consequences with the signifying use of other signs internally connected with the manifest signifying phenomena of the language-game. That arrives to clarifying that nothing is hidden absolutely, nor given as "private", or "subjective awareness", nor as "objective" – except by the operationally shared consequences and rules which unfold and get structured historically, in the context of a language-game, which is a context elucidated as a cross-section, as internally connected with manifest signifying phenomena, life in manifest as such.

Missing of that awareness about the picture is manifested by the form of expressions of imagination the reactions and habits of which are prompted by the varying pictures of historical language-games, which vary by the varying form of expressions of narratives which changes from mythological to scientific theories, depending on the changes of historical culture of the language-game. Thus, one is misled and separated apart by one's own operational language and thinking habits in reaction to and with pictures of language away from the manifest sense of life, Existence as such. And hence, away from the sense of "*ontopoiesis*", from the sense of the "unity-of-everything-there-is-alive..." as expressed by the form of expressions of Anna-Teresa Tymienicka.

This separation is deepened and hardened by the development of instrumental, operational pictures of language to the point of excluding the reality of manifest dream experience from wakeful experience of reality which appears to be subject to our willing and controlling with our operational habits with the use of pictures of language. The Other, which is not experienced as subject to habitual control and operational use, belongs to the manifest of life, and remains the Other of our life experience, as our life experience seems to be experienced as subject to empirical operational habits, the sense of the Other seems to be threatening, intriguing, unwelcome, irksome, mysterious. It is therefore covered, repressed and transformed by the pictures of narratives that order and explain them in accordance with the pictures that describe and order our operational activities of wakeful life experience.

It is only by considering the form of expressions shared we compare the concept of a "dream experience" with a concept of "real experience", as the form of expressions shared expresses our sense of experiencing reality, in comparison to a form of expression that differs from it. That means to say, our sense of experience of the reality of life is learned to be experienced and expressed as a shared intentionality so to speak in phenomenological terms. The terms "subjectivity" and "inter-subjectivity" are in need of phenomenological elucidation here, considering that there is no subjectivity prior to expressive signifying phenomena, as the terms "subjective" and "objective" are concepts, the different and opposed senses of which are internally connected with manifest signifying phenomena.

Husserl's Phenomenology started by taking consciousness always as a consciousness of something, as an intentional structure, and went to dig up the historical layers of it to come across the signifying *phenomena* with signifying relations, intersecting with other significations in virtuality. The term "Lebenswelt" (Lifeworld) refers to this virtuality rather than the world represented, objectified as a pole of a historically structured intentionality, as a pole of empirical subjectivity. Heidegger's phenomenological hermeneutics thematize this signifying field as *Dasein* and tries to elucidate it as the possibility of tracing of all the historical changes of intentionality layered by the changing forms of expressions of it, i.e., in the form of interpretations of the surrounding world-pictures objectified. Heidegger therefore needed to clarify the modes in which things exist or do not exist for us as ready to hand, or present at hand, or not present at hand in terms of the different consequences following from their being present or being absent, as i.e., the pencil's existence is presented by its being ready to my using it when the need arises for it, or conversely its absence is presented by the consequence manifested as a hindrance to my need for writing with it. He thus pointed out the manner in which the existents exist and appear with different identities and differences as part of the different operational consequences of surrounding horizons in connection with body's actions and reactions in internal connection with the signifying surrounding phenomena, which the latter assumes the appearance of a world horizon objectified as a result of an intentionality structured operationally with the representations that picture conceptual identities and differences of language and culture. Hence, Heidegger thematizes the signifying manifest field of Lifeworld (Lebenswelt) in which the intentional consciousness is characterized by phenomena expressed in signifying internal connections in manifest which unfolds and structured operationally with operating with signs. Merleau-ponty contributed to the elucidation of signifying field of virtual phenomena by tracing back the sensation into the signifying phenomena internally connected as expressive phenomena with other significations in manifest, from the internal connections of which nothing is thinkable in isolation; that is to say, no-body, no-behavior, nor the intentionality as subjectivity of bodily behavior, nor any thing can be shown, or supposed to be perceived as subject to an ostensive definition. That seemed to be possible to the analytical habits of taking objects and events in the manner in which it seemed to a naive realist, or to a logical positivist, or to a logical empiricist as expressed and betrayed by the forms of expressions expressed by gestures and gesticulations of meaning and analysing a sense data, which is then exposed by Wittgenstein's reminders that they are not essential for representation in language, like an wheel, turning nothing with itself, signifying nothing, except the fact of one's confusions, which result from a failure of understanding about how signs operate, mean, name, picture (represent) identities and differences in the actual stream of using and operating with signs of any language-game.

Anna-Teresa Tymieniecka expresses the same insight when she points out that:

To grasp life's patterning *all should be presented at once in one cross section of an image.*[9]

That requires starting always from the many aspects in simultaneous with their manifest spontaneity in bringing out the sense of ontopoiesis that expresses its internal logos as opposed to interpreting logos as external to the manifest of life. Her words: *"Unity-of-everything-there-is-alive..."* also expresses its original sense in connection with the same awareness of the virtual phenomena of life in manifest, in the signifying web of which intentionality of particularizations are expressed and structured in the form of operational habits of using signs as pictures, as means and ends of saying, showing, pointing, *mean*ing *any*thing.

Phenomenological movement of thinking seems therefore promising as long as it keeps up the good work of elucidating the intentional structure and behavior of historical consciousness with its internal connections with the eternal manifest moments of life. It is promising in bridging the gap with the original awareness that started philosophizing, with the original and different sense from the sense which assumed historically: which took on the particular shape of the intentionality of the thinking and philosophizing habits by constructing general pictures in the form of theories and arguments – the form of expressions of which are operationally structured to fluctuate in reaction to pictures and their associated images, always modifying and generalizing pictures, in the weave of which intentionality is layered and conditioned as historical consciousness and imagination, as historical modifications of intentionality, without however freedom of movement of thinking in distance to the historically modified intentional consciousness.

Such is the Anglo-American analytical way in which the language of philosophical analyses and forms of arguments shaped the way for philosophizing with its norms and journals and peer reviews, so far away from the original roots that started a movement of thinking peculiar to the insight which inspires to articulate a language to share its logos which is internal to it. Philosophy in its original sense of astonishment, which is prompted by the flash of an insight about manifest of life or nature, without the mediation of representational language, having found itself speechless given the established rules and representational tools of language and thinking, had to articulate a language to express its own sense of life in manifest. Such sense of life is expressed by the language of *apeiron* of Anaximander, by the *Flux* of Heraclitus and the unmoved full plenitude (that leaves no space for a movement) of Parmenides. That movement gave way on the other hand to another cultural development that created a language of picture constructions by analyses and syntheses; hence picturing Nature in accordance with the rules of constructing and comparing pictures of language; which ended up by misrepresenting and misunderstanding the identities and differences pictured. The misunderstandings and deep confusions of which still infects philosophy education and its industry, leaves us now with facing the problem of unifying the so deep fragmentation of human consciousness and world horizon, which has always been the deep concern of authentic philosophical insight. The inherent crisis of that fragmentation manifests with its own consequences in human life and culture, with its own fate ("karma") so to speak, as was once noted by Heraclitus: *"One's character is one's fate"*. Logos in Heraclitus sense is a "call" of awareness that is addressed to awaken our deepest intelligence, and not to thinking habits structured to operate with rules and pictures of conventional, cultural, historical language-games.

Istanbul Kultur University, Istanbul, Turkey
e-mail: erkutsezgin@hotmail.com

NOTES

[1] Heraclitus of Ephesos, John Burnet *Early Greek Philosophy* (Adam & C. Black, 1963) p. 133.
[2] Wittgenstein, *Philosophical Remarks* (Blackwell, 1975), p. 69.
[3] Wittgenstein, *Zettel*, 541.
[4] "When I say that I mean by substance that which is conceived through and in itself; and that I mean by modification or accident that which is something else, and is conceived through that wherein it is, evidently it follows that substance is by nature prior to its accidents. For without the former the latter can neither be nor be conceived. Secondly it follows that besides substances and accidents nothing exists really or externally to the intellect." Spinoza, *Correspondences*, IV.
[5] Professor Anna-Teresa Tymieniecka's philosophy of *ontopoiesis* seems to me to be such an effort to explore these significations with an eye touching and tracing their internal connections with the manifest of life of phenomena. That is an interpretative activity motivated and resonated by the same live movement in manifest.
[6] *Philosophical Remarks*, Foreword; Ed. By Rush Rhees and translated into English by Raymond Hargreaves and Roger White (Blackwell, 1975).
[7] "We fail to be struck by what, once seen, is most striking and powerful." Wittgenstein , *Phil. Inv.*, 129.
[8] The concept of "Lifeworld" ("Lebenswelt", "vivencia": Ortega y Gasset) of Phenomenology is elucidated by Wittgenstein's reminders assembled as "language-games" which present "cross strips" cut out so to speak from the signifying manifest stream of phenomena as the background internally connected with our showing, pointing, naming, giving an ostensive definition of any thing or event in physical and temporal space of acting and operating with our memory habits shared and kept by our learning and acting with the rules of the game. The internal connections of manifest signifying phenomena are elucidated as the possibililily of defining and describing external connections between signs as "word" and "object" the actual identities and differences of which are represented by their differing uses and significations objectified as objects or events ordered in physical and temporal space. The missing awareness

of the whole manifest life experience manifests by being conditioned to operate with memory habits in reaction to pictures of language, hence confusing pictures used as means of representation with pictures represented as reality, due to the missing awareness how language phenomena, forms of our expressions manifest as internally connected with the signifying phenomena in manifest.

[9] A.-T. Tymieniecka, *Logos and Life* (Kluwer, Book 4) p. 5.

CARMEN COZMA

"SOPHIA" AS "TELOS" IN THE "ONTOPOIETIC PERSPECTIVE"

ABSTRACT

Centered upon the "logos of life" and the "creative human condition", and being developed in terms of an integrator Apollonian vision about the universal harmony within the "Great Plan of Life", the original phenomenological work of Anna-Teresa Tymieniecka reveals itself as a cogent urge to retrieving the signification of a major value: sophia – in the terminology established by the ancient Greeks. Taking it in a broader semantic openness, as wisdom in general – not merely a theoretical, but also a practical one, including moderation, too – this concept appears us even like telos – in the Stoics' distinction from skopos – for human being in its self-individualization and self-accomplishment-in-existence. To approach sophia, that engages inquiry, insight, knowledge, reflection, comprehension, a complex practice of man-in-quest-of-wisdom, an elevated attitude toward the entire experience of life; seeing that sophia covers a telic oriented creative tension, an aspiration toward the attainment of an ideal situation. It represents a challenge for human becoming inscribed in a worthy movement as self-fulfillment in an aretaic horizon, which is one of sense-bestowing for the human being's participation to the logoic flux and order in the "ontopoiesis of life".

Given the situation of our time with the turmoil of science-technological advances, but no less with a general climate of disarray in which humanity "is apparently plunging into further chaos as disorientation about everything",[1] phenomenology of life with the pursuit of the *logos* – the "sense of sense" that "penetrates All" – comes to heralding a "New cultural Enlightenment".

Searching after reason, putting in act a "new critique of reason", Anna-Teresa Tymieniecka develops a philosophical/proto-phenomenological vision claimed by that she considers to be a "brewing flux of renewal, growth, and the perfecting of humanity",[2] asserting the need of turning to the wisdom. "The state of our culture prompts us to search after reason"; "it calls for philosophy to free us from our impasse and to lead on", by regaining the potential of wisdom, finally, in "our maneuvering upon the chaotic flux of life".[3]

Throughout an original work unfolded in four tomes of her fleuve-treatise *Logos and* Life,[4] continued in the recent Book 1 of *The Fullness of the Logos in the Key of* Life,[5] we face an impressive demonstration of *philosophizing* upon *life* and *human condition*, that encourages us to reconsider even the mission of this act in its roots of the Hellenic tradition concerning the value of *sophia* (σοφια)/*wisdom*.

Actually, we face a celebration of philosophizing practice in its principle vocation conducted by the human "quest for wisdom" – that seems to be the Tymienieckan reflective way at anchor in the creative human experience as part of the "Ontopoiesis of Life". This is the central nerve of an integrator and dynamic vision from the vital to the sacral levels of life, following a constructive design under a *telos* oriented schema focused on the "self-individualization of life" process. As the author states: *onto-* refers "to the 'firstness' of this process with respect to the scale of the existential formation", meaning also "the indispensable and universal character of whatever there could be in the 'objective' form proper to human reality"; and *poiesis* refers to the continuous transformation, "advance" and "qualification", respectively the mark of creativity as the intrinsic factor of "becoming".[6]

In its entirety, phenomenology of life emphasizes an Apollonian fundamental choice for philosophizing, on the orbit of trustfulness given by light, construction, order, harmony. It displays an affirmative attitude and a balanced healthy comprehension upon the whole existence, concomitantly in each dimension of "the inorganic, the organic, bios or zoe, gregarious life, social and cultural life", and in "the unity-of-All-is-alive".[7] We find an offer to re-discovering *sophia* with the opportunity to be guided in the effort of overcoming "our present decadence", in surpassing the lack of orientation of "an Alexandrian Age" – with the picture of "futility and absurd as normal", of "spiritual paralysis" and "moral confusion", of the lamentable rise of the "demotic" (a "tyranny of the common man").[8]

Acknowledged by mythology as attribute of goddess Athena – the personification of wisdom and the patroness of creative humans – *sophia* is one of the values that stand the test of time and that in nowadays – maybe, more than ever – deserves to be retrieved in its plenitude.

Sophia almost became an imperative to be incorporated in our life, to be explored in its function of orienting our discernment as regards the right things and actions, for the benefit of ourselves and others, for the common good of the societal, cultural and natural environment. *Sophia* must be cultivated like an important faculty of personality, organizing a reasonable human life on the ground of the best use of available knowledge; shaping and adjusting our behavior in accordance with what is really true, significant, lasting; helping us to choose well, to decide and to act by a responsible commitment for a positive long-term future of a meaningful life in its totality.

We approach *sophia* as *telos* within the context of phenomenology of life, being rather interested by the implied process, as active component in philo-sophizing. So, it is more fitting to the "ontopoietic" vision that is crystallizing not on beingness as a state, a finished, established one, but on beingness in progress, as a process, as beingness-in-becoming.

The term of *telos* with which we are dealing here is that used in Stoicism, tacking into account its overall meaning: the ultimate object of desire, both as an incorporeal and a corporeal, predicate and thing, stressing the first situation by priority. Without entering the complex discussion on verbal meaning about things and predicates, we just remind that Zeno, Cleanthes, Chrysippus have distinguished between the *telos* ($\tau\epsilon\lambda o\varsigma$), an operational finality toward which man is striving for, and the *skopos* ($\sigma\kappa o\pi o\varsigma$), a target or aim in actual fact.

In correspondence with the Stoic theory of the *telos: to eudaimonein/to be happy* – including: the happiness, the happy life, and being happy,[9] we try a phenomenological dis-closure of *sophia* as creative wisdom, combining actual and possible human experience unfolded on an ascending trajectory to an end. We take the concept in the sense of the wisdom exercise, as a horizon into which man directs his conduct, with the movement between transient and eternal.

Sophia as *telos* rather supposes a tension to pursuing an ideal: to be able to achieving wisdom, to be devoted to philo-sophizing, to become wisely. There is an interplay of instrumental, motivational, teleological dimensions of *sophia* to be putted in act for a full understanding of the dynamic web of life in its "ontopoietical course", articulating the unity of "the rationalities of the cosmos, life, nature, and those of human creative genius" in "an all-embracing vision . . . of the entire spread of the Logos in its manifestation".[10]

A creative wisdom, a creative way towards wisdom is at stake, in the tonality of the matrix of phenomenology of life with the question of *creativity* as the "Archimedean point" for the whole life's unfolding progress.

Sophia as *telos* manifests itself like a fundamental process of spiritual awareness supporting the creative act of human being – that in which "man is the doer and is dealing with the inner workings of nature within himself and as they relate him, to all other human beings, and all living beings", what Tymieniecka calls "the unity-of-everything-that-is-alive". It is a peculiar lore, a sage learning that defines a creative manner of philosophizing, one that goes "to the roots of human thinking and acting" in respect with man's descending "to this deepest plane on which everything is being played".[11]

Returning to a capital problem of the antique philosophy, we get *sophia* in a large sense, covering the *integrator wisdom* as a priority we should strive for in the present situation of an "anatomy of bewilderment – of the disarray humanity now finds itself in".[12]

Although Anna-Teresa Tymieniecka doesn't explicitly resort to the notion of *sophia*, the contemplative and practical potential of it emerges throughout her phenomenological discourse, sustaining a clear-sightedness and a thorough understanding for the scrutiny of the "logos of life" – "the motor and carrier of the entire ontopoietic enterprise", considered to be "not only force and shaping but also the ordering principle of life".[13]

By relating the assertion about the "quest of wisdom" with the pivotal thesis that: "to be human means to be creative", we can follow a veritable pathway of philosophizing upon the meaning of life. It re-sets and renews philosophy itself "in a major key in an age of minor variations" under the auspices of acknowledging "a deep-seated need in human nature to respond to, and preserve the mystery of life".[14]

A praise of wisdom is revealing from the phenomenology of life. Thus we turn back to *sophia* – the so highly esteemed concept by Greeks. Its exercise can be distinguished in the framework of philosophizing as that used to be shared during the early times in European culture. So, to tackle the issue of *sophia* entitles us to turn back even to the Delphic precepts, "because this was the manner of *philo-sophia* among the ancients, a kind of laconic brevity", as Socrates says in a Platonic dialogue, speaking about the "Seven Sages of Greece" ("Seven Wise Men") and their wisdom.[15]

Precisely, we refer to the in so far famous formulas: "$Γνωθι\ σεαυτον$"/"*Gnothi seauton*"/"Know thyself" and "$Μηδεν\ αγαν$"/"*Meden agan*"/"Nothing too much" ("Nothing in excess").

The meaning of both these sentences is activated in the territory of phenomenology of life.

Self-knowledge enters by necessity in the process of "self-individualization in life" – a nodal concept of Tymieniecka's effort to clarify the workings of "the logos in life" and "the life of the logos" in the "ontopoietic design of life". The "self-individualization of life" is thought like the center from whence the beingness rays; it is that "holds the vital strings of beingness"; it represents the "vehicle" of the *logos* in its constructive advance in life under the convergent action of the principles: "*impetus and equipoise*".[16]

Self-knowledge carries the development of self-consciousness and orients the praxis of man by singling out the creative human condition within the web of life. It is to be supposed that such of self-knowledge operates on the unveiling of human individualization like a process that consists "in an *in itself* but not a *for itself*"; it is a process of serving the entire system of life, with the interplay of "the singular and the whole", "within the mesh of interlocking existential ties and life-communion".[17] In completion with self-explication and self-understanding, it grounds the "self-interpretation" that – engaging the play of self and other/the self like the other in the unity sameness-distinctiveness deciphering – marks the human creativity in the functional polyvalent network of existence. According to Tymieniecka, *creativity* appears to be the "uniquely, specifically human self-explication in existence"; it means "interpretation par excellence".[18]

Such determinations of the "creative self" interest us especially from the point of view of the moral experience, of the human fulfillment for which the *wisdom* is a virtue of prime order.

"The quest for wisdom" claimed by the phenomenologist of life is in resonance with the Greco-Roman philosophy about the precedence of *sophia/sapientia* among the other virtues, orchestrating a real art of living. "Wisdom is the chief and leader; next follow temperance, ... courage and justice" – teaches us Plato, valuing the cardinal "virtues" in the "class of divine goods".[19] And Cicero, finding in *sapientia* an *ars vivendi*, he has seen wisdom as "the first of all virtues" ("*princeps omnium virtutum illa sapientia*"), explaining it like "the ability to perceive what in any given instance is true and real, what its relations are, its consequences, and its causes".[20]

In the same tonality of highlighting the creative moral experience of human being, we reconsider the second wording we have already mentioned as summary of the Ancients' philosophizing: "Nothing too much". The referential, here, is the *measure*.

The issue is re-assessed by Anna-Teresa Tymieniecka as an ethical demand of our time. She unfolds an insight of the *golden measure*[21] as key valuation stick for the "telic schema of life's constructivism" in its "cycle: *generation, fruition, accomplishment*" with its "innermost *sense of continuation, renewal*".[22]

In retort to the present situation – seen as being "much more complex and far-reaching than any crisis" –, the phenomenologist of life offers a viable way by launching "the universal call for measure"; in the value and the principle of *measure* she finds a strong factor to be used "in striving for the common good of life".[23]

Actually, Tymieniecka's conception circumscribes a rich and supple *science of right measure in all*; it enlightens upon the valences of the "art of measure": *to metrion* (το μετριον) coming from the Hellenic philosophy, expressing "that is fitting", "that is timely", "all is necessary", "everything there is living in the middle of the distance between extremes".[24]

The question of measure prompts us to rethink about wisdom as *virtue*, applying the Aristotelian definition: the "laudable mean state between excess and deficiency". We undertake the concept in its creative function of guiding and educating man towards his moral excellence, placing him on the royal area of the *metron ariston* (μετρον αριστον) of the Greeks' teleology. It is *the virtue* conceived like measure between two vices: a "too much" and a "too little", the "middle way" that "if regarded from the point of view of the highest good, or of excellence, it is a climax".[25]

Beyond the theory of Aristotle about virtue as being twofold: "partly intellectual and partly moral",[26] we aim at *virtue/arete* in general, designated here as a renewed generic *sophia*. It demands teaching and habituating, experience, time, and ability; because our concern is upon its exercise. Thus, we are mostly focused on both the theoretical and practical dimensions of wisdom and also on the value of moderation, as potentiality to making "a good condition of man" and to enabling him "to perform his proper function well";[27] respectively, to become wisely, and so to succeed in seeking and understanding "the ontopoietic intentionality – a sentient and intellective one – of life".[28]

In our sphere of interest, to exercise *sophia* implies to exercise an intellectual and moral virtue that accounts at a superior degree for the ennobling human condition and life. In an *aretaic* approach, *sophia* functions like a commendable and useful quality of character, facilitating to man a specific freedom to surpass himself and to continuously work for his moral personality's fulfillment.

Especially, facing the present "anatomy of moral disarray", we need to prize and to restore such a cardinal virtue as the wisdom has been acknowledged in the Antiquity philosophy. According to Anna-Teresa Tymieniecka, "we are challenged to enter into our depths in order to achieve a new understanding of our place in the cosmos and the web of life, to find new wisdom for charting our paths together and fresh inspiration to animate our personal conduct."[29]

We think that, for our time, a more adequate appropriation and practise of wisdom entail a mixture of the ancient Greeks' terms: *sophia* (σοφια), *phronesis* (φρονησις) and *sophrosyne* (σωφροσυνη), covering the entire action of *arete* (αρετη) as *mesotes* (μεσοτης); respectively, the virtue-"golden measure" that operates in the "ontopoietic, specifically human self-individualization", and concomitantly "in the coordination and harmonization of the whole of life".[30] Experimenting with and assuming contradictions, but aiming towards equilibrium which sustains creation, development, and preservation, through an activated wisdom as *arete-mesotes*, man can coordinate the balance of his confrontation with the given world and his creative capacity to transform it. Regaining a peculiar *sapientiality/sageness*, following his own creative *telos*, finally man completes his status of *homo sapiens* with that responsible position of "custodian of everything- and all-alive-unity", by using his inventive faculties to searching and disclosing the *logos*' rhythm and harmony in maintaining and increasing the Good and the Beauty of existence.

It is the creative human mode of becoming to put in act the moral excellence in the endeavor to ennobling the entire life; that means a victory of the *aretaic* vision that engages good reason, sensitivity and motivation, creative imagination, promotion of values of order, measure, refinement, construction, harmony; all, in an Apollonian *poietical* perspective of "man, the creator" acting under "that principle existing in individuals that Homer had named *divine image and likeness*".[31]

To become wisely, to treat *sophia* as *telos* – that appears like a significant part of to the utmost experiencing of a human moral life as a creative one. Considered as plenary functional virtue, wisdom makes sense the intertwining point of man's autonomy and his natural and social relationships. In Tymieneckan language, it facilitates the grasping of the conjunction of human commitment to "self-individualizing" and to "sharing in life" in its totality, revealed in consonance with the overall flow of the "logos of life". Establishing the centrality of the investigation of virtue, the phenomenologist of life features: "we situate the question of morality and virtue at the primogenital human plane where reason with its faculties, on the one hand, and the vital forces, on the other, emerge as partners in the creative orchestration of human functioning that forms [eventually] ... the human expansion of the schema of Nature. In short, the question of virtue lies at the heart of the life strategies of the Logos".[32]

To Anna-Teresa Tymieniecka's assertion that philosophy, respectively wisdom is "made possible owing to the creative virtualities of man",[33] we add that, in its turn, wisdom represents a creative factor within the expanse of human life; it can be explored like a function of "man, the creator" in order to rise from the initial spontaneity, going to enact and to fulfill the inventive virtualities in a specifically human individualization, guided by a "creative telos", considering the total human experience between the rootedness "in Elementary Nature" and the tendency "toward Transcendence".[34]

Phenomenology of life opens new opportunities to the *experience of wisdom*, that is bolstered by the reflective, "Apollonian intellect", with the "creative imagination" and cultural memory. At the same time, such experience makes possible a balance between these, on the one hand, and, on the other, the sentient, emotive, communicative "Dionysian" and the inventive, freeing in spiritual transcendence, "Promethean" modes of rationality, in a divisive as well as a complementary harmony of life.

The exercise of wisdom is a condition required by a creative philosophizing upon the "great plan of life" making possible the unraveling and catching of the innermost sense of life. An indispensable component of philosophizing – in the circularity: outcomes and means –, wisdom enables man to register his becoming on the route of progress, connected to the *logoic* – natural, creative and sacred – process of life.

Wisdom can fruit in the mobilization of all the human energies "for the discovery of means, ways, and materials, and for channeling them into a constructive apparatus capable of concretizing a new vision".[35] It is an open-ended, integrating and dynamic vision with the dialectics of opposites in their distinctive identities, but also in their full unity. In Tymieniecka's terms, it is a vision of comprehending life simultaneously as "timing and spacing", "flux and stability", in its "fleetingness and essence", with disruptions and continuity, by "inward-outward directions", on "hidden and obvious arteries", unveiling – beyond any contradictions – the universal harmony of the All. Such a vision is completely in accordance with the incipient uses of philosophizing, on the Pythagorean channel about "philo-sopher"/"*sophos*"/"wise man".[36]

To practise wisdom, that fortifies man – the agent of the heroism of the life struggle – to endure the most difficult trials of his worldly existence, and to find and to increase the joy of living. According to Joseph Bochenski, we observe that, like Janus the dual headed god, "wisdom has two faces; it teaches us on the one hand that all is futile, and on the other that we must enjoy life". The message to be learned is that, essentially, "there is no contradiction between the doctrine of futility and the precept of pleasure".[37]

By practising wisdom, we reach to comprehend the play of contradictions and equilibrium alike, in the individual, societal and natural life. Thus, we attain to revive the awareness of a basic truth from the Greek philosophy, that *sophia* is correlating to the principle of *harmonia* that brings order to chaos.[38]

Intimated tied to an *aretaic* culture, the experience of wisdom leads human being close by a supreme order, in a cosmic perspective, toward which man aspires from ever; metaphorically conceived from Pythagoras and Plato until Tymieniecka, it is the order of an ideal spectacle of the "music" and "dance spheres".[39]

In its intricate movement of deciphering the "logos of life" and the "creative human condition", phenomenology of Anna-Teresa Tymieniecka shows itself as a praise-demonstration of the exercise of *sophia-telos tou biou* (wisdom-purpose of life). Like process of a continuous spiritual freshness and ordering of human action, wisdom-at-work helps man to inscribe himself on an ascending route in the becoming flux of life; consequently, it helps man to understand life "in its surface phenomenal manifestation, in a formal, structural, constitutive fashion", as well as "into the depths of the energies, forces, dynamisms that carry it relentlessly onward";[40] no less, to understand his own vocation of philosophizing as style of a healthy and happy living.

"Al.I.Cuza" University of Jassy, Romania
e-mail: carmen.cozma@uaic.ro

NOTES

[1] Anna-Teresa Tymieniecka, *The Fullness of the Logos in the Key of Life*, Book I: *The Case of God in the New Enlightenment*, Springer, Dordrecht, 2009, p. xxiii.
[2] Ibidem.
[3] Ibidem, p. xxv.
[4] See Anna-Teresa Tymieniecka, *Logos and Life*, Book 1: *Creative Experience and the Critique of Reason*, Kluwer, Dordrecht, 1988; Book 2: *The Three Movements of the Soul*, Kluwer, Dordrecht, 1988; Book 3: *The Passions of the Soul and the Elements in the Ontopoiesis of Culture. The Life Significance of Literature*, Kluwer, Dordrecht, 1990; Book 4: *Impetus and Equipoise in the Life-Strategies of Reason*, Kluwer, Dordrecht/Boston/London, 2000.
[5] See Anna-Teresa Tymieniecka, *The Fullness of the Logos in the Key of Life*, Book I: *The Case of God in the New Enlightenment*, op. cit.
[6] Anna-Teresa Tymieniecka, "Measure and the Ontopoietic Self-Individualization of Life", in *Phenomenological Inquiry*, Volume 19, Belmont, Massachusetts, The World Institute for Advanced Phenomenological Research and Learning, 1995, p. 40.
[7] Anna-Teresa Tymieniecka, in Ivanka Rainova, "Interview with Anna-Teresa Tymieniecka", Moscow, August 1993, http://www.phenomenology.org/interview.html.
[8] Cf. Jacques Barzun, *From Dawn to Decadence. 500 Years of Western Cultural Life, 1500 to the Present*, Harper Perennial, 2001.
[9] Cf. Stobaeus, *Eclogae*, 2.57, in Arthur J. Pomeroy (ed.), *Arius Didymus. Epitome of Stoic Ethics. Texts and Translations*, Society of Biblical Literature, Atlanta, GA, 1999.
[10] Anna-Teresa Tymieniecka, *Logos and Life*, Book 4: *Impetus and Equipoise in the Life-Strategies of Reason*, op. cit., pp. 189, 187.
[11] Anna-Teresa Tymieniecka , in Ivanka Rainova, op. cit.
[11] Anna-Teresa Tymieniecka, "Measure and the Ontopoietic Self-Individualization of Life", op. cit., pp. 26–27.
[13] Anna-Teresa Tymieniecka, *The Fullness of the Logos in the Key of Life*, Book I: *The Case of God in the New Enlightenment*, op. cit., pp. 29; 113.
[14] Lawrence Kimmel, "Logos: Anna-Teresa Tymieniecka's Celebration of Life in Search of Wisdom", in Gary Backhaus (ed.), *Thinking through Anna-Teresa Tymieniecka's «Logos and Life»*, *Phenomenological Inquiry*, Volume 27, Hanover, New Hampshire, The World Institute for Advanced Phenomenological Research and Learning, 2003, pp. 20; 21.
[15] Plato, *Protagoras*, 342e–343b.
[16] Anna-Teresa Tymieniecka, *Logos and Life*, Book 4: *Impetus and Equipoise in the Life-Strategies of Reason*, op. cit., pp. 5, 6.
[17] Anna-Teresa Tymieniecka, *The Fullness of the Logos in the Key of Life*, Book I: *The Case of God in the New Enlightenment*, op. cit., p. 89.
[18] Anna-Teresa Tymieniecka, "The Creative Self and the Other in Man's Self-Interpretation", in Anna-Teresa Tymieniecka (ed.), *Analecta Husserliana*, Volume VI, D. Reidel, Dordrecht, 1977, pp. 189; 168.
[19] Plato, *The Laws*, I, 631.
[20] Marcus Tullius Cicero, *De Finibus Bonorum et Malorum*, I.42 ; *De Officiis*, I ; II.5.
[21] See Anna-Teresa Tymieniecka, *Logos and Life*, Book 4: *Impetus and Equipoise in the Life-Strategies of Reason*, op. cit., pp. 613–639.
[22] Anna-Teresa Tymieniecka, *The Fullness of the Logos in the Key of Life*, Book I: *The Case of God in the New Enlightenment*, op. cit., pp. 109, 110.
[23] Anna-Teresa Tymieniecka, "Measure and the Ontopoietic Self-Individualization of Life", op. cit., p. 28.
[24] Plato, *Political Man*, 286d, 284e.
[25] Aristotle, *Nicomachean Ethics*, 1106b 20; 1107a 5.
[26] Ibidem, 1103a 5.
[27] Ibidem, 1106a 15–25.
[28] Cf. Anna-Teresa Tymieniecka, *The Fullness of the Logos in the Key of Life*, Book I: *The Case of God in the New Enlightenment*, op. cit., pp. 140–141.

[29] Anna-Teresa Tymieniecka, "Measure and the Ontopoietic Self-Individualization of Life", op. cit., p. 26.
[30] Ibidem, p. 36.
[31] Plato, *The Republic*, 501b.
[32] Anna-Teresa Tymieniecka, *Logos and Life*, Book 4: *Impetus and Equipoise in the Life-Strategies of Reason*, op. cit., p. 598.
[33] Ibidem, p. 313.
[34] Ibidem, p. 485.
[35] Ibidem, p. 469.
[36] Diogenes Laertius, *Lives and Doctrines of Eminent Philosophers*, I, 12.
[37] Cf. Joseph Bochenski, *Manuel de sagesse du monde ordinaire*, Editions de L'Aire, 2002.
[38] See Diogenes Laertius, op. cit., VIII, 33.
[39] See Anna-Teresa Tymieniecka, *Logos and Life*, Book 4: *Impetus and Equipoise in the Life-Strategies of Reason*, op. cit., pp. 651–657.
[40] Anna-Teresa Tymieniecka, *The Fullness of the Logos in the Key of Life*, Book I: *The Case of God in the New Enlightenment*, op. cit., p. 35.

OLGA LOUCHAKOVA-SCHWARTZ

THE SELF AND THE WORLD: VEDANTA, SUFISM, AND THE PRESOCRATICS IN A PHENOMENOLOGICAL VIEW

ABSTRACT

This article examines two opposing perspectives in the formulation of phenomenological analysis which take as a starting point either the self, or the world. The phenomenologically grounded (based on direct intuition) focus on self-knowledge in ancient philosophy and esotericism emerges out of its apparent epistemological counter, the world. A similar dialectical synthesis can be traced in Tymieniecka's Philosophy of Life, with its emphasis on the primacy of the world in analysis. To that end, the article examines phenomenological reduction of the self in Western egology, a more holistic approach of the practical philosophy of Advaita Vedanta, the function of the Heart and the mutual mirroring of world and self in Sufism, and the shaping of the self by the world in Greek doxographical traditions. It suggests that the human condition has the possibility of an awareness which encompasses the world, as in the Sufi notion of the Heart, and that the positioning of the self in the entire context of life brings one closer to "things as they are" in Tymieniecka's philosophy of life.

At the 55th International Phenomenology Congress in Nijmegen, 2007, I witnessed a historical dispute between Anna-Teresa Tymieniecka and Angela Ales Bello, which summed up the two opposing perspectives in the formulation of phenomenology: Where does one begin the analysis—in the self, or in the world? In this paper, I will compare self-world relations in several systems of knowledge, such as Western phenomenology, Indian Vedānta, Sufism, doxographic Greek philosophical tradition, and, finally, in Tymieniecka's Phenomenology of Life. The knowledge in these philosophical or mystical philosophical systems is largely obtained through the direct intuition of the inner contents of consciousness, which is also characteristic of phenomenology. In Vedanta, presocratic Greek philosophy, and Sufism, all of which begin with distinction between the self and the world in the natural attitude, the experiential dichotomy between the self and the world resolves into the recognition of their foundational ontological unity. However, the paths uncovering this principal unity are compassed in a very different manner, and it is this difference that I will examine.

PHENOMENOLOGICAL EGOLOGY

In egological investigations, the self and the world often appear as two dialectically connected opposites. When the systematic egological investigations reach the bottom-line within human subjectivity, the residue of pure awareness comes out of anonymity. In reflective analysis,[1] the pure subjectivity of awareness becomes fully differentiated from its objects (not so in the internal, experiential motion of reduction, but this consideration is usually dismissed). The discovery of a perceptual potentiality beyond space or time, and of an apparently limitless principle within human consciousness, is so overwhelmingly intense that the question of the existential status of pure subjectivity is pushed out of consideration. Whether this limitless and pure principle is an actually lived self-awareness, or is merely an abstraction—that is, a mental possibility conditioned on a volitional act of discrimination between a subject and an object—remains unexamined. The prototypically Cartesian subject-object differentiation is completed, and the residue of reduction of the self becomes a separately standing "thing" suitable for analysis.[2]

Pure subjectivity becomes a transcendental ego which is released into the world, either as an infinite ontological substratum of the latter (in Cartesianism), or as a constituting principle of intersubjectivity (as in Husserl's transcendental phenomenology). The manifold world of names and forms *"is"* by the *"amness"* of this pure subject-awareness and, in turn, participates in the constitution of the self which is subjected to reduction. The dialectical cycle has been completed, but the synthesis has not become fertile: it has nowhere to go, and it never transcends itself into a new emergence.

As a mental mode, reduction is different from a natural flow of experience.[3] Besides uncovering, or releasing, pure subjectivity, the operation of reduction also constitutes it.[4] Although pure awareness is a condition that is at best extremely difficult to achieve pre-reflectively, it remains experientially easily available as an idea. It is unclear not only whether such a condition as pure subjectivity of awareness can be lived,[5] but also whether without reflection which bridges the perceptual gap between the self and the world the dialectical synthesis outlined above would even be possible.

In practical philosophies which are concerned with human fulfillment,[6] the question of where does self-knowledge begin, in the self or in the world, is even more tangible: as opposed to thinking what we are, we become what we think.

VEDĀNTA: SELF IS THE ONLY ONE REMAINING

In comparison with the pure philosophical agendas of western phenomenology, the goal of the Indian Advaita Vedānta is much more practical: it is liberation (Sanskrit *Moksa*) from limitations associated with the worldly existence of the body. In positive terms, it is an attainment of full, flawless happiness (Sanskrit *Ānanda*). As practical philosophy, Vedānta not only frees the self from the tenets of individuality, particularity and separateness, but goes even further to turn this condition into a natural state. The method of Vedānta is self-inquiry with systematic reduction of self-experience, combined with metaphysical ruminations on the nature of self-awareness.

FIRST MOMENT OF VEDANTIC SELF-REALIZATION: DISCOVERY OF THE SELF

As an example, the following verses (Lakṣmīdhara, trans. and commentaries Berliner 1990, verse 2) begin with a discovery and examination of the formless and limitless Self:

> Always I am; always I shine;
> Never am I an object of dislike to myself.
> Therefore it is established
> That I am Brahman,[7]
> Who is of the Nature of Existence, Awareness and Fullness.

To find "Always I am; always I shine;/Never am I an object of dislike to myself" in one's lived experience, one performs a systematic negation of all cognition and perception. Known in Vedānta as "differentiation between the seer and the seen" (Sanskrit *Dṛk dṛśya viveka*)[8] or "not this, not that" (*neti, neti*), the process differentiates a subject who has the nature of awareness, signified as "I," from objects of awareness that are labeled as "this" or "mine," and do not have inherent awareness. This reduction happens not only within the sphere of pure ideas but involves the whole self-awareness, and the whole of bodily perception.

The cascade of reduction-based switches in self-identity culminates in the realization that who one is pure, limitless, self-subsistent, self-aware subjective awareness identical only to itself—the good old Transcendental Ego. Detached from changing objects, one's identity has to be firmly associated with this formless awareness. This central step, disidentification with names and forms, and identification with the limitless principle of awareness, means that one is not a separate person and an individual, but is instead an all-encompassing, one-in-existence entity (Sanskrit *Brahman*) with no boundaries.

Psychologically, reduction alone is an insufficient means to construct such a universal self-identity, to cancel the existential reality of being a separate individual, and to attain the Vedantic goal of liberation from suffering. After the dialectical loop by which the principle of individual awareness is realized as universal ("Therefore it is established/That I am Brahman"), next must come the mental changes responsible for the cessation of suffering. Unless this is accomplished, liberation/happiness remains a purely theoretical, and not an existentially realized value.[9] Both the stability of the universal self-identity and the cessation of suffering are conditioned on the radical reconstitution of the self-world relations.

In order to get rid of the persistent impression of re-emerging individuality, the teaching methodology focuses on the recognition of the ontological primacy of pure awareness:

> In me, in the space of awareness,
> Rises this celestial city called the world.
> Therefore how am I not Brahman,
> Who is all-knower and the cause of all?[10]

The verse emphasizes that awareness only appears finite due to the illusion of the senses. In reality, awareness is beingness, which is clearly indivisible and present in every element of the world. Experientially, one cannot distinguish between the beingness of the subject and of the object. Thus, pure awareness/transcendental ego acquires the quality of a universal substance which is indivisible because of its transcendent nature.

> As a man regards the food he has eaten as one with himself, the Adept Yogin sees the Universe as one with his Self...[11]

Identification with this awareness-substance is believed to change the mental processes: having discarded all identifications with the phenomenal field the seeker ceases to experience suffering. However, the pressure of intentionality continues even though awareness is experientially differentiated from its objects: bodily participation in the world continues, and the influences which constitute individuality cannot be canceled. One quickly discovers that Self-Realization does not hold by itself and that the world keeps imposing itself on the seeker, reconstituting the individuality undermined by Vedāntic self-exploration.

As expected according to Western phenomenological philosophy, this world-attachment of the self is precisely what reduction is expected to reveal.[12] However, Vedānta intends not only to examine, but to modify Husserlian intentionality. Besides deconstructing the reality of the world in its theory of Māyā-vāda, Vedānta introduces the complementary methodological counterpart to the analysis of the self, the analysis of the Self-of-the world, or *Iśwara*. The transcendental ego is extracted from a less-than-real individual psyche and is bestowed onto the less-than-real world. As a result, the world acquires a sentient Self. Paradoxical self-world relations are contained within the notion of different levels of reality, as in the commonly cited in Vedānta circles famous verses from Shankara's work Manīṣāpañcakam:

> On the level of the body I am your servant.
> On the level of the soul I am your lover.
> On the level of the Self I am you.[13]

The method of reduction persists on many levels as the means of establishing self-world unity. The self is reduced to pure awareness, i.e. Self; the world is reduced to the Self, and even when analysis begins with the world, the world has to be interpreted as a self in order to give space to the same reduction. So Self remains as a reigning entity, never questioned in its causal positioning. "It is the ego as transcendental, i.e. as having abstained from granting the validity of the world's existence, including that part of the world that comprises its own psychophysical being, that bears the responsibility for the entire sense and the existential status of the objective world".[14] Is this true, that only reduction can rescue consciousness from the perceptual illusion of isolation and establish an otherwise unreachable self-world unity? To answer this question, I will turn to the systems of knowledge that, instead of imposing reduction on lived self-experience, follow the natural dynamics of the phenomenal fields of the self and of the world. If we are, indeed, always "confronted with the process of lived self-acquaintance whose distinctive feature is its non-reflective character, and which must be understood as an immediate expression of life itself",[15] then there may be something in our experience that lies deeper than the analytically recognized self-world unity creates a ground for a gestalt of unity as different from the understanding rooted in reduction and logic.

Such direct intuitive epistemologies of Islamic or Buddhist mysticism are not egalitarian in a sense of being available to every kind of mind. On the contrary, they correspond with the developmental maturity of the mind's capacity to know.[16] These capacities can increase spontaneously[17] or can be trained.[18] These developmentally available faculties form epistemologies rooted in the direct, unmediated awareness of "things themselves," as in Islamic mystical philosophy, with its epistemology of knowledge by presence.[19]

SUFISM: THE SELF IS THE LIGHT IN THE MIRROR OF THE WORLD

The Sufi approach to knowledge fully incorporated Plato's "repeated insistence that what to a superficial person is just 'myth' may have all decisive attributes of a logos for someone whose perception runs deeper".[20] Islamic philosophy establishes the limits of reason, and shifts the emphasis of knowledge-giving practices to heart-intellect with its direct intuition of the contents of consciousness. The central figure of Islamic philosophy, Abu Hamid Muhammad al-Ghazālī, in his autobiography *al-Munqidh min al-Dalāl*,[21] describes his search for true knowledge as a progression from a radical doubt in the truth of sense perception, to a deep study of contemporary scientific-philosophical systems, and finally to the mystic discipline of experiential cognition.

In contrast to systematic reduction in Cartesianism, early Husserl's phenomenology, or Vedānta, Islamic mystical philosophy relies on individual revelatory perception where knowledge by presence[22] plays a major role. Human life is viewed in Sufism as a journey of knowledge leading Sufi Gnostics,[23,24] to the escape from existential alienation through the recovery of a primordial state of mystical Union:

> Hear the reed as it makes its lament
> telling a tale of how it was rent
> from its root. . . . (Mesnevi, line 1)
> All who have wandered far from their source
> make their beginning the end of their course (Mesnevi, line 4).[25]

The journey is created by the dynamics of the three participating principles: the self of the gnostic, the world, and God. Thus, Sufism investigates a different kind of self: the self which is always relational, possessing of all its potentialities, and inclusive of the full spectrum of perceptual possibilities available in the human condition. These possibilities are dependent on the psychological maturity and spiritual development of the practitioner.[26] The majority of Sufi authorities do not use systematic reduction of experience in order for the knowledge (Arabic *ma'rifa*) to emerge because (a) the Real (Arabic *al-Haqq*, "the Truth") is revealed in the advanced states of perception, and (b) the relationship between the human being, God and the world are mutually pervasive and paradoxical:

> He who affirms the duality (of God and the world) falls in the error of associating something with God; and he who affirms the singularity of God commits the fault of confining Him to a (rational) unity... though wilt see Him in the essence of things, sovereign and conditioned at the same time.[27]

Systematic reduction, therefore, would distort the picture.

At the beginning of the journey, knowledge is incomplete. Both the self and the world are present in the beginner's perception, but God is hidden. Progress from the state of veiling to the states of direct perception of God corresponds to the increase of self-knowledge. "Who knoweth himself knoweth his Lord" is a recurring theme in the collection of Hadīth Qudsi.[28] The whole process is orchestrated by God who is the teleological force behind self-knowledge.

> God (Arabic *al-Haqq*, 'Reality') wanted to see the essences of his most perfect Names whose number is infinite – and if you like you can equally well say: God wanted to see His own Essence in *the global object* [the world, italics mine] which having been blessed with existence summarized the Divine Order so that He could manifest His mystery to Himself.... As the vision that a being has of himself is not the same as that which another reality procures for him, and which he uses for himself as a mirror...[29]

In this schema of things, the existence of the world is conditioned on the existence of God (who is the ultimate subject/carrier of consciousness), and the world is also necessary for God's self-knowledge. This creates a very definite role for the human being (Adam) as a vehicle of divine self-knowledge:

> God first created the entire world as something amorphous and without grace, comparable to a mirror not yet polished... For the entire reality from its beginning to its end comes from God alone, and it is to Him that it returns. So then, Divine Order requires the clarification of the mirror of the world; and Adam became the light itself of this mirror and the spirit of this form.[30]

Hence, philosophizing in Islam begins with the posited unity of God, world and self, but the starting platform of uncovering the experiential correlates of this unity is the self. The latter is, of course, similar to the natural state in phenomenology,[31] except that the self is not isolated as the only relevant subject of analysis. On the contrary, the self in Sufism is viewed in the context of indivisible unity with the world, and via a shared medium of an all-encompassing, transcendent and self-subsistent God.

Because the world is the place of Divine self-disclosure, and because human awareness is the vehicle of Divine self-knowledge, Islamic mysticism pays very close attention to the actual givenness of experience. The latter, according to Sufis, naturally fluctuates between the two poles of self-transcendence, the inner pole, which is the pure subject, and the outer pole, which is the world. Eventually, gnostic experiences a variety of states, all of which carry existential value. Fluctuations in spontaneously expanded perceptual states occur on the axis of meaning God—the self—the world, with a shifting focus of identification (Sam Goldberger, personal communication, 1997). The fluctuations are reflected in Sufi concepts of *fanā'* and *baqā'*[32] and in the maps of the states of knowledge.[33]

If the fluctuations of the states of internal spiritual union resolve in an experience of oblivion, the Sufi tradition suggests a return to the awareness of external existences. Return to the world leads gnostic to a greater knowledge than that the knowledge in the state of ecstatic extinction in God, because God, knowledge of whom the gnostic is seeking, is "sovereign and conditioned"[34] at the same time. Eventually, the introspective self-transcendent and external self-transcendent modes of awareness are indistinguishable.[35]

In this ebb and flow of revelatory experience, various forms of reduction may emerge naturally; however, there is no need to take a path of systematic reduction to arrive at self-world unity. The certainty of knowledge of unity of the self and the world is discovered perceptually, within the medium of the *embodied* lived experience, as opposed to an abstracted mental effort in phenomenology of pure ideas. The cosmological insights of Sufi Gnostic are connected with the embodied structure of self-awareness known as the Spiritual Heart.

THE HEART

The attitude towards the world in esoteric Islam varies from a straightforward rejection of it as a place of pollution of the soul in Tirmīdhi,[36] to much more nuanced perspectives on the self-world relationship in Niffari[37] or Ibn 'Arabi.[38] The attitude towards the world influences the understanding of what is true knowledge. The shift in the kind of knowledge is "from discursive to spiritual; [the shift] in the *subject* of knowledge [is] from the mind to the heart; and... [the shift] in the *object* of knowledge, [is]from discrete, formal data, to the essential principles of Reality as such".[39]

Discrimination between mental knowledge and the knowledge of the Heart is widespread in folk-theories of cognition. The post-Husserlian studies of constitution of the self, such as Merleau-Ponty's phenomenology of perception,[40] Rosen's formulation of topographic phenomenology,[41] and Gendlin's work on the embodiment of meaning,[42] create a phenomenological framework for this common sense distinction. The primary constitution of knowledge always engages both the meaning and the body; consequently, the distinction between the discursive mind and the heart-intellect will be related to the constitutional horizon within which the knowledge is viewed. In the constitution of knowledge, the absence (discursive mind) or presence (heart-intellect) of a consciously highlighted somatic self-awareness creates a major difference. Paradoxically, it is not the domain of pure ideas, but the somatic self-awareness, i.e. the Sufi "heart", that is associated with the function of imagination and carries potential for all the possibilities of consciousness available in human condition. The "heart" is both outwardly, and inwardly aware, as in Brentano's notion of inner consciousness, except for that this consciousness in Sufism is necessarily egological. Sufism uses the practices of *Dhikr* (Arabic "Rememberance")[43] to activate the direct awareness of embodied imagination. A process of conscious self-reconstitution, rooted in the awareness if this inward embodied realms of imagination, consists of the sequence of the states of Jam' (Arabic "Gathering").[44] Jam integrates into a conscious psychic unity all the elements of consciousness which had been previously anonymous or "veiled", and later rendered transparent by *Dhikr*. This form of awareness leads to the gestalt of unity of the self and the world.

Within the phenomenological field of the Heart, Sufi maps of consciousness identify several domains (Arabic *Laṭā 'if* "graces, subtleties")[45] which correspond to various forms of identity, from the individual self, to the self "beyond" the ego boundaries. The "self-in-the-Heart" is always a self-in-relationship,[46] always in a self-transcendent mode. As was both in psychological research[47] and in Sufi gnostic explorations, the modes of knowledge of the world, and the capacity to penetrate the world's internal organization, depend on the configuration of the self of the knower. Hence, the Heart presents an array of possibilities of knowledge in regard to the self-world tandem.

The internal contents of the heart-awareness is the infinite world of imagination,[48] with the logoic, ontopoietic hierarchical ordering.[49] Even without a phenomenological analysis, one easily discovers correspondences between the constitution of the heart-self, the naïve perception of the world, and the mythological cosmologies of creation.

Direct intuitive apperception of the interiority of the Heart-self, i.e. the embodied core of self-consciousness, creates the conditions of awareness necessary for experiential realization of the unity of its individual and cosmic aspects. There follows the unification of paradoxical conditions, such as in the statement of Ibn 'Arabi, partially quoted above: "You are Him, and you are not Him; you'll find Him in the nature of things, sovereign and conditioned at the same time".[50] Thus, the self-world dialectics embedded in Sufi gnosis is not a sequential flow of ideas resolving into a synthesis. This is the ontopoietic time,[51] in which Heart-consciousness manifests its characteristic topological flip-flops ("You are Him, and you are not Him" at the same time),[52] and creates gestalt of unity out of paradoxes of ordinary awareness. This awareness is not "altered", but expanded and sober.[53] It simply highlights the moments in consciousness which were previously anonymous.

The ultimate condition of the gnostics is the condition of a Universal Man, when the "Real is identical with them, while they do not exist".[54] The Real (Arabic *al-Ḥaqq*, "real, truth"), the perennial subject-awareness, generates the world as its global object, therefore, the perceptual condition of the Universal Man is open to complete awareness of everything there is within his own "inner consciousness". Thus, in one's perception the intersubjective is included in the intrasubjective:

> Each individual of the human species contains the others entirely, without any lack, his [her] own limitation being but accidental... For as far as the accidental conditions do not intervene, individuals are, then, like opposing mirrors, in which one fully reflects the other...[55]

If the inner consciousness transforms towards integration of the intra- and inter- subjective domains, the reflective self-awareness and the relationship with the world will also change. The self and the world in Sufic practice are connected by this perception, and not by the zigzags of Hegelian dialectics. In Sufic awareness, the transcendent function of consciousness is highlighted both inwardly and outwardly, unifying the self and the world through the experience of self-transcendence.

The inward-outward self-transcendence constitutes the two mirroring worlds: a world of imagination within the bodily egological heart-self, and a "real" world outside the body. In this system, the forms of awareness fluctuate, from egological awareness, to the awareness beyond the ego. There emerges one indivisible structure, a unity which encompasses both the intersubjective and intrasubjective domains located in the spectrum of perceptual possibilities between the polarities of awareness.

Where is the "beginning" of this knowledge? If it is in the self, then the degree of uncertainty of knowledge skyrockets, as the forms of subjective awareness are in constant flux. The angles of interpretation, created by the positioning of awareness, create a variety of phenomenological systems; from standpoint of Islamic metaphysics, the whole of cultural history is nothing but the dynamics of the modes of witnessing. The analysis in the present paper began with the self-based/reduction-based approach in egological phenomenology, proceeded to a more holistic view of the self in Vedanta, and demonstrated the self-world as a system in phenomenological epistemology of Sufism. I will now examine the possibilities which open when the world is taken as the beginning platform for the analysis. For that, I will turn to where things had begun, to the "Golden Age" of philosophy before the dawn of reduction, i.e. to the pre-Socratic Greek philosophy in the doxographic tradition.

ANCIENT GREEKS: THE WORLD'S SHAPING OF THE SELF

Kingsley's meticulous analysis of early Greek philosophy leaves no doubt that there is much more to the Greeks explorations of the world than a simple natural philosophy. It seems quite plausible that Empedocles and Parmenides practiced some kind of esoteric disciplines[56] which might have refined their intuition of internal consciousness. Even though the Greeks did not leave a record of formalized introspective contemplative practice which would lead to the self-knowledge by presence, one cannot exclude the possibility that the oral Greek tradition included such practices.[57] If this is so, Greek cosmology can be at least partially rooted in the phenomenological datum of inner self-consciousness. While such similarities can be (and generally are) ascribed to diffusion, the similarities between Greek and Sufi eschatology also support an assumption that in this case, the diffusion of ideas could've been accompanied by the transmission of introspective practices. The latter will lead to the similar constitution of the *pre-reflective* level of consciousness, reflexively presented in the similarities of mythological and philosophical ideas.

There are striking similarities between the central philosophical-mystical ideas in the doxographic tradition, attributed to the influence of South Italian landscapes,[58] and the essential topographic structures of the internal landscape of the self. The format of the article limits me to only a brief outline of the cycle of ideas "at the roots of classical Greek and Roman mythology...[which] appears in many forms, and ... exist(s) within themes and motifs in oriental mythology, both Near Eastern and Indo-European".[59] This cycle includes the idea of a fiery essence contained in the heart of the matter, the idea of the sun coming in and out of the underworld, the idea that one needs to visit hell before rising to heaven, the idea of an axis connecting heaven and earth, and the whole script of the post-mortem travels of the soul, with the spiritual "descent" rather than "ascent". For a practitioner of egological esoteric introspective practice involving the spiritual heart, such as the practice of *Dhikr*, these ideas will be perceived as metaphorical descriptions of the internal impressions emerging in the process of the advanced practice. Tirmīdhi,[60] not only mentions the perception of light emerging from the darkness at the core of one's being, and the sun-like luminosity rising inside one's body, but believes these impressions to be the signs of spiritual advancement. Corbin, in his comparative analysis of Iranian mystical philosophical texts, provides the detailed descriptions of the types and occurrences of the internal lights.[61] References to internal luminosity appear not only in the Sufism or pre-Islamic Iranian philosophy, but also in Hindu scriptures,[62] in Buddhist cosmologies,[63] and in Christian descriptions of the effects of internal somatic attentional focus in prayer.[64] Perception of the internal movement up and down the central axis of the body connecting the internal imaginal worlds and the descent into the limitless spaces of spiritual darkness within the inner space of the chest appears in the practices of *Dhikr*, Prayer of the Heart, and Kundalini Yoga—as well as in the processes associated with the spontaneous spiritual transformation described in Tantra.[65] Descent into darkness, and the annihilation of the personal identity akin to death and the following opening of the spiritual ascent are also the typical elements of inner practice.[66]

These stable impressions create a topography of the internal universe and serve as the landmarks by which the aspirant defines his or her progress.[67] Some systems, such as tantric Laya Yoga, Indian Sufism, or Taoist Alchemy, formalized the inner topographies as the system of chakras, Lata'if (subtle centers of the Heart-consciousness), channels, centers or meridians.[68] In tantric yoga, the detailing of the inner landscape reaches a high degree of elaboration reminiscent of the real topographic maps.[69] The temporal, spatial, meaningful and hyletic relations within these internal structures scaffold the flow of introspective self-experience. Whether these structures are "uncovered" in practice, or constituted, is not clear.[70] For example, Suhrawardi[71] describes the dynamics of lights without any reference to a corresponding spiritual practice. These lights, internal movements, and the overall internal landscape belong to the spatial, perceptual/somatic, topological and hyletic structural elements of the inner organization of the self, as opposed to the meaning-related constitution described earlier by Merleau-Ponty or Gendlin. The ideas of the above mentioned cycle are likely to form as a reflection of both the external, and of this internal landscape.

The similarities between the inner-imaginal and the outer-real landscapes is more than coincidence. The stable correspondencies between the cycle of mythological ideas, and the shape of the internal landscape point out to the complicated constitutive process including observations and interaction with the world, coupled with mythical thinking, languaging and introspection.[72] Neuroscietific studies of the correlates of consciousness showed that constitution includes various cognitive-perceptual processes, mirroring being one of them.[73] Mirroring theory suggests that the living, embodied self uses for its constitution the reflection of things external to the self. Indeed, "all things are required for fire and fire for all things as goods for gold and gold for goods".[74] In this schema of things, the structures of consciousness appearing in the internal landscape of the meditating mystics, are the hyletic and spatial blueprint of the external world, in a manner on a negative, one sees the image consisting of the white light outlines, as compared to the fully fleshed figures seen on the positive.

Ancient Greeks write without using the personal pronouns. Unless it is simply a writing convention, their written language reflects the way of thinking. The whole human being, the body and the soul, is viewed not in isolation, but within a logoic, natural flow of things. Thus, both on the pre-reflective, and on the ciphering level, the knowledge of the world becomes a key to the knowledge of the self; the focus on the world incorporates the phenomenological urge to self-knowledge, in that it pushes philosophers such as Empedocles, Parmenides or Heraclitus to examine whether there is, in the nature of being, anything that endures the flux of things. From here, the two possibilities emerge. One is to isolate the self from the flow of things, and to examine it reflectively,—this is the path of reduction. The other path aspires to understand of consciousness, reflective as well as pre-reflective, in its reflexivity with the world within the phenomenological order of things. Which leads us, inadvertently, to Tymieniecka's phenomenology of life.

TYMIENIECKA: SUBTERRANIAN CURRENTS OF LIFE

In the early stages of her philosophy, Tymieniecka follows the traditional phenomenological focus on the self. However, after the self-individualizing-in—existence of life, and the human condition are articulated, Tymieniecka develops the focus on the primacy of the world in the overall formulation of her philosophy. Self emerges out of this matrix gradually, as the evolution of the "...crucial specific existential/ontological device [the soul, i.e., the principle of consciousness or measuring observer] that differentiates all life from non-life, that is, through the inward/outward oriented central 'agency' of the individualized beingness, that life's ontopoietic processes are carried out".[75]

Tymieniecka switches the angle of phenomenological awareness from the segmented horizons of knowledge, to the foundational nature of the phenomenon of life: life is a primary given, and it has to be phenomenologically studied before anything else in order to give the raise to the true understanding of things.[76] Since Tymieniecka's whole categorization bears fidelity to life per se, the primacy of the world is dictated by the logic of her interrogation. She breaks the millennia-long philosophical trance in which the self always "feels itself as a fulcrum, center of our experience's furthers horizon".[77] While "the persistent care of the self is built into the very life stream",[78] Tymieniecka's focus on the world as a beginning analytical platform is the direct outcome of her primary thematization of life as compared to the more traditional thematization of being and knowing, and of the radical spirit of her phenomenological approach. She describes the dynamics in the ontopoietic Logos of life, and fully follows the flow of this dynamics in her interrogation. In the same motion she breaks free from of the monumental pressure of the logoic flow by establishing herself and her philosophizing as the locus of logoic self-reflection. Tymieniecka's turn to the world as a point of departure is both fulfilling the logoic individuation, and defeating the trance inherent in it.

The world is the *primum mobile* in Tymieniecka's philosophy, creating a context for the rest of the analysis. In its principle methodological positioning in the whole discourse, the strategy of referencing the insights against the non-reducible presence of the world is analogous only to a pervasive motion of reduction in Husserl's. The turn to the world does not put the self into the oblivion, on the contrary, it creates a condition of expanded and hightened awareness which propels one from simply living to living as a personal developmental practice. This is because of this move of expanded awareness, and the resulting aesthetic freedom coupled with a phenomenological descriptiveness of Tymieniecka's method, the categorical apparatus of Tymieniecka's philosophy such as life, sharing-in-life, unity-of-everything-there-is-alive etc. comes as close to things themselves as it is at all possible in relations between a signifier and the signified. Her analysis is never an inference, or an opinion, but it is always a penetrating vision and an essential description. Designs in the self-individuating givenness of life become visible without alienation or superimposed modifications of their operating dynamic principles. Tymieniecka effectively controls "the destructurting [of life's] constitutive efforts inherent to the logos of philosophical interrogations",[79] as her phenomenological "gaze" is both the expression of self-reflexivity of life,[80] and of the "symbiotic empathy" embedded in the matrix of life.[81] Thus, self and the world are both positioned within the overarching principle of life, interconnected, as it were, within a network of myriads phenomenologically present interactions and/or dynamic bonds.

In the phenomenology of life, the phenomenological explications of life happen within the phenomenological modus vivendi,[82] i.e. the systematic outlook at life as a phenomenon given in the first person experience. Experiential states present themselves as self-luminous, i.e., self-generated and self-posted, i.e., both *creative*, and *creating* virtualities of life. The strength of Tymieniecka's approach, as it seems to me, consists exactly in the extreme intimacy with and high appreciation of this direct givenness of the self-luminous life. The attraction to her philosophy is nearly sensate; she simultaneously eidetically expands, and meditatively grounds in the body the attention of the reader. In this manner, she remains true to the logos of her interrogation, overcoming the centuries of philosophical "dissociation" from life. She manages to stay with ontopoiesis as it happens, in the world and in its' mirror, the self, in the moment-by-moment unfolding of "subterranean" currents of life. In this specific positioning of awareness, I believe, lies the reason for the success of Tymieniecka's philosophy.

Institute of Transpersonal Psychology, Palo Alto, CA, USA
e-mail: olouchakova@gmail.com

NOTES

[1] T. Toadvine, Leaving Husserl's Cave? The Philosopher's Shadow Revisited, in *Merleau-Ponty's Readings of Husserl*, eds. T. Toadvine and L. Embree, Netherlands: Kluwer, 2002, pp. 71–94.

[2] For the examples, see Fergus Kerr, The Modern Philosophy of Self in Recent Theology, in *Neuroscience and the Person. Scientific Perspectives on Divine Action*, eds. Robert John Russell, N. Murphy, T.C. Meyering, and M.A. Arbib. Berkeley: Center for Theology and Natural Science, 2002, pp. 23–40.

[3] D. Zahavi, *Subjectivity and Selfhood: Investigating the First Person Perspective*, Cambridge: 2008, p. 79.

[4] For the description of a phenomenological constitution of pure subjectivity, see O. Louchakova, Ontopoiesis and Union in the Prayer of the Heart: Contributions to Psychotherapy and Learning, in *Logos of Phenomenology and Phenomenology of the Logos. Book Four: the Logos of Scientific Interrogation. Participating in Nature – Life-Sharing in Life*, ed. A.-T. Tymieniecka. Analecta Husserliana, V. 91, Dordrecht: Kluwer, 2005, pp. 289–311.

[5] See debates in *The Problem of Pure Consciousness*, ed. R.K.C. Forman, New York: Oxford University Press, 1990.

[6] See explanation of the practical philosophies in D.B. SenSharma, *The Philosophy of Sadhana: With Special Reference to the Trika Philosophy of Kashmir*, SUNY Series in Tantric Studies, Albany: State University of New York Press, 1990.

[7] i.e., "I am limitless".

[8] *Drg-Drśya-Viveka*. op. cit., 1976.

[9] Similar questions are discussed in the preface by Iyer, *Drg-Drśya-Viveka*. op. cit., p. iv.

[10] Laksmīdhara, *Advaita Makaranda*. Trans. and commentary A. Berliner, Bombay: Asia Publishing House, 1990/15th century.

[11] Śrī Śaṅkarācārya, *Dakshinamurti Stotra*, with Sri Sureshwaracharya's *Mānasallāsa*, Trans. A.M. Sastry, Madras: Samata Books, 1978, ch. 1, 14–15, p. 8.

[12] Toadvine. op. cit., 2002, p. 81.

[13] Modified from the translation by Śri S. V. Ganesan "In the form of body I am your servant. // In the form of life, O three-eyed one, I am part of yourself. // In the form of soul, you are within me and in every other soul."

[14] Toadvine, Op. cit., 2000, p. 76.

[15] Zahavi, op. cit., 2008, pp. 59–63.

[16] For the developmental dynamics of awareness over the lifespan, see A. R. Arasteh, *Final Integration in the Adult Personality. A measure for health, social change, and leadership*, Leiden: E.J. Brill, 1965. Also, for the description of the training of the faculties of the mind necessary for mystical philosophical enquiry, see O. Louchakova, On advantages of the clear mind: Spiritual practices in the training of phenomenological researcher. *The Humanistic Psychologist*, 33(2), 87–112.

[17] As shown in O. Louchakova, Ontopoiesis and Spiritual Emergence: Bridging Tymieniecka's Phenomenology of Life and Transpersonal Psychology, in *From the Animal Soul to the Human Mind, Analecta Husserliana*, ed. A.-T. Tymieniecka, Vol. 94, Dordrecht: Kluwer, 2007c, pp. 43–68, 50–51.

[18] M. Kozhevnikov, O. Louchakova Z. Josipovic & M.A. Motes. The enhancement of visuospatial processing efficiency through Buddhist deity meditation. *Psychological Science*, 20(5), 645–653. 2009.

[19] M.H. Yazdi, *The Principles of Epistemology in Islamic Philosophy: Knowledge by Presence*, Albany: State University of New York Press, 1992.

[20] P. Kingsley, *Ancient Philosophy, Mystery and Magic: Empedocles and Pythagorean Tradition*, Oxford: Clarendon Press, 1995, ch. 7, p. 80.

[21] *Al-Ghazali's Path to Sufism: His Deliverance from Error (al-Munqidh min al-Dalal) and Five Key Texts by Abu Hamid Muhammad al-Ghazali*, R.J. McCarthy SJ, David Burrell CSC, and William A. Graham, Louisville: Fons Vitae, 2000.

[22] Yazdi, op. cit., 1992.

[23] For knowledge in Islam, see R. Shah-Kazemi, "The Notion and Significance of Ma'rifa in Sufism," *Journal of Islamic Studies*; 13(2) (May 2002), p. 155; W.C. Chittick, *The Sufi Path of Knowledge: Ibn 'Arabi's Metaphysics of Imagination*, New York: State University of New York Press, 1989.

[24] Sufism can be viewed as gnosticism in it's primodial, non-sectarian sense, as a knowledge of the inner realities of consciousness, S.H. Nasr, General Introduction, in *An Anthology of Philosophy in Persia*, eds. S.H. Nasr and M. Aminrazavi, Vol. 2, I.B., London: Tauris Publishers, 2008.

[25] The following book was used *The Essence of Rumi's Mesnevi* by Prof Dr. Erkan Turkmen, Konya, 1995. Quoted verses are translated by Martin Schwartz.

[26] Shah-Kazemi, op. cit., 2002.

[27] M. Ibn 'Arabi, al-Futūhāt al-Makkiyya: Texts choisis/Selected texts. Trans. M. Chodkiewicz, W.C. Chittick, C. Chodkiewicz, D. Gril and J. Morris, Paris: Sindbad, 1989, p. 34.

[28] Hadith is a collection of revelational statements attributed to the Prophet Mohammed. *Forty Hadith Qudsi*, Trans. E. Ibrahim and D. Johnson-Davies, Beirut/Damascus: Dar ai-Koran al-Karaeem, 1980, 104, no. 25.

[29] M. Ibn 'Arabi, *The Wisdom of the Prophets (Fusus al-Hikam)*. French Trans. T. Burckhardt. Trans. from French to English A. Culme-Seymour. Beshara Publications. Circa 1229/1975, pp. 8–9.

[30] Ibid., pp. 9–10.

[31] For definition of the natural state, see Zahavi op. cit., 2008, p. 63.

[32] W.C. Chittick, *The Self-Disclosure of God: Principles of Ibn 'Arabī's Cosmology*, Albany: SUNY, 1998, p. 84.

[33] For the Sufic states, see Muhammad ibn' Abdi'l-Jabbár al- Nīffari. *The Mawáqif and Mukhátabát with other fragments*. Trans. A.J. Arberry, Cambridge: E.J.W. Gibb Memorial, 1935; A.H. Abdel-Kader. *The life, personality and writings of Al-Junayd*. London: E.J.W. Gibb Memorial, 1976; M. Chodkiewicz. *The spiritual writings of Amir 'Abd-al Kader*. Trans. J. Chrestensen and T. Manning. Albany: SUNY, 1995.

[34] M. Ibn 'Arabi, *The Wisdom of the Prophets*, op. cit., p. 34.

[35] Shah-Kazemi, op.cit., 2008.

[36] For perspective on the world as a place of pollution of the soul see Al-Hakim Al-Tirmīdhi. *The concept of sainthood in early Islamic mysticism*, Trans. B. Radtke and J. O'Kane, Richmond, Surrey: Curzon Press, 1996.

[37] For nuanced perspectives on the world in Sufism, see Muhammad ibn' Abdi'l-Jabbár al- Nīffari. *The Mawáqif and Mukhátabát with other fragments*. Trans. A. J. Arberry, Cambridge: E.J.W. Gibb Memorial, 1935.

[38] For Ibn 'Arabi perspectives on the "world", see Chittick, op. cit., 1998.

[39] Shah-Kazemi, op.cit., 2008, p. 160.

[40] M. Merleau-Ponty, *Phenomenology of Perception*, Trans. Colin Smith. Routledge, 1962.

[41] S.M. Rosen, *Topologies of the Flesh: A Multidimensional Exploration of the Lifeworld*, Athens, OH, University Press, 2006.

[42] E. Gendlin, *Experiencing and the Creation of Meaning: A Philosophical and Psychological Approach to the Subjective*, Evanston, IL: Northwestern University Press, 1997.

[43] On Dhikr, see A. Hammarlund, Introduction. In eds. A. Hammarlund and E. Özdagla *Sufism, Music and Society in Turkey and the Middle East* (Swedish Research Institute in Istanbul Transactions Volume 10). Routledge, 2001; H. Algar, Silent and Vocal Dhikr in the Naqshbandi Order, in *Akten des VII Kongresses für Arabistik und Islamwissenschaft*, Göttingen: Vandenhoeck & Ruprecht, pp. 39–46, 15–22 August 1974, 1976.

[44] As in Arasteh, op. cit., 1965; Chittick, op.cit., 1998, p. 179.

[45] *Lata'if* described in T. Dahnhardt, *Change and continuity in Indian Sufism: A Naqshbandi-Mujaddidi branch in the Hindu environment.* Islamic Heritage in Cross-Cultural Perspectives Series, No. 3. New Delhi: D.K. Printworld (P), Ltd., 2002.
[46] O. Louchakova, op. cit., 2005.
[47] S. M. Johnson, *Humanizing the Narcissistic Style*, Norton Professional Books, 1987.
[48] H. Corbin, *Creative Imagination in the Sufism of Ibn 'Arabi*, Princeton, NJ: Princeton University Press, 1969; H. Corbin, *Swedenborg and Esoteric Islam*, West Chester, PA: Swedenborg Foundation, 1995.
[49] O. Louchakova, op. cit., 2005.
[50] Ibn 'Arabi, The Wisdom of the Prophets (Fusus al-Hikam) op. cit. Circa 1229/1975, p. 34.
[51] For kaironic, ontopoietic time, see A.-T. Tymieniecka. Life's primogenital timing. Time projected by the dynamic articulation of the onto-genesis, in *Life. Phenomenology of Life as the Starting Point of Philosophy*, Book III, ed. A.T. Tymieniecka, Analecta Husserliana, Vol. L, Dordrecht: Kluwer, 1997, pp. 3–22.
[52] For topological flip-flops, see Rosen, op.cit., 2006.
[53] For the "sober" awareness in Sufism, see A. H. Abdel-Kader, op. cit., 1976.
[54] M. Ibn 'Arabi, op. cit., 1989. Ch. 219, 512, 9.
[55] 'Abd al-Kerim Jili, Universal man, Trans. T. Burckhardt, U.K: Beshara Publications, 1983/14th cent., p. xxv.
[56] P. Kingsley. *Ancient Philosophy, Mystery and Magic: Empedocles and Pythagorean Tradition*, Oxford: Clarendon Press, 1995.
[57] P. Kingsley, *In the Dark Places of Wisdom*, The Golden Sufi Center, Inverness, 1999.
[58] P. Kingsley, op. cit., 1995.
[59] Ibid., p. 54.
[60] Al-Hakim Al-Tirmīdhi. op. cit., 1996.
[61] H. Corbin, *The Man of Light in Iranian Sufism*, Omega Publications, 1994.
[62] For inner lights in Hinduism, see F.B.J. Kuiper, The bliss of Asha, in Ancient Indian Cosmology, essays selected and introduced by J. Irwin, Vikas publishing house, 1983, pp. 56–89; J. Gonda, *The Vision of the Vedic Poets*, Mouton and Co. Netherlands, 1963, ch. xII.
[63] D. Goleman, *The Meditative Mind*, New York: Perigree, 1977.
[64] O. Louchakova, op. cit., 2005.
[65] J.S. Harrigan, *Kundalini Vidya: The Science of Spiritual Transformation*, Knoxville, TN, 2005.
[66] As in Ibn 'Arabi, *The Tarjumán al-Ashwáq*, Trans. R.A. Nicholson. London: Theosophical Publishing House, 1911.
[67] O. Louchakova & A. Warner, Via Kundalini: Psychosomatic excursions in transpersonal psychology. *The Humanistic Psychologist*, 31(2–3), 115–158, 2003.
[68] Ibid., 2003.
[69] For an example of a tantric yoga map of the internal world see Harrigan, op. cit., 2005.
[70] For constitution of the internal structures of awareness in the process of spiritual practice, see Louchakova, op. cit., 2005,
[71] On internal lights, see Shihab al-Din Suhrawardi, *The Philosophy of Illumination*, Trans. J. Walbridge & H. Ziai, Provo, UT: Bringham Young University Press, 1999. (Original work written 1185).
[72] For constitution of mythological consciousness, see E. Cassirer, *Language and Myth*, Trans. S, Langer, Harper and Brothers, 1946.
[73] S. Gallagher, Neural Simulation and Social Cognition, in *Mirror Neuron Systems: The Role of Mirroring Processes in Social Cognition*, ed. J.A. Pineda, Totowa, NJ: Humana Press, 2008, pp. 355–371.
[74] C.H. Kahn, *The Art and Thought of Heraclites*, New York: Cambridge University Press, 1979, fragment XL, p. 47.
[75] A.-T. Tymieniecka, op. cit., 1997, p. 7.
[76] M. Groth and A.E. Gunn, Life. Phenomenology of Life as the startying point of philosophy, *The Review of Metaphysics*, Vol. 52, 1998. Retreived 7/6/10 from http://www.questia.com/googleScholar. qst;jsessionid=MynHXTDrxYQLpf2wCChJ2td3nLL4dZmLlyHSPcGNMvjJJ5RJm8QL!122306455!547733517?docId=5001410055.

[77] Tymieniecka, op. cit., 1997, p. xiii.
[78] Zahavi, op.cit., 2008, p. 79.
[79] A.T. Tymieniecka, The Human Condition Within the Unity-of-Everything-There-is-Alive and Its Logoic Network, in *Logos of Phenomenology and Phenomenology of the Logos, Book Two. The Human Condition. Analecta Husserliana*, Vol. LXXXIX, ed. A.-T. Tymieniecka, Dordrecht: Kluwer, pp. XIII–XXXIII, 2007, p. XXXI.
[80] Ibid., p. xxx: Logos reversing its course.
[81] Ibid., p. xv.
[82] Michel Henry, *La Barbarie*, Paris: Grasset, 1987 (réédition, Paris, PUF, avec une préface inédit, 2001).

THOMAS RYBA

TRINITARIAN APPROPRIATIONS OF THE TRANSCENDENTALS: GIVENNESS AND INTENTIONALITY IN LEVINAS, MARION, AND TYMIENIECKA

ABSTRACT

Consciousness' contribution to the constitution of objects is a central concern for Husserlian phenomenology and its heirs. The philosophies of Levinas, Marion and Tymieniecka problematize this concern in the form of two commonly shared questions: "What is given in object-constituting consciousness?" and "How—if at all—is the given co-constituted by realities which transcend consciousness and are (in some sense) prior to it?" Though problematizing the issues similarly, the answers Levinas, Marion, and Tymieniecka give to these questions seem very different. For Levinas, the Other is given constitutively in all of its particularity in a relationless relation. Because it is outside of the totality of the Transcendental ego, it resists possession, is not an object of enjoyment but shows a freedom which is both a call to responsibility and to obligation. For Marion the phenomenology of givenness is tied to the kind of reduction accomplished: the reduction of the to-whom-given, the reduction of the to-which-given, the reduction of the how-given and the reduction of the how-far-given. For Marion it is the third kind of givenness—givenness as gift (Charity/Agapē) that delivers the Being of beings. For Tymieniecka, the New is poetically constituted in chaotic deconstruction and reconstruction of perceptual givens, after the incipient phase of awareness. In this paper, my intention is to argue that all three formulations of givenness, constitution and transcendence are complementary and tantamount to a rediscovery of three transcendental modes of being. For Levinas, it is the Other's ingression into constitution of consciousness that points to the givenness of the Ethical (or the Good), for Marion it is Love that delivers Truth, and for Tymieniecka it is the New's constitution out of given features of aisthēsis that points to the Beautiful. I will, further, suggest that each of these transcendentals may be understood theologically as Trinitarian appropriations belonging, respectively, to the Father, Son and Holy Spirit. Finally, I will also relate these different respective notions of givenness as descriptions of the Trinitarian personalization of revelation.

INTRODUCTION

The paper is an exercise in resourcement theology, an approach to generative and speculative theology in Roman Catholicism which some have cynically termed the "back to the future" approach to theological innovation, a view which might suggest that it is associated with a conservatism in theological outlook relative to the RC tradition. However, a quick inquiry into the identities of its practitioners on both sides of the *Communio* and *Concilium* divide, dispels the charge of conservatism. One finds among its many practitioners many luminaries of 20th century theology—luminaries such as Henri de Lubac, Eric Przywara, Ives Congar, Bernard Lonergan, Karl Rahner, and Hans Urs von Balthasar, among others, luminaries whose theological writings made them suspect radicals relative to the conservatism of the Roman church, though that same church eventually caught up with and even sanctioned many of their ideas.

"Resourcement Theology" defies simplistic reduction because it is a "cluster concept" covering a variety of theological orientations and methods. Even so, across its many varieties there are some common features: First, in intention it parallels the Husserlian epigram "Back to the sources," "sources" being understood by the resource theologians as a reactivation of ideas internal and external to the Tradition, ideas whose meaning had become ignored, neglected, or ossified. Second, methodologically, it attempts the recovery of this meaning using the most up-to-date historical approaches and assumptions, a use which—early-on—made it susceptible to its scholastic opponents' unjust characterization as a form of modernism. Third, it attempts to reactivate these ideas in reference to the experiences/realities that ground them in order to ask the question whether those experiences are still living (or ought to be living) today.[1]

This paper is an attempt, within the scope of resourcement theology, to make an argument for the recovery and updating of the notion of the transcendentals in such a way to expand contemporary Trinitology, especially with respect to the notion of Trinitarian appropraitions.

Although there are many classic treatments of Trinitarian appropriations (and a variety of different schemas by which various Neoplatonic transcendentals are correlated with the Trinitarian persons), there is no one standard, universally accepted, set of appropriations. In the parlance of theologians, there is no formulation of these appropriations which has achieved *doctrinal* or *dogmatic* status. For this reason, theological speculation about Trinitarian appropriation of the transcendentals lies squarely in the realm of *theological opinion*, a status which has been stable for about fifteen centuries.[2] In this paper, it is my intention to revisit this speculative reflection in order to suggest a new formulation of Trinitarian appropriation based upon the thought of three contemporary philosophers: Emmanuel Levinas, Jean-Luc Marion, and A.-T. Tymieniecka.

Because each of these thinkers formulates the relationship between transcendentals and giveness in such a way to make transcendental object and epistemological process inseparable, it is possible to claim that the accessibility of each transcendental is connected to and defined by a unique epistemological mediation. This possibility has important consequences for the Christian understanding of revelation inasmuch as it provides an important new theological inflection. When the idea of revelation has been broached in connection with speculation about Trinitarian appropriations, the tendency has been to ignore the phenomenological evidences in the Scriptures that suggest revelation is pluriform in favor of a view which makes it homogenous. It is this tendency toward homogenization I would like to challenge in this essay, suggesting, instead, that if it is true that the Trinitarian appropriations are inseparable from accompanying modes of communication, then the respective revelations of the Father, Son, and Holy Spirit are qualitatively different *qua* communications and, more specifically, that they are also different as communications *of* the respective appropriated transcendentals. This does not contradict the orthodox view that the historical (or economic) missions of the Trinitarian persons are distinctive (but cooperative), yet it goes beyond the standard formulation to suggest that correlative with these unique missions are distinctive personal communications, the *idiosyncratic, economic* revelations of Father, Son, and Holy Spirit.[3]

To argue for the meaning and significance of my claims, I would, first, like to walk you quickly through some theological preliminaries connected with the notion of Trinitarian appropriations. Second, I will move to a discussion of the relationship between the *noematic* and *noetic* components of giveness according to Levinas, Marion, and Tymieniecka. Third and finally, I will conclude with an explanation of how the adaptation of the ideas can positively re-shape Christian thinking about the Trinity and revelation.

THE SIGNIFICANCE OF TRANSCENDENTALS IN THE WESTERN INTELLECTUAL TRADITION

Theories of transcendentals have been proposed since the time of Plato in order to explain (1) the overarching perfections found across beings and (2) the analogicity of language in saying something meaningful about beings from diverse realms of things. It is a part of our common human experience to speak—often in rapid succession, without skipping a beat—of good, true, or beautiful: physical theories, mathematical formulas, poems, dogs, women and men, architectural designs, virtues, shotguns, and so on. Thomas Aquinas uses an intentionally startling comparison to illustrate the second reason for the transcendentals when he describes God and a horse's unrine both as being good (*in some sense*).[4] It is the recognition that there is some common term (or measure) to such analogies—that these analogies are somehow, in some way, saying the same thing—which lies at the heart of the classic notion of the transcendentals.

Plato, whose influence on later kataphatic and apophatic theological ontologies cannot be underestimated, described the One (*henas*) and the Good (*agathon*) as beyond being (*huperousia*)—or transcendental—and *relatively* beyond knowing.[5] Plato and the later Neoplatonists suggest that the epistemic inaccessibility of the One and the Good is only *relative* because other avenues of knowing, apart from apodictic knowledge, are possible, such as: induction to the eminent source, analogy, negation, and mystical union.[6] These methods of access are later more rigorously developed by Christian and Pagan thinkers of the 1st millennium and by the Christian theologians of the Middle Ages.

Aristotle described a similar set of properties as transcending the established categories of things by being universal to them all, thus defying classification according to exclusive schemas of genera or species and substance or accidents. Among these he included being, oneness, and goodness. To capture the illusive but universal distribution of these features across reality, Aristotle, however, never resorted to the Greek noun *huperousia* nor verbs available to him and employed in other contexts, words such as *huperbainein* (stretch beyond) or *huperballein* (exceed).[7] This omission indicated a view of metaphysics confident that being circumscribed reality. For Aristotle, these properties were neither Platonic forms, nor hyper-essential. Rather, these universal features were susceptible to description and comparison analogically, their meaning neither univocal nor equivocal across reality but sharing some common intelligible measure (*metron*) or rationale (*logos/ratio*) explicable within the natural horizon of things.

The preparatory period in the consideration of the transcendental was complete when, during the Patristic period, the list of the preeminent transcendentals reached traditional theological stabilization in the triad of Goodness, Wisdom (or Truth), and Beauty in the thought of Proclus.[8] Accepting an emanational ontology, especially the Pagan Neoplatonists proposed that this triad also represented a prolation of being, the One pouring itself out in relative differentiation into the Good, the Good into the True, and the True into the Beautiful.

The medievals later propose alternative lists of transcendentals, recognizing the Neoplatonic formulations, while reintroducing Aristotelian recognitions. Thomas Aquinas draws on the prior rich development of the idea in order to make it bear theological value. In his thought, the transcendentals are correlated with the Trinitarian persons in a most original way.

For Thomas, a long list of transcendental reflects the traditional views of both the Platonic and Aristotelian philosophies, and in the *Disputed Questions on Truth*, Thomas shows his reliance on both schools by making the observation that the transcendentals may be distributed conceptually over diverse genera of things or may be assigned special metaphysical properties (Ver. Q.1, A.10). Thomas' short list of transcendentals includes oneness (*unum*), being (*ens*), thing (*res*), otherness (*aliquid*), truth (*veritas*), and goodness (*bonum*) and beauty (*pulchrum*). When considered conceptually, the transcendental are analogously present in a multitude of beings; when considered metaphysically, the properties of individual created beings are graduated and depend (by eminence and cause) on the transcendental perfections of the Godhead.

THE MEANING OF THE TRINITARIAN APPROPRIATION OF TRANSCENDENTALS

In his theory of the Trinity, Thomas Aquinas follows the Patristic distinction between the *economic* and *immanent* features of the Trinity. Thomas, following Augustine, Hilary and Richard of St. Victor, is careful to maintain that the Trinity, as God, is conjointly involved in every act. In Augustine, the missions of the Trinitarian persons are treated in some confusion with the transcendental appropriations—wisdom, for example—but it is clear that confusion is not Augustine's intent, because he keeps clear the essential features of the godhead from the Trinity's contingent actions in history. Thomas' own formulation goes a long way to clarify Augustine's treatment.

For Thomas, the *immanent* features include what has been called the "one, two, three, four, five schema." Internally, constitutive of God are five notions, one act of being (or essence), two asymmetric relations (the relation of generation of the Son, and the relation of the generation of the Spirit), three persons (constituted by the relations), four processions, and five notions. The *economic* features include the salvific givenness of the Trinity in history. These modes of givenness have the peculiar character of being (1) freely chosen, (2) technically appropriate, expedient, and efficient, and (3) personally expressive. Two of these modes of givenness—that of the Son and Holy Spirit—are more appropriately designated *missions* because they are sent into history by the Father, while the Father is not sent but is the origin of his own giving (Ia, Q. 43, A.4, Res. 1–3).

The personal expressiveness of *givenness*, requires special explanation. Thomas, following Hilary, Augustine, and Richard of St. Victor, is careful to maintain that the Trinity, as God, is conjointly involved in every act. But different Trinitarian persons assume the foreground or background relative to the modes of givenness as is appropriate to them, so that it is possible to say that the giving of the Father in the Old Testament uniquely expresses his personality, the mission of the Son in the New Testament (up to his ascension) uniquely expresses his personality, and the mission of the Holy Spirit in the New Testament and from Pentecost to present, uniquely expresses his personality.[9]

Analogous to the Trinitarian missions is the idea that each Trinitarian persons assumes or *appropriates* a transcendental which is most expressive of his personality. In using the appropriate verb [Lat.: *approprio*], Thomas indicates that each Trinitarian person takes to its own possession or self-ascription a term or quality expressing its idiomatic features with respect either to its shared Godhead or what it causes.[10] As he puts it, "To appropriate means nothing else than *to contract* something common, making it something proper" (Ver. Q.7, A.3, Res.). Here, Thomas' intentional conflation of legal and metaphysical senses of the verb *contractio* should not be underestimated. Just as, by contract, one of the parties might assume greater responsibility for the powers which all parties share, one of the Trinitarian persons can similarly take on a transcendental. But, perhaps, more telling is the metaphysical meaning of *contractio*, implying a "contraction, compression, delimitation, or focusing" of the transcendental through a *hupostasis* or person.[11]

"Now what is common to the entire Trinity cannot be appropriate to a single Person on the grounds that this belongs more to this Person than it does another," but it may be made on the grounds that "what is common has a greater resemblance to what is proper to one person than to what is proper to another" (Ver. Q.7, A.3, Res.). This being so, the appropriations are accidents, though following the language of Porphyry in the *Isagoge*, it may not be without justification to think of them as *necessary* accidents related to the economic functions of the Trinitarian persons.[12]

With the idea of transcendental appropriation comes a particular semantic procedure. "Appropriation means the making known of divine persons *by means of* essential attributes" according to the *via affimativa* and the *via negativa*, that is through kataphatic or apophatic discourse (Ia, Q. 39, A. 7, Res). Here, the transcendentals are prior conceptually in the natural order of being, but knowing the Trinitarian person in supernatural revelation illuminates the transcendental in the order of experience so that one may also come to know the transcendental through the person (Ia, Q. 39, A. 7, Res. 3).[13]

Thomas says that because it is impossible to know the divine persons and the coordinate personal properties *via* natural reason, philosophers, prior to Christianity, at best could only know the essential attributes of God—such as power, wisdom, goodness, etc.—which are the appropriations of the Father, Son, and Holy Spirit (1a, O. 32, Res. 1.). This distinction speaks to the classical division between natural and supernatural revelation indicating the complementarity of what may be known through reason and what may be known through revelation and how both kinds of knowledge reinforce one another. This distinction will have special relevance, as we shall soon see.

Thomas is aware that it is possible to assign transcendental appropriations to the Trinitarian persons in various ways. This follows from the fact that no single transcendental is a proper essential predicate of any person, but they all share them equally. In the *Summa*, so that he may bring some order to the diverse preceding views on the matter (including his own), he organizes the appropriations proposed by his predecessors according to whether they are conceptualized according to those questions which govern the investigation of all things: (1) What is it in itself? (2) How is it one? (3) How does it act and cause? And, (4) How does it stand in relation to its effects? (ST, 1a, 39, A. 8, Res.).

Corresponding to the first question, Hillary's formulation makes the assignment of transcendentals eternity-beauty-joy because eternity expresses underivability (just as the Father is the first), because beauty expresses integrity, harmony, and clarity (just as the Son is the very icon of the Father) and because joy expresses the love and enjoyment of the Father and Son for each other (just as the Holy Spirit is the procession of this love).

Corresponding to the second question, one of Augustine's formulations makes the assignment of the transcendentals unity-equality-connection because unity expresses the absolute independence of the Father as first principle, because equality expresses the oneness of the Son but in reference to the Father, and because connection expresses the unity of the relationship between Father and Son which is the Holy Spirit.

Corresponding to the third question, Hugh of St. Victor's formulation assigns power to the Father owing to the Father's primacy which *is not* flagging though he is supreme patriarch, assigns wisdom to the Son owing to the Son's being the Logos but which *is not* diminished by being an offspring of the Father, and goodness to the Holy Spirit because it is the motive and object of love, but which *is not* tainted by acquisitive violence.[14]

Corresponding to the fourth, and final, question, another formulation of Augustine assigns efficiency to the Father because it is he who is *like* the originating (kinetic) cause of all that is, assigns instrumentality to the Son because it is he who is like is the principle (organic cause) of a principle (the Father) through whom all things came into being, and assigns finality to the Holy Spirit because it he who (like the telic cause) brings the Trinitarian processions to an "end" within the mutual loving enjoyment of the Father and the Son.

Recognizing, as he does, multiple ways of assigning appropriations, Thomas does not finally settle on any particular configuration. It is my purpose, in the conclusion of this paper, to suggest an assignment which will be generally satisfactory, especially when considered in relation to the Trinitarian missions.

Although they possess many of the same accidental and economic features, Thomas never draws his discussion of the Trinitarian missions into relation to his discussion of the appropriation of transcendentals. He is precluded by this, in fact, because he thinks the distinctive appropriations are merely *notional*, that is, they are merely human conceptual distinctions (Ver. Q.1, A.1, Res. Con. 5). However, if there are analogies between each that make one transcendental more appropriate as a descriptor for a mission, I would argue that the mission becomes a vehicle for it. In other words, the mission or mode of givenness of each Trinitarian person is related to the transcendental that is most prominent in the completion of the mission. This would make the appropriation of the transcendentals something more than notional because transcendentals would be emblematic *of* and given *through* the economic Trinitarian missions. By the accepted Rahnerian principle that the economic Trinity *is* the immanent Trinity—and by thematizing the word "is" here according to the Clintonian principle that its meaning "depends on what *is* is"—then the economic missions must tell us something about intra-Trinitarian structures. Note that this is not a denial that all Trinitarian persons are involved in every mission or that all Trinitarian persons properly—that is in essence—claim all transcendentals.[15] Even so, my proposal draws the Trinitarian missions into closer relationship to the Trinitarian appropriations than Thomas seems willing to do.

But which appropriations of the transcendentals are to be thus related? Obviously, Thomas recognizes that many assignments of the transcendentals are possible, to the point of making them dependent upon the erotetic approach that one undertakes with respect to them, and that would seem to multiply them, indefinitely. In the *Disputed Questions on Truth*, he does seem to settle on a particular configuration for the purposes of discussion, but without ultimately valorizing it. That is the set of appropriations that associates the essence of the Trinity with Being, the Father with Oneness, the Son with Truth, and the Holy Spirit with Goodness.

My question is whether there isn't a supervening appropriation which both (a) corresponds to the economic mission of the person and (b) best represents its unique mode of givenness. I would argue that the supervening assignment of the Trinitarian appropriations is that of Goodness for the Father, Truth for the Son, and Beauty for the Holy Spirit and that it is this assignment that is borne out by the recent phenomenology of givenness in the thought of Levinas, Marion, and Tymieniecka.

Because of the limitations of time and preparation, I can only give a sketch of this demonstration. I hope that this sketch will be sufficiently detailed to be understood.

LEVINAS, MARION, AND TYMIENIECKA ON TRANSCENDENTALS AND GIVENNESS

The phenomenologies of Levinas, Marion, and Tymieniecka share three important features that have particular significance for the remainder of this paper. First, all of them are critiques of the limitations of the Husserlian formulation of intentionality. Second, all of them result in the re-interpretation of the givenness of experience. Third, all result in theological repercussions, some recognized and some unrecognized by their authors.

Although Husserl's philosophy itself, may be viewed as a resistance to the absorption of the world into consciousness, Levinas, Marion and Tymieniecka dispute, in various ways, the effectiveness of this refusal in order to propose alternative, more radical refusals (Levinas 1998a, 86). In the case of Levinas, the result is the reductive thematization of a specific kind of givenness as the primordially and existentially constitutive of subjecthood. In the case of Marion, the result is an attempt to accomplish an eidetic reduction of the varieties of givenness in order to produce a general description of the phenomena. And finally, in the case of A.-T. Tymieniecka, the result is to establish an ontology of what is given in the synthetic apperception of sense on the basis of a phenomenological reduction.

In a way, the three views of these philosophers are a practical object-lesson emblematic of the Marion's assertion of the implicit Husserlian principle of principles: "So much givenness, so many [phenomenological] reductions" (Marion, 2002, 14; 1998, 203). But simply because there is a plurality of phenomenological reductions does not mean that the notion of reduction itself is flawed. In fact, I would suggest that the investigations of all three phenomenologists are compatible, if some key assumptions are appropriately adjusted. But that it another story. My purpose, here, is to mine the discoveries of Levinas, Marion, and Tymieniecka in order to alloy a new interpretation of Trinitarian appropriation.

GIVENNESS IN LEVINAS

"The liveliness of life is the incessant bursting of identification. As if, dazzling or burning, life were, beyond seeing, already the pain of the eye overwhelmed by light; beyond contact, already the igniting of the skin that touches—but does not touch—the ungraspable" (Levinas 1998a, 166). Life is not an ekstasis but an enthusiasm, the awakening of the Self-same to the Other (Ibid.). This process is a sobering up from the intoxication with being, an intoxication which imagines that the self is in Being and may absorb the other in the same homogeneity of Being (Ibid.).

In Levinas' this reveille, this sobering up, is tantamount to a "living reason" which is no longer judged in Husserlian terms as the lucidity of self-evidence (Ibid.). A "living reason" is not the kind of reason that seeks repose in the "the Same," in knowledge facilely understood in terms of being; it is not the kind of reason that seeks "repose, conciliation, appeasement" in the ultimacy of the Same (Ibid., 167). Nevertheless, Husserl opened the door, particularly in *Ideas* 1, to a "transcendence in immanence," a differentiation and rupture in the midst of presence to self in the recognition that the other constitutes a fissure in consciousness deeper than its unity (Ibid., 176). Unfortunately, Husserl promptly closed the very door he opened on this possibility, when, according to Levinas, he attempted to derive the consciousness of the other through an egological reduction. Levinas radicalizes the openness to the heterogeneity of the Other by suggesting that it lies at the basis of the constitution of the self.

The "Other wrenches me from my hypostasis, from the here," where at the heart of my being and the center of my world, I posit myself (Ibid., 177). This wrenching out of self is the root of the philosophically deep awareness of my identity. This wrenching confers on me an awareness of the alterity of myself, the very alterity by which I wrench the other to a similar rupture in his solipsistic identity (Ibid., 177). As this happens, "the *here* and the *there*" are "inverted" and the I, once comfortably installed, moves into the background. Here, I see myself denuded before the other to whom I am obliged "to render account" (Ibid.). Facing the other is a sobering trauma which awakens the ego from its dogmatic slumbers; it awakens the ego to its freedom from itself but also to the realization that the Other cannot be assimilated to it (Ibid.).

The living reason described by Levinas is not the contemplative equanimity of the egological constitution and absorption of the other, nor is it the subsumption of both ego and other and their homogenization within the categories of being. Rather, living reason is the reckoning of living transcendence: This is the "[t]ranscendence in which, perhaps, the distinction between transcendence toward the other [hu]man and transcendence toward God should not be made too quickly" (Ibid., 178).

The transcendence which Levinas describes constitutes a salvation from the solitary self, a self which on one hand must face the horrific anonymity and homogeneity of the "there-is" (*il y a*), the self's alienation, and the self's false constructions of the being of the world around it (Purcell 2006, 98). It is the encounter with the other which brings with it immediate responsibility for that other, a responsibility which is also transformative for the self. In its "unanticipatable alterity," in its incomprehensibility and inexhaustibility, I confront the other as unlimited in him/herself and as an opening to the infinite which stands behind it (Levinas 1969, 34). Here, what is established in Husserlian terms is an intentional relation in which the noematic content is totally at the disposal of the other and infinitely beyond my ability to accomplish a concept of it in the intentional act (noesis) (Levinas 1996, 54). I am in complete passivity to it. Though lost in forgetfulness, the encounter with the wholly other grounds human self-hood and responsibility. This means that the Infinite has primacy as the noematic principle which is constitutive of self-hood. All philosophies which deny the irreducible nature of this confrontation are self-deceptive.

According to Levinas, the unsettling confrontation with a person in his otherness, as it were, bears with it the question, "What has he to do with me?" It bears with it the irreducible and undelimitable datum of responsibility, infinite responsibility for the other. "My responsibility is an exceptional relationship in which the Same can be concerned by the Other without the Other being assimilated to the Same" (Levinas 1998b, 13). Opposite the face of the other, I find that there is no limit to what I must demand of myself (Ibid.) Opposite the face of the other, infinite responsibility is given to me. Levinas calls this relationship a relationship of Deaconship [!] of absolute service to the other, a service where I lose myself in responsibility to the other. (Levinas 2000, 161).

GOD IN LEVINAS

The reduction to being which is impossible for the other is intensified when God assumes the place of the other beyond otherness. To speak of God as the eminent—one of Levinas' favorite epithets—is not adequate so long as that God is understood as an inhabitant of the house of being. For Levinas, God transcends being and every idea that would attempt to think him in terms of being.[16] A more apt, though hyperbolic, expression is that God is the pre-eminent. The only adequate approach to God is one which recognizes that: "This idea of God surpasses every capacity, its 'objective reality' as a *cogitatum* causes the 'formal reality' of the *cogitatio* to break apart. This claim overturns—in advance—the universal validity and the original character of intentionality. ... [T]he idea of God causes the breakup of the thinking that—as investment, synopsia, and synthesis—merely encloses in a presence, re-presents, brings back to presence, or lets be" (Levinas 1998b, 63). Nevertheless, the Infinite finds a paradoxical place within consciousness but one which transcends thought "which is structured as a comprehension of the *cogitatum* by the *cogitatio*." The place of the Infinite within consciousness is the placing-in consciousness a passivity beyond all passivity which may be actively assumed by the subject, and in this it is rather like the notion of an obediential potency to be passive, a passivity impossible without the Infinite's awakening one to that passivity (Ibid.). This non-intentional modality does not constitute an orientation in the subject directed to the fulfillment of evidences in the Husserlian sense. As Levinas puts it, this non-intentional modality is the disproportionate transcendence which makes the subject "hostage" to measureless responsibility for the human other but also to the Infinite other (Levinas 2000, 137–138). This relationship is tantamount to a saying whose interpellative force is not in *what is said* but in *the saying*. Its illocutionary force—the illocutionary force of the summons—is absolute and without a sedimentation of specifics (Levinas 2000, 161).

The relationship of radical responsibility for the other demands that Levinas characterize his movement beyond ontotheology as the development of a transcendental ethics. Simply put, ethics thus becomes for Levinas first philosophy and substitutes for that part of theology traditionally called "the doctrine of God." This means that God is thought no longer primarily in terms of being but, rather, that pride of place (as far as first philosophy is concerned) goes to the transcendental, the Good. God is not longer to be considered pre-eminently being but pre-eminently, *the Good*. Here, Levinas admits, he follows some hints in Plato but especially the themes of some of the Neoplatonists, and primarily Plotinus.

The indivisibility of the Good and its identity with the One, the way in which the Good brings the subject into responsible unity with the other, the way the Good elects me to responsibility in a way that is prevenient to a choice I might make in favor of irresponsibility, the way *the* Good is superior to any idolized descriptor of God as a lesser good or in terms of the characteristics of beings—all of these are the reasons that Levinas gives for preferring to consider God as the Good beyond ontotheology (Levinas 2000, 176–179). This is a God who is transcendent to the point of absence; this is a God whose saying can never be circumscribed in the ossification of what is said (Ibid., 204). This is a Good *who* is "otherwise than being."

GIVENNESS IN MARION

Based on his formulation of the implicit Husserlian foundation principle ("So many reductions, so much givenness."), Marion has approached the idea of givenness by following a path of phenomenological reductions designed to render the essences of the gift and the essence of its givenness. This investigation follows the analogy between the noematic and noetic correlates (in intentional relations) and the gift and the given.

Marion discovers three features of the gift. First, the giving of the gift is always accompanied by the withdrawal of the giver and the creation of a distance, a distance between the benefactor and the beneficiary and between the intention of the gift and its appropriation. No matter how gratuitous the gift may seem to be, once given, the temptation of the beneficiary is an appropriation of it. This appropriation is an appropriation of control and one of the forms of that control may be the interpretation of the meaning of the gift itself. The danger of the appropriation is that it occludes the communication of significance which is contained in the giving of the gift.

Second, even though the distance of withdrawal makes the appropriation of the gift a live possibility, it is also what preserves the giver as its origin, and because the givenness of the gift cannot be occluded—though its significance, might—the subject from whom or the direction from which it issued cannot entirely be effaced.

Third, in its purity, the giving of a gift is not fundamentally the giving of a thing. The thing given is an excuse for the gesture—or, better, is the instrument for the gesture, and it is the gesture which is what is supremely important in the purity of the giving. In its purity as an act of giving—in its sheer gratuity—the gift of *the* giving is the gift of *giving*. In other words, pure giving is perichoretic. It produces a spontaneous reciprocity by in which the giving fans out to others and, if possible, returns to the primary giver. In a community of generosity, generosity is returned.

When he take up the essential (intrinsic or noetic) aspects of the giving, Marion finds that there are three requirements and five conditions or determinations for the giving of a gift. The requirements speak to the empirical reality of the giving. That it be recognized as empirical, the giving, first, must be thematized simply in its givenness and not as the preapprehension of its relation to its giver; it must show its givenness in the immanence of consciousness. Second, it must be thematized as irrevocably given. Third, it must have its whole significance as givenness (Marion 2002, 119–120). These requirements characterize the purity of the phenomenon.

But the pure phenomenon also possesses modalities or determinations which are unique, once its empirical nature is established. (1) A gift is given anamorphically, because its significance is revealed only when a particular perspective is achieved; (2) it is given contingently and unexpectedly; (3) it is given as an indisputable fact whose efficacy cannot be disputed; (4) it affects me with a suddenness and excessive profundity which is irreducible to its form; and (5) the giving of the gift is the giving whose definitive character is that of an event which precedes its cause and, in fact, cannot have an adequate cause (Marion 2002, 123–170). Emancipated from classic notions of causality, three *notae*—or distinguishing marks—may be abstracted from the five determinations: the phenomenon of any particular givenness is unique and unrepeatable; the phenomenon of givenness is excessive because it defies all expectations; and the phenomenon of givenness is supercharged with possibility because it exceeds the limitations essence imposes on potentiality (170–173).

GOD IN MARION

For Marion, the thinking of the being of God is, as in the thought of Levinas, the audacious attempt to limit God to an enclosed horizon of being which is preapprehended and capable of being thought. But God may not be thus limited. Like Levinas, Marion opts for a rejection of ontotheology with the hope of transcending totality.

What this means is that the only way God can be thought is not in subordination to humanly established categories of being, but only on God's terms. God himself dictates the conditions for the possibility of his own conceptualization. These conditions are freely achieved in the way God gives himself as love, that is as preeminently a gift that dictates the terms of its reception (Marion 1995, 49). There is a remote similarity between Levinas' attempt to think the goodness of God—which is a matter of the relationship between the summoned and the Summoner—and Marion's understanding how the gift of love establishes the ground rules for thinking a God without being—which is a matter of the relationship between the beneficiary and the Benefactor. Both seem to opt for an *analogia relationis* over an *analogia entis*.

Marion deviates from the radicality of Levinas' call to an anti-idolatry of thinking the Good, especially where Levinas treads very close to the acceptance of the relationship between the death of God and the death of metaphysics. For Marion, the death of metaphysics is not equivalent to the death of God. To make it such is to succumb to idolatry. To understand this claim, one must understand that Marion views idolatrous conceptions of God as those which in their ability to dazzle us are reflective of the crass values we hold in common with the carvers of the idols. The idol is an idol because it is reflective of the idolater's self-idolatry. The attempt philosophically to take on these false idols and to best them by installing some other ultimate in their place is to make more idols. Moreover, even those who take the critical leg of this dialectic of the idea as having finality without accepting the newly installed idol, succumb to idolatry, indirectly. Thus, even Levinas strays very close to idolatry by not being sufficiently critical of the critics of the idea of God. For Marion, the death of metaphysics does not mean the death of God; it is merely a call to a new way of thinking God.

The iconic provides a preferable way of describing how the Infinite might be expressed through the finite. Unlike the idol, the icon is not reflective of the worshippers own gaze. Instead, the icon becomes translucent to the divine and draws the Infinite through the finite to the worshipper. The worshipper's gaze does not stop at the finitude of the transmitting image but transects the finite surface to intend the Infinite that it presents. In the process, the gaze of the worshipper crosses the gaze of the Worshipped. In the relation of the worshipper loving the Worshipped "the weight of the other's unsubstitutable gaze as it crosses ... [the] intentional aim [of the worshipper]" (Horner 2005, 70). As Robyn Horner has observed, this is a quasi-Levanasian intentionality of love, in which the weight of Worshipped's "gaze is experienced as an always-prior injunction that exposes and obliges me" (Ibid.) Unlike the idol the eikon is not *auto-representational*; it is *hetero-presentational*. The notion of the iconic can be broadened beyond religious artifacts and include the possibility that religious experiences and even religious ideas might function iconically. In considering the relevance of the iconic to religious experience, Marion sometimes ties together the themes of his phenomenology of givenness and his analysis of the eikon by suggesting that the experience of love can be bound up inseparably with both.

Though Marion's Christology is surprisingly underdeveloped, it is clear that Jesus Christ is the example of love, gift, and eikon par excellence. As eikon, Jesus is the very expression of the Father's infinitude in the form of finitude. In his coming into the world, Jesus' incarnation has all of the idiomatic features of the gift enumerated above. As the gift of the love of the Father, and as a lover himself, he stands as another subject, the knowledge of whom I have access to only on the condition that I accept his acceptance of me and that I return the gift of charity to him, that is, on the condition that I accept him in his otherness as he already accepts me in mine, so that he does not become a constitutable object (Horner 2005, 71; Marion 2002, 160–167).

Connected with the braid of themes—gift, eikon, love—is one of the most controversial claims of Marion is that supernatural revelation is an example of (what he calls) the "saturated phenomenon." A saturated phenomenon is a phenomenon which is not destructive of intentionality, inasmuch as a relationship between intention and fulfillment is preserved, but it is a phenomenon which gives itself in its fulfillment so excessively that the corresponding intention is overwhelmed (Marion 2004, 112). The icon is an example of such a phenomenon; so is divine revelation, and so, one would assume, is mystical experience. In maintaining that revelation is so terrifically under-determined—Marion calls is the saturation of a saturation because it possesses so many modalities as not to be unitary—he installs an incredible difficulty in its interpretation, at least as a present experience and not as a sedimented literature.[17] This is compounded also by his various statements to the effect that the saturated phenomenon does away with its noetic and noematic horizons (Marion 2002, 200–211; Horner 2005, 113–114).[18]

GIVENNESS IN TYMIENIECKA

Of the three thinkers considered, here, Tymieniecka represents the philosopher whose thought remains closest to the ambit of the Husserlian philosophy in how it approaches the problem of givenness and what is given. The direction of Tymieniecka's thought is in a direction different from Levinas and Marion. Her concern is not the with the apophatic transcendence of the hyperessential; her philosophical project is directed to the transcendence of being *within* the horizon of what she calls "beingness," a beingness that is rooted in an unconditional ground, "the God of all creation, Who announces himself as ... 'I am Who Am' " (Tymieniecka 2009, xxix). If the thought of Levinas and Marion is directed to the hyper-transcendence of a summons or a gift that come from beyond being, her thought is directed to the transcendence that is possible within the horizon of beings that share *esse commune*. Nevertheless, she recognizes that any such science of being must be rooted in a God who is Beingness, itself. This constitutes a return to metaphysics that more nearly approaches kataphatic philosophies of being. (I will say more about this later.)

Like the thought of Levinas and Marion, Tymieniecka begins with a critique of Husserlian philosophy. Like Levinas and Marion, Tymieniecka does not believe that Husserl's reductions reach the bed-rock of reality, a foundation which is equivalent to the processes of the emergence and individuation of beings (Ibid., xxviii). What is required according to Tymieniecka is "a new critique of reason" which seeks a break from its narrow traditional framework to deal with the dynamic "currents of existence" and generates additional criteria of "validity, predictability, prospects, and measure" (Ibid., xxiv).

The phenomenological approach that Tymieniecka has in mind involves "the dissolution of traditional forms of seeing reality" with the purpose of reaching "life's generative routes, the paths of the logos carrying the individualization of beingness" (xxvii). "It is the logos of life in its ... laying down of its course that gives us the access to the very becoming of beingness" (Ibid. xxvii). This—in contrast to Levinas—is what protophenomenology is about, this is the true first-philosophy. Relative to Levinas, Tymieniecka's is a drawing away from existential phenomenology back to phenomenology as a science of being.

According to Tymieniecka, humans recognize themselves as subjects "not by a cognitive act but by 'being alive'—by experiencing ... [themselves] within ... [their] milieu of beingness, directing ... [their] instincts and appetites, recognizing the elements of the circumambient world in their vital relatedness to ... [themselves], but ... [above all] by recognizing that [each is an] ... acting center of the universe ..., a self-sustaining agent who directs ... his own course and who ... endows that course with moral and aesthetic values ... and ... seeks to understand" the reasons for all that is (xxxi-xxxii). Tymieniecka's phenomenology is directed to the beingness of life, a beingness characterized by "constructivism, energy, metamorphic versatility," and the force which prompts growth and dissolution "in the regenerative fonts of the Unconditioned" (Tymieniecka 2009, xxvi). "Life is the conveyor of beingness. It partakes of its fullness" (Ibid., 3). What is sought to make sense of beingness is the "sense of sense," the Logos. The Logos must be disocluded, however. It must be rediscovered from the maze of data. It must be shown that it is that within which all reality is and is made possible (Ibid.).

Tymieniecka is in agreement with the Husserlian search for an approach which would establish phenomenology as first science, but also with his intention to explore the "logos of interrogation" which in the Cartesian Meditations should function as a "Phenomenology of phenomenological reduction" (12). We have seen that both Levinas and Marion tried their hand at the latter. Tymieniecka also has some distinctive ideas about how this is to be accomplished.

Tymieniecka's complaint about the deficiencies of Husserlian phenomenology go to the heart of the idea of intentionality. She disputes that intentionality ought to be presupposed as the "exclusive and dominating function in the human constitution of reality" (18). Undoubtedly, intentionality is "the key" to the functioning of consciousness, but it is not the complete process. Tymieniecka gives uncharacteristically succinct expression to the Husserlian omission that makes an opening to a phenomenology of life in the following passage:

The intentionality of consciousness is, indeed, the key to its functioning. As we all know ... it orients the act of consciousness in a triangular setup (the ego pole, the acts streaming from the flux, the being directed toward an objective aim); it organizes the cognitive context as the constitutive context of the objects, a context that establishes our reality. Husserl famously distinguished noetic and noematic sides of this ... act of aiming at an objective grasp. That means that the logos whose objective intention carries the act splits into subjective and objective sides, one representing the side of active performance and [19] the other that of objective shaping. Yet it is the "same" logos as it proceeds in its intimately correlated twofold way to bring forth the *presencing* of phenomena. IS IT NOT EXTRAORDINARY HOW THE LOGOS ACCOMMODATES THE 'EXTERIOR' TO CONSCIOUS ACTS THAT BY 'INTERIOR' ACTIVITY PRESENCE THEMSELVES TO THE LIVING SUBJECT THROUGH ITS OWN PERSONAL MECHANISM? [My emphasis.] The intentionality of consciousness acquires in the Husserlian schema this unique role of operating simultaneously a distinction, an operative split, such that the logos carries out its work of constituting human reality within and without, first by promoting the flux of acts, and second by endowing them with three-directional orientation to be acts of and "for" the self ... and shaping a presentational content. In this conception of presencing reality through consciousness, Husserl introduces a distinction between conscious but empirical acts, which presence reality in its changeable, fleeting appearances, and intrinsic "pure" intentional acts ..., in which the noetico-noematically revealed phenomena emerge. ... Consciousness' noetico-noematic structurations assume the character of necessity ... [and] the eidetic findings are now seen to be in relation to their formation within consciousness. (19).

But, according to Tymieniecka, this account of "the intentional shaping of reality ... does not suffice to account for it" (Ibid.) It does not address the logos in it "incipient phase" and it has been relatively unsuccessful in achieving a reduction of the empirical and the hyletic (20). Moreover, the account of the constitution of reality is incomplete and hangs "in thin air" (21). It is Tymieniecka's purpose to supply the last and grounding reduction, the one that will provide an explication of the logos which undergirds the coordination of the interior and exterior of the constituting acts. This is equivalent to a reorientation of phenomenology away from an obsession with intentionality to a consideration of the role creativity plays in the constitution of givennness. This straightaway blurs the line between a phenomenology which would remain description aloof from the creativity of life and a phenomenology that recognizes that this creativity lies at the heart of even phenomenological procedure (26). The Logos thus reveals itself as a shaping force, but it is also sentience.

The assertion that the Logos of Life is sentient means that it is not a disorganized aleatory force; it is a shaping force (30). At all levels of life's diversity, sentience is the characteristic of its organization, "an essential element in *all life*" (30). In this way, Tymieniecka grounds the Husserlian description of the constitution of experience. She maintains that the coordinating logos between what is external to consciousness and what is internal to consciousness is a constructive sentience already at work at all levels of human experience. This means that with respect to all creativity and all individuation there is no outside to the Logos of Life. It is at play in everything that is. "It is not only that in its innumerable guises sentience pervades all elements, factors, and levels of life's diversification, form the amoeba to the angels, it is also that sentience enters into life's animus, bursting forth as its essential factor" (30). Logoic sentience is the "quintessential core of life" (Ibid., xxix). Sentience runs its dianoic thread through all of life.

The mode of givenness in the Tymieniekian idea of the Logos of Life is twofold: It is the givenness of what is necessary to make sense of the Husserlian formulation of the constitution of experience, including the grounding of intentionality, itself, and it is the givenness of what is overlooked in the classic Husserlian approach because it is difficult or impossible to thematize as an accessible object.

GOD IN TYMIENIECKA

As is with Levinas, though less so with Marion whose philosophy is pitched—despite its denials—to the solution of some theological problems—Tymieniecka's philosophy does not bear its theological affinities in plain view. Even so, it is possible to mine, with some great effort, the theological gold that is buried, there. Like Levinas' account of the divine, Tymieniecka's account is more closely connected with what has been called natural theology; its extension to the realm of Trinitarian thought must, therefore, be by way of correlation.

The context for the development of Tymieniecka's philosophy is her personal concern with artistic creation and the way creativity is operative at every level within the house of being. Although her philosophy, as much as that of Levinas, claims to be a universal philosophy, it is clear that just as much as Levinas is intent upon thinking a postmodern philosophy under the themes of the Good and the Ethical, Tymieniecka is intent upon thinking a postmodern philosophy in terms of ontopoiesis and—I would argue, though I would be hard-put to find her say it in her own words—the Beautiful. Apart from her constant evocation of creativity, which occurs on virtually every page of her corpus, there are two especially important discussions in which this thematization occurs. One is the discussion of the notion of *animus*, the other is how she characterizes the virtuous development of the individual.

Tymieniecka associates the decisive factor by which the Logos of Life manifests itself and spreads itself in vast circuits as "*animus.*" It is the full glory of the Logos of Life the emergence to which the apparatus of individualization is directed. Unthinkable in itself, and eluding objectification, the animus is "in-grown" into all beings which have purpose and direction. It is not epiphenomenal but the heart of the individual's beingness (Tymieniecka 2009, 5–9).

Animus has five characteristic features. Animus is especially manifest in the unity of a living being's teleological orientation, particularly in the way it harmonizes all of its operations, all of its tensile forces, to achieve its end (7). Second, it harmonizes these various constitutive "streamlets of life" in a unique configuration with proportions and correspondences unique to the individuated being. Third, it is operative in all features of the being's nucleus so that even in interaction with other beings it is engaged in the construction of its resident being's unique identity. Fourth, its reach is through all of the capacities connected to the survival and striving of the individual. Fifth, it has as many modalities as there are living beings, "from the vegetal ... to the most complex" and "stands for the reacting, sensitive, sentient, emotive factors of life's becoming" (8). In her description of animus, Tymieniecka comes close to the classical descriptions of what created beauty is, only she has provided a description of its operative agency in the world and has named it soul or spirit.

Secondly, the progress of the soul in its sacral development shows its development in relation to the Divine. At its first stage, persons are constituted in their nascent humanity "in a vital network" shared with all living beings, that is at "all [common] levels of self individuation" (221). Individuation is pitched—at this early stage—toward the soul's interiorization of "vital-cosmic" interconnections with the "unknown, mysterious, incomprehensible, marvelous" (221). Here a distinction is postulated between the profane and the sacred. Transcendence is interpreted in terms of the experience of the uncanny other forces. At the second stage, the constitution of persons is in connection with the sociocultural world and intersubjectivity of others. This is Husserl's "community of human consciousness" which transcends the basic vitality, singularity and quotidian existence of individuals toward the development of a common spirituality, in a specific sense (222). Here, the enrichment of the human soul is a result the sacred shared with others and transcendence is a transcendence of the individual toward the group's spirit. Neither vital-cosmic transcendence nor communitarian transcendence is an "authentically religious experience of the Divine," however (222). Both are turned outward in their transcendence while the third, authentic, experience of the sacred is turned inward to the human creative act and its "transcendental-intentional" clarification (Ibid.).

According to Tymieniecka, it is the inward sacred which is the "deep work" of the soul. That deep work does not issue from the outside but from the inside of human experience. She puts her thesis in the interrogative mode: "How [is it that] revelation of the divine could be acknowledged as such and accepted" if God radically transcends the lifeworld, the human, and everything which radically encloses humankind in its "finite intentional circle"? How is it that humans can " 'listen' and get in touch with Him"? (223).

Of course, each question is answered in a different way by Levinas and Marion. Tymieniecka's answer is in a Witness whose presence is radically other "because he cannot identify himself with any living being, with anything known and with [239] nothing that could be known, because he ... [is] radically other than ... to all that is present, but also to all that which is possible" (239). At this point, Tymieniecka makes the meontic turn (in her own way) by defining the Witness as beyond possibility, possibility as defined in terms of the horizon of beingness. He is a Witness "that completely penetrates us, that participates in all our movements, that inspires all our being though his own presence"; he is who Augustine identifies "God within us and us outside of ourselves" (239).

Authentic sacrality requires that exceptional conditions be fulfilled for any message to be received by the soul since the Witness is not a part of the ontopoietic process. It requires a confiding of transformation to the Witness, and it results in a path of moral and spiritual development not determined by the Logos of Life (242). It cannot be accomplished from the side of human intentionality. An inner transformation is required by which self-detachment, self-sacrifice, and an inner communion with all beings opens "the horizon of hope for the blessedness of peace in communion with the Witness" (241). As its growth in identification with the Witness occurs, the soul comes progressively closer to a fulfillment in which is the repairing of our inadequacies and harmony with all creation and in which the divine instantiates us according to a unique measure, in the horizon of the Logos of Life (253).

REVELATION AND THE GIVENNESS OF THE GOOD, THE TRUE, AND THE BEAUTIFUL

Levinas and Marion agree that the problem with ontotheology is that it does not take the transcendence of God seriously, that is, its epigones imagine that Being is rationally de-limitable, that it could be reduced to rational descriptions and evidences. But we should be careful not to imagine that all theology is simplistic in this assumption. This is not what the idea of the analogicity of Being entails—at least not in Thomas' understanding. Thomas puts the font of Being—God—not on a continuous scale with his creation, but following Pseudo-Dionysius the Areopagite, he puts one end of the scale in created finitude and the other at infinity, but with a lacuna between the remotest end of the scale of created being and its infinite endpoint. In the theology of Thomas, all that is claimed for understanding is its adequacy to the purpose for which humans were created, not the rational circumscription of the infinite. How far we are to expand this adequacy of the knower to the known—whether we are to associate it with the Marxist dictum in the *Theses on Feuerbach*: "That humankind cannot raise any question to which it cannot find an answer", or whether it means something much more modest—is not entirely clear. Thomas' mode of operations, however, gives us some clue. His view is synoptic and one which pushes for reasonable description, whenever it is possible to give it.

Hans Urs von Balthasar has well made the point that though the *analogia entis* was made present in Christ this *does not* license the theologian to imagine that the intra-worldly scale of being (and the transcendentals which accompany it) and the extra-worldly hyperessential being of God and his accompanying freedom of revelation (with its perfusion of transcendentals) are a continuous scale. The preservation of the *analogia entis* entails recognition of its limitations, that it must not entail a projection of the features of a finite being onto the infinite, a projection of the distinction between essence and existence into God. To do otherwise is to succumb to Heidegger's condemnation of ontotheology (Urs von Balthasar 1991, 91–92). The "solution" proposed by Von Balthasar is one that takes up some of the Levinasian themes, though in a less radical form:

The real "identity" of God that unfolds the vitality of the transcendentals in its . . . inconceivable way in God's threefold personality lies—to speak with Plato and then later with Gregory of Nyssa and Dionysius the Areopagite—*epekeina tou ontos* (the far side of being [or otherwise than being]), above and beyond what we can still make out as "the 'to be' of beings" [*Seiendsein*]. Only from this "above and beyond" that points beyond all [worldly] order and law-likeness . . . to God's . . . sovereign freedom [by which he can] make use of the most comprehensive reality of all that we know. Being, not so as to define himself ([as] "I am who am"), but in order to characterize his inconceivable free turning to us ([as] "I will always be who I will always be"), in contrast to the idols that are always "identities woven whole cloth out of human thinking". This transcendence over what we think of as identical . . . is revealed in Jesus Christ. Only in this way is God's perfect freedom unveiled as an inner vitality in which the transcendentals are identified with his identity. There is no possibility of separating the life of the three Persons from God's essence. The essence is no fourth el- [93.] ement, something common to the three persons. Rather, it is in their eternal life itself in their processions. This is why God's "Being" (thought of as a substance) does not manifest itself in the true-good-beautiful. On the contrary, the manifestation of the inner divine life (the processions) is as such identical with the transcendentals, which are identical to each other (Ibid., 92–93).

In the above passage, Von Balthasar establishes several points which will figure into my conclusion. First, the thinking of God's being is not proportioned to human understanding; the adequacy of intellection does not extend to it. Thus, in a real sense the Trinity is a God beyond being, if being is understood as *esse commune*. Second, the manifestation of the inner divine life of the persons is identical with the transcendentals. Third, there is a sense in which something which approximates the Barthian *analogia relationis* provides a more direct avenue of approach to the revelation of God than the *analogia entis*. This last point is established, in the above passage, by Von Baltahsar's allusion to the covenant as the revelation of God's *relation* to humans. Similarly, relation (not being) is primordial in God's revelation according to Levinas, it is equivalent to the summons. It is possible to translate this privileging of relation over being into Husserlian terms by saying that revelation is a noesis without a clear and distinct noema. However, even if relation is seen to be the modality under which God manifests himself to humankind, it is not inconceivable that relationality could provide a mediating category by which God's being could be intuited. Here, the right question is whether God in his relations with humans reveals truthfully what he is in himself. The mediation of what God is by relation should not be understood as a path around the impasse of the impossibility of the finite fully comprehending the infinite, however.

The previous respective formulations of givenness in relation to the transcendentals has relevance to the intratrinitarian life.

The personal differentiations which are defined by the Father's generation of the Son and the spiration of the Holy Spirit by Father and Son are examples of the emergence of the hupostases and the relations of opposition which exist in God. For the Christian, the otherness of the Father, Son, and Holy Spirit in the Trinity is the metaphysical symmetric ground of all personal otherness in creation. The actualized infinite responsibility of the Father for the Son and the Son for the Father and the Father and the Son for the Holy Spirit and the Holy Spirit for the Father and the Son is realized in the perichoresis which enacts the mutual kenosis of one person for the other. Here, the infinite mutual responsibility of one for the other is actually achieved—each fully gives himself to the other—in a way that cannot be achieved in the responsible actions of humans. More importantly, each philosopher's conception of givenness has repercussions for the reformulation of the Trinitarian appropriations and missions in the economic Trinity.

THE FATHER AND THE APPROPRIATION OF THE GOOD ACCORDING TO LEVINAS

When Levinas describes the summons of the individual to infinite responsibility for the other, he apparently thinks that this summons points to the Infinite but only through the other human. This is the equivalent of the commandment to love one's neighbor as oneself. Levinas seems to think that the summons of the Infinite can be derived from this responsibility, alone. But this ignores both the summons God makes directly to the individual and the call to infinite responsibility for the supreme other that this other kind of summons entails. Even though Levinas would prefer to arrive at the idea of the Infinite through a kind of methodological atheism which brackets revelatory phenomena other than what is awakened by the faces of other humans—this being a more persuasive demonstration of its truth—Levinas would be hard put to reduce all of the revelatory modes of the Hebrew scriptures to the encounter with other humans. There are scriptural passages where what faces the individual, if not the face of God, is at least God's manifestation. Thus, apparently left out of the summons of the Infinite is the commandment to love God above all things. A Trinitarian theology sensitive to the communication of God in the Old Testament must uphold the re-instauration of the second aspect of this infinite summons; it ignores this re-instauration to its own peril.

Levinas' analysis of infinite responsibility is a description of God's call through the natural order, but he would have to admit an additional qualification to square with the data of the Hebrew scriptures. Required is the qualification that Levinas must make a place for the way for God to address his people which is characteristic of the supernatural or paranormal order. Required is that we think supernatural revelation within the philosophy of Levinas.

What one then discovers is that in the form of supernatural revelation, the content of God's summons is nonetheless likewise attenuated because of its noematic excess. This means that those to whom it is revealed often succumb or are constantly tempted to succumb to a premature limitation of its meaning. The infinite summons is constantly submitted to the distorting finitude of the individual. Just as the idolatry of being is an attempt to equate God with the finitude of being, so too the idolization of the summons is the attempt to equate the summons of God with a finite meaning which carries something less than infinite responsibility. Judaic and Christian histories are narratives replete with the idolatrous finitization of the infinite summons. This is shown in the development of religious law in the Judeo-Christian tradition where the expansion of the humanistic interpretations of the law saves it from its potential ossification. How it is saved in this process is by a kind of triangulation, the faces of humans provide the corrective to the idolatrous ossification of the law. In its infinite call to responsibility, it is made clear that the Law was made for humankind not humankind for the Law. A Trinitarian theology sensitive to the communication of God in the Old and New Testaments will recognize the incommensurability between the summons and what concretely one is summoned to, or in the words of Levinas, between the "the saying" and "the said."

Having made these stipulations, I would argue that corresponding to the mode of the Father's givenness, Levinas' characterization of it as an ethical summons means that in this revelatory givenness the transcendental of the Good is economically conveyed. Thus, the Father is first encountered in his summons to the Good. And, as Levinas has put it, the Good and the One are an inseparable unity in that revelation. The givenness of the Father is the Good.

THE SON AND THE APPROPRIATION OF THE TRUTH ACCORDING TO MARION

In light of the notions of givenness in Levinas and Marion, the economic mission of the Son and his appropriation of the Truth may now be addressed. Jesus Christ is the incarnation of God, the eikon or very image of God, the fullness of the infinite summons in the finite. This means that Jesus is perfect finite enactment of the summons but that he may also, it the first person, claim to be the very expression of the Father to whom is handed all the Father's power, so that his power to address the summons to others and to interpret the summons without its idolization falls to him. Jesus is the living law. The sinlessness which the Gospels claim for Jesus means that in all of his actions he never deviated once from the perfect enactment of the summons, he never fell short of what he was called to. His redemptive death was the actualization of infinite responsibility for everything in a finite death. His sharing of the hyperessential unity of the Father makes him unique as the only-begotten Son, but he is unique in his awareness of his unity with the Father, unique in his ability as subject to assert that unity, not as an identity with the person of the Father, but as having hyperessential unity with the Father, and being the perfect presentation of the Father, in Trinitarian immanence, and the adequate presentation of the Father incarnationally, in his earthly mission.

Jesus, in his economic mission, appropriates Truth as his transcendental because, in the classical sense, Truth is the adequation of being and concept; it is the confidence that the concept is adequate to convey the reality. Jesus, as the Son and eikon of the Father, is the Truth because he is the perfect fleshly—the perfect human adequation—of both the summons and the Summoner. And he is the Levinasian *other*, both as human and divine, both as other and otherwise than the other. He is the Truth because he establishes a commensurability between the infinite and the finite where the Infinite stands to reality as the human nature stands to concept. Hence the claim of the *Gospel of John* that he is the Logos. In the first person, he thus elicits infinite solicitude and responsibility to both great commandments. And yet, his incarnation of the Infinite and the summons is not a circumscription. Like Marion's eikon, he is merely the most *adequate* expression of the Infinite proportioned to our finite receptivity. Here, truth as adequation is reintroduced, but the very notion of adequation suggests two consequences (a) adequacy is always adequacy to some purpose and (b) the very conception of Truth as adequacy points beyond itself to deeper reality which stands behind it. Even as the incarnation of the Infinite, Jesus did not do all that he might have said, or say all that he might have done, but what he did say and did do was sufficient to establish that he was the model of perfect obedience to the summons. His actions and teachings, like the summons of the Father, is open to infinite expansion, but— also like the summons—they are in peril of idolatrous reduction to something less than the fullness of what they convey.

The givenness of the Son as the adequation of the Infinite takes the transcendental Truth as its economic appropriation.

THE HOLY SPIRIT AND THE APPROPRIATION OF BEAUTY ACCORDING TO TYMIENIECKA

In light of the notion of givenness and especially what is given according to the thought of A.-T. Tymieniecka, the appropriation of the Holy Spirit may now be adequately described as that of Beauty. In order to understand how this association may be made we must have recourse to Thomas Aquinas' classic description of beauty. In the *Summa Theologiae* (ST 1a, Q. 39, Art. 8) he provides his most expansive description of the essence of the beautiful as consisting of integrity (*integritas*), or completeness or perfection (*perfectio*), due proportion (*debita proportio*) or harmony (*harmonia*) and clarity or splendor (*claritas*). Tymieniecka's account of the ontopoietic fashioning of the individual is supercharged with these predicates, each of which is repeated numerous times but expanded according to a modern appreciation of the dynamism of life. A single passage will suffice, here, to remind us how beauty enters into her every discussion, even her discussion of other transcendentals. "No wonder that truth, in the experience of its crucial significance as the vortex of all measures, proportions, calculations, harmonies, and disjunctions in all the ontopoietic horizons from the vita to the sacral, possesses the deepest fascination and pervades all we undertake, aim at, thirst for, and enjoy as human beings, one equal only to that of the all-encompassing ecstasies of the Glory of the Fullness" (Tymieniecka 2009, xxxii). This passage is notable because it is reflective also of the title of her late friend—John Paul the 2nd's great encyclical, where—making use of Von Balthasar's own ruminations on the transcendentals—he calls the ethical presence of Christians in the world, the Splendor of Truth or, better, the Beauty of Truth.

Even though it is ostensibly a statement about Truth, in this passage, Tymieniecka associates aesthetic language with Truth as ordering being, an ordering which is experienced as a fascination felt in the dynamic measuredness of creation but also in relation to the glory of the infinite plentitude. By its description, this passage indicates the aesthetic component to the contemplation of being's order; it is also a worthy description of what Beauty is as a transcendental. This tendency to consider all features of being through the lens of beauty allows me to argue that Tymieniecka's phenomenology and her understanding of givenness are inseparable connected with the thematization of being as beautiful.

Because the Holy Spirit has been theologically associated naturally with the creative informing of reality, Tymieniecka's discussion of the Logos of Life bears affinities with it from the philosophical side. But when she described the way in which the Witness is constitutive of the human trans-natural destiny, she also speaks of a greater possible harmonization of the divine and the natural whose purpose is perfection and whose individual unit is the human soul. This is a perfection that has repercussions for sacred life in community and bears resemblance to the supernatural individuation of the members of the Church, who in the development of their unique spiritual gifts, are well-ordered constituents of the body of Christ, a body that Christ directs through his harmonizing Spirit. In her depiction of the integrity, harmony, and clarity of the operations of the Logos of Life and the complementary guidance of the Witness, Tymieniecka has identified two sides of the same coin. These are the two aspects of the mission of the Holy Spirit as well as his identifying transcendental—beauty.

Let me conclude this "little sketch" with a very brief recollection: In this paper, I have attempted to provide an insight into how the notion of the Trinitarian appropriations might be helped and updated through a reflection on contemporary phenomenological analyses of givenness and intentionality. In other words, I have *appropriated* the philosophy of Levinas, Marion, and Tymieniecka to make the notion of Trinitarian appropriations stronger and to suggest that they be more directly linked to the missions of the Trinitarian persons in the world.

As to whether all of the presuppositions of the respective philosophies considered, here, are themselves compatible, I have not ventured an answer.

Notre Dame Theologian in Residence, St. Thomas Aquinas Center, 535 W. State St., West Lafayette, IN 47906, USA
e-mail:ryba@purdue.edu

NOTES

[1] Jürgen Mettepenningen. *Nouvelle Théologie—New Theology: Inheritor of Modernism, Precursor of Vatican II*. London: T. & T. Clark, 2010, pp. 141–145.

[2] This distinction has only been viable in the Roman Catholic tradition since the 16th century. According to *Ad Tuendam Fidem*, dogma is distinguished from doctrine.

[3] This is not to preclude their cooperation in this revelation but merely to suggest that each person has a distinctive mode of revelation or givenness, both naturally and supernaturally.

[4] ST 1a, Q. 13 Art. 5.

5   The relevant passages are in the Parmenides (141d–142a), the Republic (509b), the Timaeus (28c), and Epistle (7, 341b–d).
6   These are all taken up later, by Christians, as approaches to God.
7   Despite the claim sometimes made—always without references—Aristotle never uses the word "*huperbainein*" in connection with the transcendentals. He never names the transcendentals using technical terms, either nouns or verbs.
8   R.T. Wallis. *Neoplatonism*. London: Duckworth, 1972, p. 154. Wallis says that the origin of this shortlist and their elevation to the status of supreme principles is based upon their treatment in the *Chaldean Oracles*. It becomes popular in Christian philosophy and theology, later.
9   Thomas also considers (what he calls) the invisible missions of the persons.
10 See "approprio, appropriate, and appropriation" in Roy J. Deferrari, *Lexicon of St. Thomas Aquinas*, 77b–78a.
11 Roy J. Deferrari. Lexicon of St. Thomas Aquinas. 233b.
12 Here, the necessity of the appropriation would not be causative from the side of creation – that is conditioned by it – but rather would be the result of a Trinitarian person invariantly choosing that transcendental for its self expression because of its expedience in doing so. This follows from the principle that God does not violate the nature of the created being but only enlarges it according to its potentiality.
13 Thomas is so terse in his explanation that this is the best rendering I am able to give.
14 In this example, the kataphatic and apophatic approaches to the description of the appropriations are both in effect.
15 Thomas denial that the Trinitarian persons possess the transcendentals as proper (*proprius*) or distinctively characteristic qualities is a denial that they are not common essential possessions of the other Trinitarian persons (ST 1a, Q. 39, A. 8, Res. 1–2). But simply because they cannot be considered to qualify the persons uniquely does not mean that they cannot be assumed as unique and distinctive features of the Trinitarian missions, both ontologically by way of some similitude to the immanent missions of the persons and economically because of they are modes of the missions of the persons in creation. See also; Roy J. Deferrrari, *Lexicon of St. Thomas Aquinas*, 906b–907a.
16 Levinas' rejection of being-talk is not, however, absolute. He recognizes the necessity of it with respect to sciences of immanence, sciences which have achieved amazing technical results.
17 Authority (or the authority of a tradition) guided by the Holy Spirit understood as the context of revelation helps dissolve its equivocity as a saturated phenomenon, but that, then, raises the problematic as to how authority is given. If authority is itself a matter of revelation, then one has a circle of dependent saturated phenomena.
18 To my mind, Marion treats the importance of the horizon(s) of the phenomenon with insufficient appreciation of its/their necessity. By my understanding, Husserl had a topological analogy in mind, an analogy whose features are the geometry of a sphere, an individual's position on it, and the global nature of being as bringing surprises but none which would absolute deviate from one's expectation of what is beyond the horizon. Marion's claim about the supersaturation of the horizon might be more clearly explicated by a shift in the kind of horizonality in the analogy, a shift from a spherical horizon to a hyperbolic horizon, for example. The horizons of the noematic and noetic correlates are topologically distorted. They no longer bear the contours of a totality but open hyperbolically into infinity.

REFERENCES

Horner, Robyn. 2005. *Jean-Luc Marion: A theological introduction*. Farnham: Ashgate Publications.
Levinas, Emmanuel. 1969. *Totality and infinity: An essay on exteriority*. Pittsburgh, PA: Duquesne University Press.
Levinas, Emmanuel. 1996. *Basic philosophical writings*. Bloomington, IN: Indiana University Press.
Levinas, Emmanuel. 1998a. *Discovering existence with Husserl*. Evanston, IL: Northwestern University Press.
Levinas, Emmanuel. 1998b. *Of God who comes to mind*. Stanford, CA: Stanford University Press.
Levinas, Emmanuel. 2000. *God, death, and time*. Stanford, CA: Stanford University Press.

Marion, Jean-Luc. 1995. *God without being: Hors-Texte*. Chicago: University of Chicago Press.
Marion, Jean-Luc. 1998. *Reduction and givenness: Investigations of Husserl, Heidegger, and phenomenology*. Evanston, IL: Northwestern University Press.
Marion, Jean-Luc. 2002. *Being given: Toward a phenomenology of givenness*. Stanford, CA: Stanford University Press.
Marion, Jean-Luc. 2004. *In excess: Studies of saturated phenomena*. New York: Fordham University Press.
Purcell, Michael. 2006. *Levinas and theology*. Cambridge: Cambridge University Press.
Tymieniecka, A.-T. 2009. *The fullness of the logos in the key of life. Book 1: The case of god in the new enlightenment*. Dordrecht: Springer.
Urs von Balthasar, Hans. 1991. *Epilogue*. San Francisco, CA: Ignatius Press.

SECTION VIII
CREATIVITY AND THE ONTOPOIETIC LOGOS

WILLIAM D. MELANEY

BLANCHOT'S INAUGURAL POETICS: VISIBILITY AND THE INFINITE CONVERSATION

ABSTRACT

This chapter argues that Maurice Blanchot made a distinctive contribution to philosophical thought that is irreducible to the question of influence but cannot be fully understood apart from his relationship to the phenomenological tradition. The chapter compares Blanchot's conception of reversal to what can be found in Heidegger, but also emphasizes the role that classical myth and modern literature perform in his phenomenological approach to texts. Blanchot's poetics is discussed in terms of Merleau-Ponty's notion of visibility and then contrasted to the classicism that underlies Gadamer's hermeneutics. The conclusion argues that Blanchot's poetics contains a view of beginnings that clarifies the value of phenomenology as a method of inquiry.

Maurice Blanchot's reputation as a literary essayist, who worked in the Continental tradition in criticism, has largely overshadowed his contribution to philosophy in contrast to his unique encounter with literature itself. In this chapter, we shall argue that Blanchot renews some of the basic insights of phenomenology in both engaging and surpassing early hermeneutics, while opening up a new conception of the literary work of art through the myth of Orpheus. Our discussion will proceed in four stages: first, we shall explore how Blanchot offers an account of the work of art that suggests but departs from Martin Heidegger's notion of a phenomenological reversal, just as it clarifies a new poetics of literature; second, we need to examine Blanchot's use of myth and literature as a means for clarifying the role of visibility in the process of reversal, as phenomenologically informed; third, Blanchot's approach to this reversal will be contrasted to Hans-Georg Gadamer's appropriation of classicism as a cultural and philosophical option; and, in our final remarks, we shall discuss how Blanchot invites us to envision his inaugural poetics in a manner that is both indebted to phenomenology and casts light on phenomenology as a reflective procedure.

I

The possibility of approaching literature through philosophical resources performs an implicit, if not entirely explicit, role in Blanchot's early masterwork, *L'Espace littéraire*, originally published in 1955. While clearly concerned with the concept of the work of art that often emerges in early hermeneutics, Blanchot profoundly modifies the role of this concept in describing the experience of literature in terms of a radical reversal. Such a reversal is conceived as a divestment of the self, rather than as a triumph of the subject, just as it opens up an infinite space that cannot be represented. It might be argued that this singular reversal is already anticipated by Heidegger, whose attempt to limit the power of subjectivity constitutes a radical critique of Cartesian priorities through an unprecedented renewal of ontology. We know that, in *Sein und Zeit*, Heidegger thematically opposes his own project to the Cartesian and more strongly Kantian turn toward the human subject as the primary locus of epistemological concern. The disclosure of categorical intuition is often assigned a special role in enabling Heidegger to employ the tools of Husserlian phenomenology in overcoming the subjectivism inherent in modern thought.[1] Insofar as Blanchot's reversal occurs exclusively in the privileged domain of literature, the philosophical significance of his critical reflections would be difficult to extend beyond a narrow field of cultural reflection. Hence, if connected to discrete realms of inquiry, Blanchot and Heidegger would be unlike in a way that derives from the simple difference between literature and philosophy.

This suspicion is reinforced when we consider Heidegger's work as an ontological quest that originally does not appear to privilege literature as a special source of insight. *Sein und Zeit* already suggests that the question of Being is not merely a late concern but integral to an on-going problematic that begins as soon as the situation of *Dasein* acquires a hermeneutical meaning.[2] The exact moment when Heidegger shifted from a worldly, *Dasein*-based problematic to a more ontologically diffuse undertaking may never be determined. Partly for this reason, we might consider the possibility that Heidegger's articulation of an essential reversal is already crucial to the argument of *Zein und Seit*, where the challenge to traditional metaphysics is worked out in detail on the basis of an ontological inquiry. The idea that a fundamental reversal occurs later is not always borne out by what is explored in the earlier context, which sustains a more thematic relationship to phenomenology. What this also means is that a more forward-looking reading of Heidegger points toward interpretive options that may not have been fully explored in response to his early masterwork but acquire significance in retrospect. At the same time, this does not mean that literature as such had to perform an essential role in reversal as Heidegger conceived of it.

Moreover, the role of *language* in Heidegger's basic conception of truth could easily uphold, instead of qualifying, the more strictly philosophical nature of his enterprise. The notion that Heidegger's reversal can be found early in his work is certainly compatible with the idea that the movement away from subjectivity occurs at the precise juncture that language acquires a special significance in a hermeneutical project. From this standpoint, Heidegger's middle and late apotheosis of poetic dwelling develops the linguistic clue that was already crucial to the argument of *Sein und Zeit*, where the role of language in the expression of truth performed a unique and widely acknowledged role.[3] The early philosophical works that culminate in this crucial argument do not privilege literature in the narrow sense nor do they offer us a precise way of discussing the possibility that *writing* contains a key to what would later emerge thematically as an alternative to ontological oblivion. We might even argue that the studies of individual poets and the complex discourse on poetic thinking that unfold in Heidegger's so-called middle and late periods are indebted to a philosophy of language that barely articulates the emergence of writing *as* writing.

It is only by *reading* Heidegger as a philosopher whose *use* of writing is ultimately inseparable from the nature of his philosophical activity that we can begin to question the way that his work is generally separated from literature as a crucial resource. In this sense, Heidegger's remarkable essay, "Der Ursprung des Kunstwerkes" (originally composed during the 1935–36 period), clearly demonstrates how the space of writing as a site for ontological disclosure is opened up when an experience of "world" is introduced through a specifically *poetic* discourse. From the perspective of a thematic of writing that is suggested but never developed, this thoughtfully constructed discussion of art, language and truth acquires an inaugural status in demonstrating how a poetic *text* can perform a crucial role in carrying us beyond a basically physical relationship to an external environment. *Sein und Zeit* already provides the explicit statement of how the phenomenological conception of world differs from that of Descartes, particularly when it provides a positive version of "world" on the basis of spatiality as a non-subjective mode of being.[4] However, Heidegger's crucial essay on art engages the reader in an ontological quest that can be inferred on the basis of a written description of a work of art, rather than through the example of a work that simply refers to an external situation on a mimetic basis.

Thus, in an attempt to retrieve the work of art as a thing that bears the world within it, Heidegger employs one of Van Gogh's paintings of peasant shoes to evoke the wearer, an ordinary peasant woman who belongs to a distinctive place but also alters the rural landscape of which she is a part. On one level, we might read Heidegger in this context as merely continuing the project of *Zein und Seit*, which already explained how the world comes into focus at the critical moment when an instrumental complex breaks down and forces us to reexamine our immediate environment as somehow integrated, if not entirely familiar to us. And yet, "Der Ursprung des Kunstwerkes" provides us with a way of understanding Heidegger's world-concept that is different from what the more systematic treatise provides in demonstrating the imaginary aspect of the world that it evokes. Van Gogh's painting does not provide any precise information concerning the wearer of the shoes that are depicted. Heidegger, in contrast, takes us from the things depicted to the life-world of an imaginary woman who might have occupied the empty shoes themselves:

Under the shoes slides the loneliness of the field-path as evening falls. In the shoes vibrates the silent call of the earth, its quiet gift of the ripening grain and its unexplained self-refusal in the fallow desolation of the wintry field. The equipment is pervaded by uncomplaining anxiety as to the certainty of bread, the wordless joy of having once more withstood want, and trembling before the impending childbed and shivering at the surviving menace of death.[5]

On one level, this simple description evokes a silent landscape that somehow "speaks" to us through the agency of poetic reflection. What is perhaps more plausible is that the woman has entered into the texture of the wintry landscape, just as the landscape – which otherwise would lack the features that have been worked over it – has been transformed through the persistent activity of a human host into a site of both need and withdrawal. In interpreting this passage, we would be doing the philosopher a disservice to simply remark on the practical recourse to prose poetry. The "world" of the peasant woman is evoked through a visual image that involves a *written* response to what would have remained inexpressible in a purely philosophical discourse.

The ironic aspect of Heidegger's description becomes evident when we juxtapose poetic language and visual image in the narrative of a "world" that contains visible and invisible at once. The verbal elaboration of "world" requires two media, namely, painting and poetry, in order to unify a reality that may be sundered. Nonetheless, this entire account is also a description of a certain Van Gogh painting that Heidegger has already mentioned to underscore the relative stability of the work of art as an anti-subjective thesis. While Heidegger's reversal can be traced back to *Sein und Zeit*, we might also interpret this thesis quite differently as announcing "the possibility of impossibility" (Lévinas) that emerges in the forlorn mood that the work expresses.[6] Such a possibility would not only draw upon Heidegger's earlier analyses of *Dasein* because it now involves a written account of a world that provides no heroic options to a peasant laborer who has survived many hardships. The work of art now brings to light something that cannot be seen and deepens the meaning of reversal to involve the possible collapse of human subjectivity and measurable time.

From this broader perspective, Heidegger's discourse on finitude can be interpreted as an instance of severe ontological limitation, which prevents the truth of being from coinciding with timeless presence. Moreover, this very discourse can even be related to a critique of the natural attitude that was always central to Husserlian phenomenology, especially to the degree that it requires the secondary elaboration in which a certain *resistance* to the natural world makes itself felt in poetic terms. The orientation toward "being" (rather than subjectivity) that underlies this discourse is at least suggested in Husserl's assertion that human accomplishments can be anonymous.[7] This special assertion, nonetheless, indicates why all cultural objects contain within themselves the potential for variations in meaning that qualify the objective significance of the creative work itself.[8]

It is at this point that Blanchot's conception of reversal offers a mode of access to the work of art in a way that is neither ontological nor strongly personalist in its deeper implications. Heidegger already argued that openness to the work can function as an alternative to modern subjectivism. However, when anonymous existence is philosophically demoted, we are at a disadvantage to distinguish fundamental ontology from normative concerns. While it would be problematic to read early Heidegger in overtly ethical terms, we would be hard-pressed to deny that his analysis of "everydayness" is anything other than value-laden in its tenor and implications. Perhaps more consistently, Blanchot newly appropriates phenomenological anonymity as a neutral term that describes in a formal idiom the impersonal aspect of intentional life. Hence, Blanchot's view of *the writer* specifies linguistic displacements that challenge traditional subject-based criticism: "The writer belongs to a language that no one speaks, which is addressed to no one, which has no center, and which reveals nothing."[9] Moreover, in a manner that looks forward to the early criticism of Roland Barthes, Blanchot discusses how the anonymous site of creativity often coincides with the construction of third person narratives from which the author is entirely absent.[10]

Blanchot's understanding of reversal is phenomenological in negotiating a new sense of aesthetic appearance that is irreducible to the Heideggerian problematic. The figure of Orpheus performs a crucial role in enabling Blanchot to specify how the reversal carries us from a centered notion of the human subject to a process-oriented *event* of aesthetic ambiguity. To the degree that Orpheus gazes directly on Eurydice, he ruins the work and loses what he seeks to master. However, in simply refusing to observe his approaching lover, Orpheus would demonstrate infidelity to the profound impulse to possess her as an ineluctable other. Heidegger wrote "Der Ursprung des Kunstwerkes" in the attempt to move beyond the constraints of philosophical aesthetics, which adopts the subject as its starting-point and prevents art from making serious contributions to our access to truth. Is there a sense in which Blanchot would revive aesthetics in a new guise without returning to an older conception of the subject that would ground experience in an a priori conception of its own activity? In order to answer this question, we need to examine how myth and allusion can be employed in suggesting a realm of appearance that provides insight into aspects of the human condition that are irreducible to either direct perception or conceptual understanding.

II

Blanchot's interpretation of the Orpheus myth provides a key to the meaning of *visibility* as a quasi-aesthetic category that clarifies the way in which literary texts can be read as testimonies to a unique order of experience. In discussing Heidegger's approach to the work of art, we encountered a discussion of "world" that was built out of a mysterious conjunction between person and place, but the nature of this conjunction remained unclear, perhaps because the whole notion of being-in-the-world occluded the movement *between* two zones of contact. Maurice Merleau-Ponty provides an eloquent critique of Bergson in which he explains that my encounter with the visible world pervades the structure of experience itself: "There is an experience of the visible thing as pre-existing my vision, but this experience is not a fusion, a coincidence," so that I am already within the world with which I make contact. Moreover, the visibility that is woven into my experience of things allows me to discover "a Being of which my vision is a part, a visibility older than my perceptions or my acts."[11] Hence, instead of arguing that subject and object achieve a sort of higher synthesis that invalidates self-reflectivity, Merleau-Ponty identifies the space in which I move and experience the world as one that allows me to enter into the domain of the things themselves, just as it allows the things to enter into my state of consciousness as other to myself. This dual movement is called "double reference" because of the way that it preserves the condition of being lived through and also sustains the sense of distance that prevents co-mingling from becoming a simple act of coinciding.[12]

To return to the myth of Orpheus, we might relate this analysis to Blanchot's appropriation of a classical narrative that seems to partake more strongly of the imaginary but also indicates how "double reference" pervades an aesthetic framework that suggests how the artist's gaze both responds to an appearance as an appearance and also accepts the fading of an apparition into a distance that cannot be mastered. Orpheus cannot remain indifferent to an appearance that haunts him just as he is deflected away from the special task of guiding Eurydice without observing her. And yet, the visibility that is momentarily achieved through his gaze is suddenly lost in the abyss of night. Blanchot reveals the paradoxical nature of this unveiling when he recounts the significance of the classical narrative in terms of the work of art. The Greek myth clearly demonstrates that the work cannot be pursued directly: Orpheus turns back, ruins the work, and Eurydice returns to Hades. And yet, this fateful movement becomes unavoidable as soon as Orpheus begins to understand that "not to turn toward Eurydice would be no less untrue."[13] Fidelity to what is immeasurable and to the force of circumstances require that a risk be taken, but the truth of the matter is that "only in song does Orpheus have power over Eurydice."[14] But this power is strictly limited. Eurydice has ceased to be present in the voice of the poet, while her mode of appearance cannot be separated from an encounter that *once took place* and continues to inform the memory of what now appears only as pure song.

Blanchot's interpretation of the Orpheus myth can be related to the dual nature of aesthetic appearance and also invites us to question what sort of work actually emerges through the vehicle of the artistic gaze. The gaze of Orpheus is said to be an "ultimate gift to the work," no less than it is the moment when the work is lost.[15] Heidegger places the origin of the work of art in art, rather than in the artist, and provides an alternative to aesthetic experience in reminding us that nothing can be accomplished in a creative vacuum. Blanchot, in contrast, identifies ontological instability with the transformation of the work of art into a "text" that lacks continuous presence and bears a kinship to evanescent appearances.[16] Moreover, while Heidegger provides examples of how the work of art projects a "world" that discloses truth, Blanchot anticipates Jean-Luc Nancy in discussing how the world of sense dissolves when the artist undergoes temporal displacement in an experience of solitude.[17] At the same time, Blanchot's recourse to a certain mode of appearance when describing the impossibility of the work exposes him to the criticisms that Heidegger's approach was designed to counteract. How does the opening of the work as "text" provide a sort of ground that provides a degree of stability that the dissolution of the world cannot revoke?

The dissolution of the world does not undermine the possibility of creativity itself to the precise degree that the artist is always already related to an alterity that prevents him from being assimilated to everyday self-sameness. Blanchot specifically refers to a "radical reversal" in which the artist perceives a certain object as "the point through which the work's requirements pass," thereby effacing all notions of value and utility in world loss.[18] This procedure includes two aspects that prevent the loss of world from resulting in subjective chaos. First, the artist in producing the work of art remains in contact with a quasi-subject that *stands out* and allows him to view the ordinary world in a new way. For this reason, the artist never simply rises from the ordinary world to the sphere of art but invariably enters into a negative relationship to everyday life before providing a different perspective on his goals and values. At the same time, Blanchot does not merely describe how this process occurs but offers a kind of explanation for the artist's capacity to move beyond the given world and alter our understanding of the familiar. Hence, the second aspect of this process combines with the first in bringing about a compelling transition into another realm: "It is because he already belongs to another time, to time's other, and because he has abandoned time's labor to expose himself to the trial of the essential solitude where fascination reigns" that the artist can emerge relatively unscathed from the experience of world loss and can include what is unlike in his account of existence.[19]

The movement away for the familiar world is therefore anything but a Romantic escape into subjective inwardness. Blanchot employs the literature of Franz Kafka to cast light on the artist's exile but also to demonstrate the artist's ability to pass beyond the limits of his own consciousness. Kafka is the writer who feels banished from any homeland and ultimately discovers that literature alone can offer him something that cannot even be identified with the notion of a stable world. Art is a sign of an "unhappy consciousness" and an antidote to the illusory satisfactions that are the refuge of weak souls. Blanchot identifies Kafka with one of the basic traits of art itself, which is the capacity to link us "to what is 'outside' the world, and it expresses the profundity of this outside bereft of intimacy and of repose," so that the life of the artist can seem like a perpetual misfortune.[20] The experience of being cast out can be related to a singular discovery. The choice between the homeland before us and the desert beyond does not permit a third option. Kafka understood that his own options remains limited to this stark choice, but he also knew that the artist he wished to be could not even provide him with one world that might shelter him from the condition of banishment. The artist is the "poet" for whom this one world has ceased to exist: "For there exists for him only the outside, the glistening flow of the eternal outside."[21]

While insisting that art provides access to an outside that is irreducible to inner experience, Blanchot also emphases how the artist promotes an encounter with death that assumes many forms in a general economy of creative expression. The example of Stephen Mallarmé serves the purpose of highlighting the role of death as well as absence and negativity in artistic production. The poet who remarked on the power of words to make physical things absent was also the author of *Igitur*, a verse drama in which the protagonist confronts the Midnight of freely chosen death. Blanchot notes that the final version of Mallarmé's verse drama assumes the form of a soliloquy in which the protagonist, like another Hamlet, becomes a speaking presence who directs us to the ordeals of consciousness.[22] The opposition between pure consciousness and a Midnight that threatens to obliterate all thought does not admit of a possible resolution. The problem is that Igitur has never known chance. The dice are only cast at Midnight, which is also the hour that does not arrive. Blanchot keenly observes that the successor poem of *Igitur* is necessarily *Un Coup de dés*, a literary work that gives chance its due. The first poem passes beyond the nothingness of pure consciousness to become a game of chance that compares to an inconclusive narrative, whereas the work that remains evokes the element of uncertainty and risk that inheres in all uses of language. The play between the visible and the invisible only achieves stillness when the poem itself emerges as a literary object that shines forth in the portals of being.

Blanchot's approach to Rainer Maria Rilke is consistent with a concern for the relationship between death and writing that pervades his reading of Kafka and Mallarmé, but it also provides a coda to the way that the visible passes into the invisible without ceasing to inform our sense of the poem. Mallarmé's poetry brought us to the brink of death in the consciousness of Igitur and in the transformation of the work into a site of dispersal and a mark of limits. Rilke's early attitude towards death is perhaps similar to what can be found in Nietzsche when read as a precursor to existentialism. A well-stated abhorrence for the modern depersonalization of death is a constant theme in the poet's only novel, *Die Aufzeichnunger des Malte Laurids Brigge*. And yet, Rilke's late poetry commemorates "the fruition of the visible in the invisible for which we are responsible," just as it epitomizes "the very task of dying."[23] This task is analogous to the translation of external things into verbal realities that takes place in the silent world of poetry. Blanchot contrasts the role of change in life and its more profound role in art as memorialized in Rilke's *Duineser Elegien*, the testament of the poet's final years:

> In the world things are *transformed* into objects in order to be grasped, utilized, made more certain in the distinct rigor of their limits and the affirmation of a homogeneous and divisible space. But in imaginary space things are *transformed* into that which cannot be grasped. Out of use, beyond wear, they are not in our possession but are the movement of dispossession which releases us both from them and from ourselves. They are not certain but are joined in the intimacy of the risk where we are, rather, introduced, utterly without reserve, into a place where nothing retains us at all.[24]

The space that provides the basis for this change both exceeds and occasions the things that change. Death provides one way of understanding the appearance of the visible in the invisible. The poem's space occupies the site of this change within the sphere of literature, which reconciles the world of things and the language of non-being.

Blanchot's meditation on literature therefore assigns the poem the task of constituting a unique space that allows the passage between the realm of the visible and the invisible. The possibility of this passage occurs in the space of the Open, which should not be confused with a personal site that the poet may occupy in composing the poem: "This is the Orphic space to which the poet doubtless has no access, where he can penetrate only to disappear," so that any intimacy that he brings to this opening is only achieved at the cost of silence.[25] The disruption of the world that occurs in the creation of the work of art opens a "space" in which things can newly appear, because "absence is also the presence of things" in their being.[26] And yet, the work of art radiates a "being" that is not the being of things but contains inside and outside at once, just as it refers to a space that is "prior" to everyday life experience. Blanchot is less interested in placing the work before us as the setting for truth than in foregrounding the Open as the productive space in which the work of art quietly unfolds: "The Open is the work, but the work is origin."[27]

III

Blanchot's account of art and literature allows us to assess the broader implications of a hermeneutical theory that challenges received notions of modern culture. The current exposition remains phenomenological in the precise sense of requiring a reflective component in order to clarify Blanchot's unique task, but this component is not primarily traceable to the older notion of categorial intuition or even to reflection on the question of being as first broached in Plato and Aristotle. The situation of sustaining a literary dialogue with phenomenology has a somewhat different meaning when related to the status of the reader in Blanchot's theory and criticism than it would if it were only related to the interpretation of texts. Hence, in the present discussion, we hope to enlarge upon phenomenology to include various hermeneutical motifs that foreground the interactive nature of text, reader and community in terms of the opening of the work as a gateway to time and alterity. In this part of our discussion, we will be concerned with how Blanchot anticipates but also surpasses the position of Hans-Georg Gadamer, whose major work, *Wahrheit und Methode*, develops modern hermeneutics in a systematic form largely as a response to Heidegger's ontological intervention.

First, Blanchot's emphasis on the reader in "constituting" the work of art might be placed alongside Gadamer's concept of how a "fusion of horizons" mediates between the perspectives of reader and author in literary reception.[28] Without denying that a text possesses hermeneutical value that cannot be revealed through a strictly historical analysis, Gadamer argues that interpretation can occur somewhere *between* the intentions of an author and the motivations of a reader who approaches the text in a contemporary setting. Subsequent to Gadamer's elaboration of this important concept, Hans Robert Jauss develops a more historically oriented approach to literary reception that allows us to study a given text in terms of the history of readings that transform its meaning in time. Roman Ingarden had previously demonstrated in detailed analyses that literary reception is temporally layered but allows us to correlate the reader's motivations with the production of the literary work of art as harmonious structure, which requires aesthetic criteria in order to be fully appreciated. Blanchot's contribution to the problem of reception is even more strongly anti-historicist and anticipates the thought of Barthes, Foucault and Derrida, whose poststructuralist thematic derives from Sausurrean linguistics. For Blanchot, the act of reading does not primarily establish contact with sedimented meanings but liberates us from original intentions: "The reader does not add himself to the book, but tends primarily to relieve it of an author."[29] Hence, rather than conceive of reading according to a model of co-constitution that would conceive of literary meaning as negotiated in a middle zone that mediates original intentions with contemporary directives, Blanchot conceives of the literary text as an impersonal manifestation in which writing appears *as* writing.

The author therefore "dies" in a precise sense when the reader constitutes a work that no longer coincides with the intentions of the author who produced it. On this basis, Blanchot "affirms the new lightness of the book" and displaces the role of the author in the reception of meaning. But does this imply that the reader can construe *any* meaning in disregarding the real or apparent intentions of an imputed author? Blanchot implicitly answers this question when he compares the role of the reader to the process of shaping a sculptural work: "Reading gives to the book the abrupt existence which the sculpture 'seems' to get from the chisel alone."[30] This does not mean that the book would cease to exist if it went unread, but that, like the sculpture shaped from stone, the book acquires standing existence when reading isolates it from the flow of meanings that might allow us to situate the work in the past and thus to finalize interpretation contextually. And yet, Blanchot does not conceive of the literary work as an ideal object that can be grasped as either a timeless mental entity or as the concretization of universal schemata. Blanchot posits the radical difference between a work that is always partially concealed but contains limited meanings and *a work to come* where "everything which does have meaning returns as towards its origin."[31] For Blanchot, literary reception is less of a "fusion of horizons" than a twisting free from sedimented meanings that are no longer part of an on-going interpretation.

By detaching the literary work from the intentions of the author, the reader can join the origin of the work with the movement into the future that carries us beyond the meanings that are initially evident. Because this act of detachment is possible, Blanchot can re-envision the literary work as capable of resituating us in a life-world that is not cut off from productive achievements having the power to alter everyday life in innumerable ways:

> The book, the written thing, enters the world and carries out its work of transformation and negation. It, too, is the future of many other things, and not only books: by the projects which it can give rise to, by the undertaking it encourages, by the totality of the world on which it is a modified reflection, it is an infinite source of new realities, and because of these new realities existence will be something it was not before.[32]

The reception of the literary work is therefore inseparable from an effort to vary the given environment precisely because the work derives from a world that is undergoing change on a continual basis. At the same time, we should not attempt to naturalize this process of change for the simple reason that Blanchot presupposes what we might call a phenomenological outlook with regard to both the poet and the poetics of origin. With reference to Mallarmé, Blanchot emphasizes how the poet undergoes a reduction in presence that corresponds to a decisive displacement: "The poet disappears beneath the pressure of the work, by the same impulse that causes natural beauty to disappear."[33] Both the poet and the natural world are transposed into a movement that occurs in language and nowhere else, since language is "the only initiator and principle: the source."[34]

By implicating the literary work in the process of change, Blanchot also helps us understand how the reader responds to art's vocation in historical terms. History provides us with the second point of possible convergence with modern hermeneutics, but once again Blanchot departs from what might have been a common meeting ground. Gadamer's stated preference for mediatory over historicist approaches to art suggests an opposition to antiquarianism that might seem to echo Blanchot's notion of the work to come. However, while Gadamer's notion of "the classical" was not intended to conflate normative and Greco-Roman conceptions of art, this same notion enshrines the past in the mode of continual presence, particularly when it argues that the canonical work can speak in a contemporary context.[35] Blanchot, in contrast, emphasizes that the fragmentary experience of history is essential to what remains true of traditional conceptions of art. He denies that the work of art is timeless in the sense of remaining the same throughout the ages. The reader, truly conceived, experiences the work's distance, but this is what allows him to perceive the work's genesis as origin. In Blanchot's account of literature, history possesses a divisive meaning as opposed to the distinct possibility of mediation.

Blanchot willingly acknowledges that art can become an enduring reality when it is interpreted according to a plurality of cultural values and across varied circumstances. The historical aspect of reception is what guarantees the integrity of an "endless conversation" that draws upon many perspectives and ceaselessly initiates a dialogue with the past. Gadamer refers to how the work is encountered in a "history of effects" that sometimes has the cumulative significance of implying a hermeneutical totality. Blanchot argues somewhat differently that the continual search for new interpretations is what gives the work its historical future. Art has a public significance, which is not predicated on the continuous presence of a past achievement that has been reaffirmed as a canonical value. The work of art in its historical manifestations is both a presence and a disappearance, which means that reception is sometime difficult to correlate to genuine appropriation. Blanchot acknowledges that the Greek dramas contain meanings that have become opaque in time, signifying a reality that is no longer accessible. The Eumenides will never speak again, but from another standpoint, "each time they speak it is the unique birth of their language that they announce."[36] Their first utterances occurred in the primeval night of myth, whereas they later became synonymous with the ascendancy of law and order. When they speak tomorrow, their words may be part of a literary work in which the language of origin has acquired a more intimate meaning.

Finally, Blanchot shows us that the work of art is an *event* in the radical sense of providing a basis for new beginnings. The notion of the work as an event constitutes the third possible area of convergence between Blanchot and modern hermeneutics as conceived in the wake of Gadamer's critique of Romantic historicism. A purely traditionalist approach to art detaches us from the process character of what comes to us from the past and reaches us in the here and now. Gadamer's critique of the Romantic approach to history as remote and inaccessible is consistent with Blanchot's suspicion of academic historicism, but, more importantly, the hermeneutical rehabilitation of art as a possible source of knowledge draws upon the notion that "the language of art is an encounter with an unfinished event and is itself part of this event."[37] Gadamer cautiously affirms Hegelian models of historical research, which privilege the present over the past, over Romantic ones that value the past only for its own sake. Blanchot also acknowledges the power of Hegel's arguments, just as he emphasizes the limited nature of Romantic conceptions of art and artist. And yet, in a somewhat different spirit, Blanchot returns to the work of art as historical in a way that is irreducible to any progressive survey that would minimize the importance of the work's origin. The trained historian may be too methodologically encumbered to grasp the event-like quality of art works, but the work itself does not lack historical resonance, since "it is an event, *the* event of history itself, and this is because its most steadfast claim is to give to the word *beginning* all of its force."[38] How can we assess the phenomenological significance of Blanchot's assessment of the work of art as an event, especially when the work of art is a *literary* event that awakens a more reflective sense of our beginnings?

IV

The role of phenomenology is Blanchot's inaugural poetics has been implicit, if not explicit, in our study of his use of myth, his approach to modern literature and in the hermeneutical implications of his view of literary reception. In the early part of our discussion, we compared Blanchot's approach to the literary work to Heidegger's understanding of the work of art, and also broached the possibility that Merleau-Ponty's notion of visibility might clarify the mode in which the space of literature opens up our sense of particular works. What we need to do now is examine the phenomenological aspects of this space, particularly in view of the problem of interpreting Blanchot as in some respects Heideggerian but also as engaged in a philosophically distinctive task that cannot be subsumed under established headings. In order to achieve these things, we first need to identify Blanchot's poetics within a general framework that is compatible with a phenomenological approach to literature. We then need to return to the question of Heidegger's own indebtedness to phenomenology, since Blanchot's version of inaugural poetics would be enriched if this indebtedness could be more clearly specified. In the final part of this discussion, we will want to examine the possibility that Blanchot's poetics of origin makes a unique contribution to the phenomenological tradition when it explores the question of beginnings in a new way.

While Blanchot's criticism cannot be understood apart from his persistent interest in modern literature, we should not assume for this reason that his inaugural poetics lacks philosophical import. We might take his frequent references to Mallarmé as an indication of his literary stance, which involves a clear rejection of representational conceptions of art and literature. Blanchot's approach to the literary text opens up the significance of writing, as opposed to a purely verbal understanding of what constitutes the literary. Timothy Clark has discussed how this approach required the development of modern poetry in order to become theoretically compelling: "The space of text, with Mallarmé, becomes no longer one of voice, but of writing, whose force is always to break away from narrowly representational constraints."[39] This notion of literary or textual space does not map onto external reality anymore than it participates in the regime of everyday speech. Blanchot refers to an "essential language" that appears when the poet occupies a space that opposes our mimetic expectations:

Sounds, rhythm, number, all that does not count in current speech, now become most important. That is because words need to be visible; they need their own reality that can intervene between what is and what they express. Their duty is to draw the gaze to themselves and turn it away from the thing of which they speak. Yet their presence is our gauge for the absence of all the rest.[40]

And yet, without depriving poetry of its visible dimension, Blanchot also emphasizes how writing itself is the crucial term that expresses "a rupture with language understood as that which *represents*," just as it breaks with the manifestations of sensible appearance.[41] Writing must be conceived in a concrete way as a kind of "other" that provides the space within which thinking can occur: "Uncontained by any system or any conceptual or empirical limit, it is a species of infinity, or, better, of infinitizing."[42] The word of the poet is thus an appearance of what no longer appears, evoking "the imaginary, the incessant, the interminable."[43]

The infinite aspect of this space requires that we reconsider the role of reflection in phenomenology as a *modern* discipline that presupposes a thematic understanding of conscious acts. Early in our discussion, Blanchot's use of the Orpheus myth as a paradigm for considering the poetic imaginary was compared to Heidegger's use of the work of art in "Der Ursprung des Kunstwerkes," which provides a model for assessing the concept of "world" in phenomenological terms. What is easy to overlook in both cases is the modern contribution to our understanding of the matter at hand. In Blanchot's case, the myth of Orpheus is elaborated in terms of Rilke's poetry rather than simply as a classical myth that had its home in ancient Greek tradition. For Heidegger, in a similar way, the "world" of the peasant women derives from reflections on Van Gogh's painting of peasant shoes, a specifically modern work of art that opens up certain hermeneutical possibilities that are expressed in language. In *Der Grundprobleme der Phenomenologie*, Heidegger directly maintains that classical ontology is inadequate because it holds to "a common conception of *Dasein*," rather than because it is unreflective. Classical ontology returns to the compartments of *Dasein* when it demonstrates an awareness of "*Dasein*'s everyday and natural self-understanding."[44] But this does not mean that Greek philosophy offers clarity with regard to the ontological problematic, or that the Greek concept of the world can be identified with a phenomenological use of the world-concept. What we need to do, therefore, is to examine the contribution that phenomenology is capable of making to our understanding of how ontological inquiry can occur.

By returning to the work of Heidegger's so-called phenomenological decade, which falls roughly between the *Habilitation* in 1916 and the publication of *Sein und Zeit* in 1927, we can begin to grasp the importance of Husserl's work to all that follows. In the Marburg University lectures on historical ontology that were given in the winter sessions of 1923–24, Heidegger discusses the kinship between Descartes and Husserl but, more importantly, offers a separate discussion to clarify their "fundamental differences" ("fundamentale Unterschiede") through which the phenomenological notions of evidence, reduction and pure consciousness acquire original meanings. Central to phenomenology is Husserl's effort to mark out a "wholly new domain" that clarifies the ways of "self-relating-towards" (Sichbeziehens-auf) and the intentional manner in which self-relating becomes present.[45] Heidegger argues on this basis that it is a error to reduce the meaning of Husserl's achievement to a phenomenology of the act or to a transcendental psychology, since phenomenological reflection is not reflection in the limited sense but what allows ontological research to proceed in a scientific manner.[46] Apart from the grounding effort that enables us to see any entity in terms of its being, we are not able to pursue ontology as a philosophical discipline.[47]

We might apply these principles to Blanchot's approach to the imaginary as the "space of meaning" that both requires meaning and goes beyond it in order to function as the frame within which literature is apprehended. On the one hand, Blanchot's recourse to the myth of Orpheus is a paradigm for interpreting the reversal that occurs when appearance passes into disappearance but also produces a work that is other than anything else in the world. This work possesses being and therefore can be reflected upon in the manner that Heidegger illuminates in his assessment of Husserl's departure from Cartesian rationalism. At the same time, the literary work of art is not grounded in the manner of a mere object but marks the entry of history into the substance of poetic achievement. Blanchot underscores the role of beginnings in history without depriving inaugural poetics of its phenomenological meaning as a space in which the past is reflectively seized upon in new ways, thus allowing the self to cross a certain threshold that cannot be anticipated on the basis of present knowledge alone.

Hannah Arendt has more clearly discussed how the possibility of beginning anew exceeds the limits of established knowledge and cannot be understood apart from the question of self-identity. This possibility is as old as the Augustinian belief in the possibility of claming a new origin in the mode of a recurrent recollection, but what Arendt emphasizes in this case is not so much the role of memory in enlivening the past as the ontological conditions that allow the beginning to be made: "This beginning is not the same as the beginning of the world; it is not the beginning of something but of somebody, who is a beginner himself."[48] Blanchot is closer to this viewpoint than he is to the Gadamerian notion of affirming the truth-value of art as an alternative to the post-Kantian tendency to relegate the aesthetic dimension to the fringes of knowledge. Blanchot understands how a poetics of beginnings requires a movement beyond knowledge in the cognitive sense as well as an openness to time that cannot be guaranteed within the framework of pure knowledge. Moreover, the possibility of achieving this new beginning is never far from "the impossibility of possibility" that reminds us of our finitude but also opens a sense of responsibility to ourselves and others. The work of art provides us with a reminder of a death all of us share, but it also unfolds in the fragile space of an infinite conversation and likewise suggests that no community is more difficult to preserve than the community to come.

The American University in Cairo, Cairo, Egypt
e-mail: wmelaney@aucegypt.edu

NOTES

[1] The role that Husserl's conception of categorical intuition performs in Heidegger's ontology is examined in Jiro Watanabe, "Categorial Intuition and the Understanding of Being in Husserl and Heidegger," *Reading Heidegger: Commemorations* (Bloomington and Indianapolis: Indiana University Press, 1993), pp. 109–117. The influence of Husserl's *Logical Investigations*, particularly the Sixth Investigation,

on Heidegger's attempt to disclose the limitations of the propositional theory of truth is crucial to the hermeneutical approach to being which grounds our understanding of beings as such.

2   Reiner Schürmann argues that the first eight sections of *Sein und Zeit* constitute a sort of prolegomena that draws upon Plato and Aristotle, rather than modern philosophy, in the effort to retrieve the question of being In Schürmann's reading of Heidegger, intuition itself is under critique to the degree that it is concerned primarily with the cognitive status of objects as opposed to the understanding of *Dasein*. We might argue somewhat differently, however, that *Sein und Zeit* remains indebted to the Husserlian breakthrough insofar as it grasps phenomenology as a series of questions about the truth of being as a whole. For details, see Reiner Schürmann, "Heidegger's *Being and Time*" in *On Heidegger's "Being and Time"* (London and New York: Routledge, 2008), pp. 56–131.

3   The status of language as everyday discourse as well as the role of language in the expressive disclosure of truth are both discussed in Martin Heidegger, *Being and Time* (Albany: State University of New York Press, 1996) I.5A, section 34, pp. 150–56; I.6 section 44, pp. 196–212.

4   After developing a comprehensive criticism of Descartes's concept of world, Heidegger works out a phenomenological understanding of world on the basis a new approach to spatiality as presented in *Being and Time* I.3, sections 22–24, pp. 194–205.

5   Martin Heidegger, "The Origin of the Work of Art," *Poetry, Language, Thought* (New York: HarperCollins, 2001), p. 33.

6   Under the heading of the radical reversal, Blanchot refers to Levinas after discussing Heidegger's "possibility of impossibility," thus inviting us to consider finitude in a new way. See Blanchot, *The Space of Literature*, p. 140. A discussion of the full implications of this shift can be found in Lars Iyer, *Blanchot's Vigilance: Literature, Phenomenology and the Ethical* (New York and London: Palgrave Macmillan, 2005), pp. 16–20.

7   The formation of the life-world by "anonymous" subjective phenomena is discussed in Edmund Husserl, *The Crisis of European Sciences and Transcendental Phenomenology* (Evanston, IL: Northwestern University Press, 1999), IIIA, section 29, pp. 111–114. From this standpoint, the task of philosophy is to investigate "anonymous" subjectivity as existing prior to what we accomplish in more limited spheres: "Before all accomplishments there has always already been a universal accomplishment, presupposed by all human *praxis* and all prescientific life." Ibid., p. 113.

8   Husserl explains that the analysis of intentionality would remain "anonymous" if it remained at the level of "a naïve devotion to the intentional object," but also that "the noetic multiplicities of consciousness and their synthetic unity" could not reveal "one intentional object" unless the phenomenologist adopted a reflective attitude toward mental acts. For details, see Edmund Husserl, *Cartesian Meditations: An Introduction to Phenomenology* (Dordrecht: Kluwer, 1999), pp. 47–49. Blanchot's view of anonymous subjectivity invites us to intervene reflectively at the precise moment when the work of art calls on us to reverse an unreflective absorption in both self and world.

9   Maurice Blanchot, *The Space of Literature* (Lincoln, NE: University of Nebraska Press, 1982), p. 16.

10   For a critical discussion of the use of third person narrative in modern novels, see Roland Barthes, *Writing Degree Zero* (New York: Hill and Wang, 1968), pp. 29–40.

11   Merleau-Ponty, Maurice. *The Visible and the Invisible* (Evanston, IL: Northwestern University Press, 1968), p. 123.

12   Ibid., 124.

13   Blanchot, *The Space of Literature*, p. 172.

14   Ibid., p. 173.

15   Ibid., p. 174.

16   See Roland Barthes, "From Work to Text," *Image Music Text* (New York: Farrar, Straus and Giroux, 1988), pp. 155–164.

17   Jean-Luc Nancy argues that, under conditions of late modernity, the world no longer has a sense because it lacks an essential relation to either another world or a divine creator. The end of the world, however, does not mean the end of sense but that the world's sense has become immanent. See Jean-Luc Nancy, "The End of the World," *The Sense of the World* (Minneapolis, MN and London: University of Minnesota Press, 1997), pp. 4–9.

18   Blanchot, *The Space of Literature*, p. 47.

19 Ibid., p. 47.
20 Ibid., p. 75.
21 Ibid., p. 83.
22 Ibid., pp. 115–116.
23 Ibid., p. 141.
24 Ibid., p. 141.
25 Ibid., p. 142.
26 Ibid., pp. 158–159.
27 Ibid., p. 142.
28 See Hans-Georg Gadamer, *Truth and Method* (New York: Crossroad, 1991), pp. 300–307. We should note here that the concept of a "fusion of horizons" presupposes two separate horizons, rather than a process whereby the past and present are completely unified.
29 Blanchot, *The Space of Literature*, p. 193.
30 Ibid., p. 193.
31 Ibid., p. 196.
32 Maurice Blanchot, *The Work of Fire* (Stanford: Stanford University Press, 1995), p. 314.
33 Maurice Blanchot, *The Book to Come* (Stanford: Stanford University, 2003), p. 228.
34 Ibid., p. 229.
35 Gadamer carefully distinguishes this conception of the classic from Greco-Roman models of normativity but also attempts to ground this notion in an experience of timelessness according to which the past is experienced in the mode of the present. See Gadamer, *Truth and Method*, pp. 285–290.
36 Blanchot, *The Space of Literature*, p. 206.
37 Gadamer, *Truth and Method*, p. 99.
38 Blanchot, *The Space of Literature*, p. 228.
39 Timothy Clark, *Derrida, Heidegger, Blanchot: Sources of Derrida's Notion and Practice of Literature* (London and New York: Cambridge University Press, 1992), p. 68.
40 Blanchot, *The Work of Fire*, pp. 31–32.
41 Maurice Blanchot, *The Infinite Conversation* (Minneapolis, MN and London: University of Minnesota Press, 1993), p. 261.
42 Clark, *Derrida, Heidegger, Blanchot*, p. 80.
43 Blanchot, *The Space of Literature*, p. 40.
44 Martin Heidegger, *The Basic Problems of Phenomenology* (Bloomington, IN and Indianapolis, IN: Indiana University Press, 1988), p. 110.
45 Martin Heidegger, *Einführung in die Phänomenologische Forschung*, III.2, *Gesamtausgabe* 17 (Frankfurt am Main: Vittorio Klostermann, 1994), p. 262.
46 Ibid., 262.
47 While offering sympathetic criticisms of Kisiel's account of this early period, Crowell develops some of these compelling suggestions on Heidegger's interface with Husserl and phenomenology that preceded the publication of *Sein und Zeit*. For details, see Steven Galt Crowell, "Heidegger's Phenomenological Decade," *Husserl, Heidegger, and the Space of Meaning: Paths Toward Transcendental Phenomenology* (Evanston, IL: Northwestern University Press, 2001), pp. 115–128.
48 Hannah Arendt, *The Human Condition* (Chicago: University and Chicago Press, 1989), p. 177.

REFERENCES AND READING

Arendt, Hannah. 1989. *The human condition*. Chicago: University and Chicago Press.
Barthes, Roland. 1988. From work to text. *Image music text* (trans: Heath, Stephen), 155–164. New York: Farrar, Straus and Giroux.
Barthes, Roland. 1968. *Writing degree zero* (trans: Lavers, Annette and Colin Smith). New York: Hill and Wang.
Blanchot, Maurice. 1982. *The space of literature* (trans: Smock, Ann). Lincoln: University of Nebraska Press.

Blanchot, Maurice. 1993. *The infinite conversation* (trans: Hanson, Susan). Minneapolis and London: University of Minnesota Press.
Blanchot, Maurice. 1995. *The work of fire* (trans: Mandell, Charlotte). Stanford: Stanford University Press.
Blanchot, Maurice. 2003. *The book to come* (trans: Mandell, Charlotte). Stanford: Stanford University.
Clark, Timothy. 1992. *Derrida, Heidegger, Blanchot: Sources of Derrida's notion and practice of literature*. London and New York: Cambridge University Press.
Crowell, Steven Galt. 2001. *Husserl, Heidegger, and the space of meaning: Paths toward transcendental phenomenology*. Evanston, IL: Northwestern University Press.
Gadamer, Hans-Georg. 1991. *Truth and method* (trans: Weinsheimer, Joel and Donald G. Marshall). New York: Crossroad.
Heidegger, Martin. 1988. *The basic problems of phenomenology* (trans: Hofstadter, Albert). Bloomington, IN: Indiana University Press.
Heidegger, Martin. 1994. *Einführung in die Phänomenologische Forschung*, III.2, *Gesamtausgabe* 17, 254–290. Frankfurt am Main: Vittorio Klostermann.
Heidegger, Martin. 1996. *Being and time* (trans: Stambaugh, Joan). Albany, NY: State University of New York Press.
Heidegger, Martin. 2001. The origin of the work of art. *Poetry, language, thought* (trans: Hofstadter, Albert). New York: HarperCollins.
Husserl, Edmund. 1970a. In *The crisis of European sciences and transcendental phenomenology*, ed. Walter Biemel. Evanston, IL: Northwestern University Press.
Husserl, Edmund. 1970b. *Logical investigations*, Books I/II (trans: Findlay, J.N.). London: Routledge and Kegan Paul.
Husserl, Edmund. 1999. *Cartesian meditations: An introduction to phenomenology* (trans: Cairns, Dorion). Dordrecht: Kluwer.
Ingarden, Roman. 1973a. *The cognition of the literary work of art* (trans: Crowley, Ruth Ann and Kenneth R. Olson). Evanston, IL: Northwestern University Press.
Ingarden, Roman. 1973b. *The literary work of art: An investigation on the borderlines of ontology, logic, and theory of literature* (trans: Grabowicz, George G.). Evanston, IL: Northwestern University Press.
Iyer, Lars. 2005. *Blanchot's vigilance: Literature, phenomenology and the ethical*. New York and London: Palgrave Macmillan.
Jauss, Hans Robert. 1981. *Toward an aesthetic of reception* (trans: Bahti, Timothy). Minneapolis: University of Minnesota Press.
Kisiel, Theodore. 1993. *The genesis of Heidegger's "being and time."* Berkeley and Los Angeles: University of California Press.
Merleau-Ponty, Maurice. 1968. *The visible and the invisible* (trans: Lingis, Alphonso). Evanston, IL: Northwestern University Press.
Nancy, Jean-Luc. 1997. *The sense of the world* (trans: Librett, Jeffrey S.). Minneapolis and London: University of Minnesota Press.
Schürmann, Reiner. 2008. Heidegger's 'being and time'. In *On Heidegger's "being and time"*, ed. Steven Levine, 56–131. London and New York: Routledge.
Watanabe, Jiro. 1993. Categorial intuition and the understanding of being in Husserl and Heidegger. In *Reading Heidegger: Commemorations*, ed. John Sallis, 109–117. Bloomington, IN: Indiana University Press.

HALIL TURAN

LOVE OF LIFE, TRAGEDY AND SOME CHARACTERS IN GREEK MYTHOLOGY

ABSTRACT

There are certain personages whom almost everyone recognizes as concrete characters, and takes as universal descriptions of types of human beings. The coherence of the traits of such figures makes them real persons with virtues or vices, with certain perspectives and inclinations for various pleasures. These creations of the poets are simple and pure in the sense that they appear to incarnate unique values and a will to determine or influence the course of events. Unlike the tragic heroes, we do not consider ordinary men and women worth aesthetical or moral attention.

I

The poetic account of the war in Troy and that of the heroes' lives gives us substantial clues for reconstructing the past of humanity. The kings, the princes, the aristocrats and those who happen to come across them in their lives as their servants, as prophets and beggars represent universal characters conceived and described by the ablest writers of all ages. No doubt, the ancient poets focused their view to the aristocrats, or on the lives of those men and women who were kings, queens, princes, and to the gods, but this choice neither blurs our sight of humanity nor represents it in a biased manner. That Homer and other great poets bring before our eyes a lively picture of the ancient men and women is clear once we consider that the descriptions of the lives and aspirations of those heroes are universal in their basic features. Although Odysseus and Hamlet owe their lives in our imagination to those who speak about them, they exist as universal human types, as real individuals, even more real than the actual ones.

The objects of the longings and the dreads of human beings do not seem to be infinitely many in kind. A given community has characteristic practices, particular laws that regulate life, an ethos, but the restraints or the encouragements concerning the common and private life appear to aim to shape the same human nature, namely the same inclinations for pleasure, health, security, comfort, wealth and honor, and the same aversion for pain, misery, insecurity and shame. Death, loss of the beloved, freedom or property; illness, disgrace are objects of aversion for all human beings regardless of culture or age; similarly a peaceful, healthy and prosperous life with those whom one loves, recognition and honor are universal objects of desire. Since the first recorded descriptions of common life, these objects of aversion and desire seem to have constituted the substance of the discourse on values: losses and gains that come from the society, or from the powers one thinks one can communicate, please, appease and convince, from the other men, women or gods is the major theme in all discourse about value in terms of motives or fears. A particular mode of action is said to lead to gain and hence to happiness, and its opposite or lack to the loss of valuable things of life.

The theme of loss and gain appear in almost all artistic, moral and juridical discourse. We may even assert that communication rests on these conceptions. Losses are either punishments or they are seen to be so. Gods, for example, are thought to have the power to punish those who do not abide by the universal laws. Again, the society is thought to be naturally holding the right to punish by depriving the individual of his or her rights of property, freedom, honor and even of life. Both restrictions and deprivations are losses. In a similar vein, gods, men and women are believed to have the power to allow the individual to enjoy the liberty to attain whatever he or she lawfully demands or by rewarding him or her by rights or objects that give pleasure and happiness.

This general nature of society seems to bring different ages and cultures together in the meaning, aim and possibilities of human happiness or misery. Without this common understanding of good and evil, of pleasure and pain, or of gain and loss, neither different generations nor different cultures could understand the tragedy and happiness of the other. Of course this does not mean that a chasm never exists; in fact, education, opportunities, and political order affect this communication in the highest degree. However, we are concerned not with the actuality, but with the possibility of universal meaning and communication.

Ancient Greek imagination still preoccupies the minds of many, and it is generally held that the modern artistic conception rests on that solid base. The ancients have laid down the foundations of knowledge in almost every science and art. Although most of the elementary knowledge in physical sciences now seems to be refuted by the paradigm change that took place in the modern period, the philosophical and artistic achievements of the Greeks preserve their value, and prove to be on a par with the other peaks of human history. Philosophy is a Greek word, and thought on existence and value is still called by this name. The mythical or fictional personages of the Greek poetry and drama still exemplify actual human characters; otherwise they would not remain concrete individuals in the contemporary, or in the universal imagination. They live as conceptions of certain types, and in the dramatic contrast or harmony with the others, serve as poetic tools to describe "facts" to which we attribute "value."

I intend to concentrate on several such universal characters. I will focus on Odysseus, his life, character, expectations and his story of victory and survival. I believe that Odysseus constitutes a universal exemplar of the individual who aims tranquility in the peaceful life of a farmer, and who, facing unfavorable circumstances employs all possible means to attain this end. Odysseus is not a man after the uncertainties of an adventurous life. Rather, he is depicted by Homer and Sophocles as one who leads an adventurous life against his will. Odysseus seeks happiness in his beloved country, in his family, in farming, in feasting and in leading people to happiness of the same sort. No doubt, Odysseus is a king, not an ordinary man, but his kingdom resembles rather a household. Desire of peace, prosperity, friendship and order in any society, be it a small or a large one, seems to be universal to humanity. A philosopher who renounces life, or a hero who looks for death for a cause are only exceptions.

The story of the Trojan War and the characters conceived by the poets give us sufficient clues to reconstruct the characters of the ancient men and women who know the value of a tranquil, prosperous and honorable life, but who experience tragic losses and see drastic changes in their stations. Indeed the concept and the form of tragedy rest on such losses, and the poets delineate the pain of loss in the personages and situations they elaborate.

The encounter of king Odysseus and his ally and later rival Ajax referring to Sophocles' famous tragedy is illustrative. Sophocles brings the two figures together to describe the universal human experience of competition and jealousy. Odysseus, who has proved himself to be one of the keenest warriors in the Achaean army, and who has therefore deserved the honor of his allies, is challenged by a less praiseworthy man, Ajax, who deems himself abler than the classical hero. Ajax's frenzy, which we learn to be a punishment by the goddess Athena, brings him a disgrace which he cannot bear. This loss and his suicide that follows his consciousness of the dishonor that his hallucination would bring are exemplars of tragic loss. Odysseus, on the other hand, appears as a man of sense resisting his commanders and insisting that Ajax should be buried in accordance with universal laws as befits an honorable warrior. Here Odysseus is depicted as a just man who consistently argues against further disgrace of his rival by the mortals. This piety or justice of the hero is coherent with Homer's interpretation: Odysseus is a reliable warrior and ally, a keen strategist, and seems to have rightfully deserved the arms of Achilles. Odysseus is never depicted by the ancient poets as demanding recognition for his virtues or as disregarding justice for illicit gain. Sophocles' portrayal of Odysseus is in line with Homer's: in *Philoctetes*, Philoctetes accuses Odysseus for being unwilling to sail with Agamemnon and Menelaus for the war. Odysseus loves peace, but he is an excellent warrior. Although he values and seeks peace and tranquility, he came to be the paragon of the adventurous. And, he is rewarded with the highest honors and stations in the army for his courage and cunning although he is against war. As we read in the *Odyssey*, he resists the temptations of a life of pleasure, even an immortal life, having nothing but the memories of his land and family in his mind.

I will now concentrate on two characters which can be contrasted with Odysseus, a character which has proved to be universal, although he is not universally seen as representing virtue. The main themes of contrast will be related to the classical virtues, namely courage, prudence, moderation and justice. I will refer to the ancient poets' descriptions of the situations and the discourses of their personages to substantiate my claim that the desire for tranquility and the desire for honor are the two dynamic powers that lead human beings to action, and are, therefore, the causes of their passions. Ancient Greek literature gives us almost a complete picture of human existence as driven by these two forces. The scenes described by the authors represent our destiny and render our desires and aversions, thoughts and acts visible in the losses or gains in life. The tragic characters that represent these accidental or willed passions, virtues and vices, power and weakness are models through which we can conceive universal values and the universally valuable.

II

Odysseus' story seems to teach us that a life of adventure is possible, endurable, and even enjoyable through one's natural inclination to tranquility. The memories of a land, a home, family, friends, objects and activities for living must be dear to one who is uncertain of his or her future. It seems that one envisages the dangers of a voyage or of a flight because of this calling of memories. An escapade, even an imaginary one as in hearing or inventing a story must become conceivable only if one has an aim or a destination. Outlaws or fugitives must be looking for their home in the memories of a yet innocent life, or must be dreaming of one in which their crimes are forgotten as in dreams. All travelers, regardless whether they are tourists or emigrants must be keeping the memories of happiness in a past and expecting their recovery. It seems that a departure's charm is either in the thought of returning home, or that of rebuilding it on a secure land.

Sophocles represents Odysseus as unwilling to participate in the Trojan expedition. Philoctetes is the Achaean warrior left on the deserted island of Lemnos. After the prophecy of Priam's son Helenus which says that Troy can only be taken by his help or with Heracles' weapons the army decides to call Philoctetes who has Heracles' bow and arrows, but Philoctetes refuses to come with Odysseus.

Sophocles makes Philoctetes say that unlike Odysseus, he was willing to come to Troy: "you sailed with them after being kidnapped and compelled, and I, the unfortunate one, had sailed of my own free will with seven ships before they, as you say, but as they say you, threw me out dishonoured."[1]

Philoctetes was left alone on Lemnos because of his stinking leg which seems to be taken as an ill omen. Apparently he was willing to join the war, which he clearly wanted for fame, but, according to one legend, was bitten by the snake at the altar of Chryse, which suggests he was overhasty and imprudent in trying to reach the honor he was looking for. Odysseus is there with the mission to bring Philoctetes and his weapons back, and was also among those who left him alone on the deserted island. Odysseus answers the bitter words of Philoctetes who blames him and the other Argives:

Where there is need of men like this, I am such a man; but where there is test for just and noble men, you will find no one more scrupulous[2] than I. But it is my nature always to desire victory.[3]

Sophocles makes Odysseus speak for himself that he is desirous of victory. The story related to the hero, namely that of Homer, however, makes Odysseus appear obedient to the words of the gods, to the unwritten laws of nature, and to the established order in the army. This seems to be reminded to those who are familiar to the legend by Sophocles as he makes Odysseus say "where there is test for just and noble men, you will find no one more scrupulous than I."

Odysseus is one of the two or three men whose analyses and strategies are esteemed at the headquarters, he is an indispensable ambassador in every important negotiation, an able warrior and scout in dangerous missions. He is depicted by Homer and Sophocles to follow the dictates of universal reason in conflicts, and to be at variance with his comrades only when he can justify himself by reference to the universal laws. He is in Lemnos on a mission to bring Philoctetes and his bow back to Troy, a mission to be accomplished at any cost, and by any means. Odysseus may appear unjust to many among the modern readers, as too cunning or even stealthy in making Achilles' inexperienced son Neoptolemus a part of his scheme, and in forcing the resistant Philoctetes by threatening him with taking his bow and leaving him on the island as a prey to animals. But Neoptolemus gives him the bow back and Odysseus puts an end to his mission which now appears to be accomplishable only by unjust means. Besides, the prophecy that had led the army to decide to call Philoctetes back also tells that Philoctetes would be cured by the ablest physicians of the time, which must be invaluable for Philoctetes who suffers from a painful chronic disease.

It is not Odysseus, but Philoctetes who appears unreasonable and unjust in the play, for gods make him accept the offer. Philoctetes was willing to participate in war, and he wanted it for honor. Hence, he appears in utter contrast with Odysseus who proved to be an excellent warrior: although he never wanted war and the honor it might bring, Odysseus has deserved the highest of honors, Achilles' arms. We also find Odysseus as responsible for the crucial mission of following the prophecy of Helenus, which all Argives believed and decided to be the only means to win the war.

Reluctance for war implies love of a tranquil life. War has a great symbolic power: it can be taken as representing competition or strife in the most general sense. Strife is not observed in work: the herdsman, the farmer, the artist do not "make war" with natural powers, they only obey them to attain their ends. Odysseus was fond of his life in his small and relatively unimportant island Ithaca, and is always represented by Homer as dreaming of an ordinary family life throughout his odyssey. He is never at home, never looks for a new home, but his own home from where he is torn apart for an unreasonable cause, namely for the war for Helen, a war which appears to have attracted many kings and princes for honor. Philoctetes was one of those who came for fame, and certainly, Ajax another.

Odysseus loves tranquility in ordinary life where one is at work, and work in its primordial form must not be strife, but a systematic activity in collaboration with the unchanging or uncontrollable forces of nature and gods. Humans worship nature, they try to predict its behavior, it is impossible to fight with it, and one can only expect to appease it, to persuade the power which makes life possible and enjoyable.

After years of absence, Odysseus, left on his island Ithaca by the Phaeacians, goes to his farmstead, where he meets his old swineherd who managed to increase production by good techniques and hard work. This farm, which does not seem be a place fit for the first meal of a king who has been absent for many years, but the poet seems to suggest that the well governed farm and the honest swineherd Eumaeus, the good farmer is the most reliable person of the land, that production is the most vital function of a country, that work for living is essential for the feeling of security and hence for happiness in an ordinary life.

Odysseus, the unwarlike, has to slay the suitors for a happy life with his wife and son. It seems that the sharp contrast between the hero's love of tranquility and his eagerness for victory for a well reasoned cause, which often appears as only a means for his ultimate aim, namely his home, makes him ever more interesting as a figure throughout the ages and the greatest hero of antiquity.

Odysseus is both a farmer and a warrior. He is a follower of reason common to all possible activities of men, in this sense in his struggle with those who deny him the liberty of tranquility he resembles an artist who plays with nature for the happiness of play and living. But this struggle is not war at any cost: it is through obedience to laws, or to those powers one can and must conceive as universal that Odysseus tries to attain his end. *Logos*, or the common reason is the most perfect tool in the hands of the person who can comprehend and use it. *Logos* must be the essence of any conceivable human product, including especially those of politics and art, activities whose objects are not as concrete as the others. Odysseus represents the human being who strives to change the course of events by employing this common and powerful tool. Neither production nor management could be possible without understanding and obeying the laws one cannot change: one can neither find food, nor create objects or bring forth examples rare and unique without a sufficient understanding of the laws that govern nature and men. It seems that common sense constitutes the principal criterion by means of which we discern the quality of human products, and that those who create or use them in a distinguished manner must have an accurate understanding of the laws discovered by common reason which make these products conceivable and essential as they are.

Mastery in understanding the common laws, especially those related to politics was Odysseus' merit, and this makes him a universal hero. Polytheistic religious practices too seem to be politics in the most general sense, for they concerns the relations between gods and men.[4] Odysseus is a negotiator, an ambassador, a statesman. He was a member of the commission sent to Priam, and his merits are admitted by the person who must have the final word, the king himself:

The basic conflict of the Trojan War was that between Agamemnon and Achilles, which Epictetus used as an example to teach that passion for a woman (Briseis) must not be a reason of conflict between those who went to Troy for war. He has a similar argument for the passion of Menelaus for Helen. The stoic ideal of morality blames Menelaus' sensitivity for honor and love. Epictetus repeatedly stresses that Helen, being a disloyal wife, is not a good cause for war. In both conflicts that arose because of passions related to sexual love and honor, we find Odysseus as a mediator. He tries to win Achilles back to war, and manages at last to bring Briseis back to Achilles.

Again, Homer shows him as an ambassador trying to take Helen back from Paris to avoid the painful war. Given his dislike for war, Odysseus appears before our eyes as a pacifist, but also as a warrior to end the war with victory. He kills or wounds because this appears to him unavoidable for survival; and this is not without a good cause: he longs for his country, his wife, his son and everything related to his modest life as a king. What should a war for Helen, for spoil, for honor in the far away land Anatolia bring to the Ithacan? He has enough for happiness on his island.

Homer shows Odysseus as excelling all Argives in reasoning and speech; but never shows him after honor, or after recognition of his merits, but only after avoiding war. Odysseus who joined the fleet with a modest naval force, is the greatest hero of antiquity. The ancients must have seen the cardinal virtues, namely justice, moderation, prudence and courage as incarnated in the personage of this invincible mortal. Moderation and prudence, as proper to the ideal farmer of Hesiod, are virtues of one who seeks peace and a tranquil life. Justice and courage, though they are inseparable from moderation and prudence, and have little value if they are found in a person who lacks the former, become visible more in one's relations with the others.

III

Sophocles depicts Odysseus in *Ajax* as the model of justice putting him on a par with Antigone, his heroine of justice and piety. Odysseus is the advocate of the hero, his rival in the contest for the arms of Achilles, which resulted in his victory. In Sophocles' interpretation, Ajax, thinking that not Odysseus, but he deserved the arms, plans to kill Agamemnon, Menelaus, Odysseus and others, but being deceived by the goddess Athena, slaughters the cattle kept for the army taking them for his enemies. Ajax, perceiving that he has been deceived, or that he has lost his mind, commits suicide. We find Odysseus arguing for the necessity of a burial ceremony for the hero, defying his seniors Agamemnon and Menelaus. Although he knows that he was one of those whom Ajax wanted to kill, he does not approve of further punishment, nor does he think that leaving his body to the wild animals would be a good example to the army, for this, even if it is ordered by rulers, would be a transgression of the eternal laws. In Sophocles' play, Odysseus says the following to the furious Agamemnon:

I beg you not to venture to cast this man [Ajax] ruthlessly, unburied. Violence must not so prevail on you that you trample justice under your foot! For me too he was once my chief enemy, ever since I became the owner of the arms of Achilles; but though he was such in regard to me, I would not so far fail to do him honour as to deny that he was the most valiant man among the Argives, of all that came to Troy, except Achilles. And so you cannot dishonour him without injustice; for you would be destroying not him, but the laws of the gods. It is unjust to injure a noble man, if he is dead, even if it happens that you hate him.[5]

Justice is generally conceived as the virtue which makes order and peace possible, and treatment after death seems to be one of the things humans have always been sensitive. Both Odysseus and Antigone appear as the guardians of the eternal laws exemplified in the religious practice of burial, they remind that this must be a right of those who once performed their duties even if they ultimately fail tragically like Ajax. Sophocles makes justice appear in full splendor in Odysseus' words in favor of Ajax, but the hero's courage, prudence and moderation are no less apparent. For, without the courage to resist his commanders, his consideration for the events that would follow if Ajax's body is left as prey to wild beasts, and without the strength of will that makes him forget his enmity, his justice would not appear as striking as it does in his resistance to the rulers of the army.

Ajax, as we read in Homer, was a distinguished warrior of the Achaean army. But, like Philoctetes, he seems to have joined the league for honor, for booty and fame. Ajax is depicted by Homer and Sophocles as too eager for dangerous missions. They both show him as one the most valiant warriors of the army. And, unlike Philoctetes, we must consider Ajax as careful, as Sophocles makes the goddess Athena say the following words to Odysseus:

> Do you see, Odysseus, how great is the power of the gods? What man was found to be more farsighted than this one, or better at doing what the occasion required?

But Ajax's farsightedness, (his quality of being *pronous* or *promēthēs*) is not prudence, *phronēsis* in the proper sense. For *phronesis* as one of the cardinal virtues, requires the presence of the other virtues in combination. And, Ajax's love of honor or self-conceit makes him deficient in moderation and prudence. He is represented by Sophocles as insolent to gods, hence as a sinner, and his valor loses its significance in the eyes of gods and men. Virtues do not have values by themselves; they are valuable if they are united in the character of the person with the other virtues. Hence, Ajax, who is depicted by Sophocles as lacking in respect for the gods, can hardly be called virtuous. As we read in the play, the messenger reports the prophecy of Calchas as follows:

> When men grow to a size too great to do good, the prophet said, they are brought down by cruel misfortunes sent by the gods, yes, each who has human nature but refuses to think only human thoughts. But he [Ajax] from the moment of his leaving home was found to be foolish when his father said to him, "wish for triumph in battle, but wish to triumph always with a god's aid!" And he replied boastfully and stupidly. "Father, together with the gods even one who amounts to nothing may win victory; but I am confident that I can grasp this glory without them." Such a boast as that he uttered; and a second time, when divine Athena urged him on and told him to direct his bloody hand against the enemy, he made answer with these dreadful and unspeakable words, "Queen, stand by the other Argives; where I am the enemy shall never break through." By such words as these he brought on himself the unappeasable anger of the goddess, through his more than mortal pride.

Odysseus pays due respect to gods, and their words. He takes prophecies seriously, as he is represented in *Philoctetes*. He is favored by Athena, the goddess of reason, but he deserves her favor as a free agent. In the ancients' imagination gods were interested in men and women, they favored or even respected certain human beings. They did fall in love with them. But respect and love are different. Odysseus appears more to be respected than loved by the goddess Athena. It may be a plausible conjecture that the ancient poets, and especially Homer, regarded the relation between the goddess and Odysseus as representing the ultimate tie between gods and humans. I do not say that they intended this as an embellishment for their stories of human tragedy, but only that they conceived this divine relationship as the highest piety. I must note that I am speaking of a dynasty of gods and a human race where there are countless conflicts and intrigues, both among gods and men. Odysseus is "pious"[6] in this sense, he fears gods, except involuntarily never acts or speaks insolently as we read Ajax to have done; he pays due attention to rituals and takes prophecies seriously.

It seems that the Greek poets saw piety as careful reasoning and moderation in thought. Hence, what makes Odysseus interesting to us is his moderation, his sound reasoning in finding expedients. There are restrictions to one's will and actions among gods and men, and Odysseus takes this as his first truth, desires only a human life, and thinks only "human thoughts." He unceasingly tries to make his way in the turmoil, thinks as a human must think, and never takes himself as something more than a mortal.

I have tried to understand the significance of character Odysseus in terms of reason and virtue. Many ancient and modern writers of ethics use these terms interchangeably. Without reason, one can hardly be virtuous, and without virtue one can hardly be seen as making use of reason properly. The anonymous tradition, and the poets and philosophers of antiquity seem to have conceived that one who is most skilled in proper reasoning is also the most courageous, moderate, prudent and just. For, without a due to consideration of the given and the means to an end, one does not deserve to be called virtuous, but is rather pitied as weak, or blamed as evil. It is true that Odysseus is not universally deemed to be virtuous, but we were speaking of the world of ancient Greeks and their conception of virtue where piety meant something different than what it means for monotheism.

Orta Doğu Teknik Üniversitesi, Felsefe Bölümü
e-mail: hturan@metu.edu.tr

NOTES

[1] Sophocles, *Philoctetes*, 1025–28; Hugh Lloyd-Jones translation (Loeb ed.), p. 359.
[2] The Greek adjective is *eusebēs*, which could also be translated as religious or pious.
[3] Ibid. 1049–51; Hugh Lloyd-Jones, p. 361.
[4] In this sense Odysseus' reverence for Athena and her providence to him are of great significance. The ancient texts show Odysseus as favored by the goddess who is recognized as equaling Zeus in the power of thought. But Odysseus is keen to observe religious practices for almost all gods, and to abide by the words of prophets who convey gods' decrees.
[5] Sophocles, *Ajax*, 1332–45; Hugh Lloyd-Jones translation (Loeb ed.), pp. 153–155.
[6] See note 2 above.

ANNA MAŁECKA

HUMOR IN THE PERSPECTIVE OF LOGOS: THE INSPIRATIONS OF ANCIENT GREEK PHILOSOPHY

ABSTRACT

The paper discusses the ancient Greek concepts of humor interpreted in the perspective of logos, as well as the inspirations of these philosophical ideas of antiquity for contemporary humor studies. Generally speaking, humor can be considered as a charming, yet paradoxical counterpart of logos, supplementing the one-sidedness of strictly discursive cognitive approach and allowing for the perception of phenomena in multifarious and contradictory planes of reference. Such seem to be the intimations of leading Greek philosophers. Thus, the philosophical humor may be presented as having its roots in the universal logos and alluding to it in an *à rebours* manner. The concepts of humor and laughter formulated by Democritus, Socrates, Plato, Aristotle and Diogenes have been reviewed using contemporary interpretation measures. In this respect, among others, Anna-Teresa Tymieniecka's concept of ontopoiesis of life with its logos-source and the resulting innumerable perspectives turns to be a useful tool of analyzing the abundance and individuation character inherent in humor, as revealed already in classical systems and reflections. Her notion of the ontopoiesis of life allows for intimating the phenomenon in terms of creativity and insight.

Strikingly opposite to logos as it may seem, humor can also be viewed as its counterpart or even paradoxical outcome. So understood, humor – and laughter that tends to accompany it – symbolically participates in the logos universality, indirectly and perversely pointing to the existential or even metaphysical essentials. It was the Greeks who first suggested this line of humor discourse.

We should remember that the notions of humor, laughter, comedy, wit, or irony have often been mixed up and confused in 2500-year-old discussions on the topic that can be traced in the philosophical literature. In the present paper we shall understand humor – referring to the distinction made by John Morreal – as arising from a pleasant cognitive shift, whereas laughter is assumed to be related to a pleasant psychological shift.[1] Thus the transformations occurring within the process of cognition and resulting in a feeling of surprise that pleases the human yearnings for novelty and weirdness – may be considered as characteristic of humoristic approach. And though the ancient philosophers often used the term "laughter" while talking about the humorous perception of experienced (comical) situations, they pointed to the very essence of the phenomenon that is, in our opinion, embedded in the fundamental cognitive paradoxes. That is why humor in its substance can be associated with – seemingly distorted – aspects of the cognizing reason itself. What is more, the philosophical humor – searching for the essences, alongside the whole business of philosophy – possesses a veiled metaphysical inclination, discovering a dialectical potential in innumerable (and as such also contradictory) emanations of the logos itself.

Philosophy has traditionally been rooted – at least until the postmodern times – in the logos. It was the ancient Greeks who built the foundations of Western thought upon the universal rational principles, and made the questions concerning the most vital and ultimate matters the core of the discipline. The gravity of such inquiries might seem profoundly inconsistent with the light spirit, absurdness and triviality that are commonly recognized as inherent to humor. Yet, in the writings of the greatest philosophers of the time we can find at least the outlines of certain theories of humor either corresponding to – even if apparently competing with – the primary task of philosophy proper.

Undoubtedly, the category of logos was the one dominating the classical thought, designating and embracing in itself the harmony of being as such as well as the rules of human cognition and conduct. It was Heraclitus who recognized logos as a factor common to all things, in particular to human beings; as the basic rational principle governing all phenomena and making for the unity of things. However, as he believed, this fact could be pronounced only by people following and accepting the necessary and omni-present nature of the logos: "Therefore it is necessary to follow the common; but although the logos is common the many live as though they had a private understanding".[2] In this context, the chaotic abundance, multifariousness, whimsical randomness and freedom, irregularity, exaggeration and ephemeral trait embedded in humor might – at first sight – be perceived as drastically different from the harmonious clearness, as well as the necessary and eternal nature of the logos.

Analyzing the fundamental and source-like character of the logos from the contemporary phenomenological perspective, Anna-Teresa Tymieniecka writes: "The classical ways proposed by philosophy – ontology, epistemology, metaphysics, aesthetics, anthropology, etc. – all have their source in this logos and yet escape from it into the labyrinths of their singular intellective approaches, getting more and more remote from the source and from each other, getting lost in endless intellectual speculation".[3] Following this idea of the author of phenomenology of life, we aim at showing how, also in the case of humor, one can speak in terms of its primordial logos roots and associations, which seem to be negated in the labyrinths of seemingly illogical and highly individual endeavors. The interpretation based on Anna-Teresa Tymieniecka's concept would help us understand the phenomenon of the universal humor – the humor that philosophy is interested in – as having its source in the universal logos, similarly to the whole "enterprise of philosophy". Referring to the quoted author's postulates, we feel the necessity of reviving this ancient and powerful idea that makes for the unity of being and experience, believing that it can be paradoxically envisaged also in the philosophically understood humor.[4]

The profound humorist standpoint is not that distant from the approach of philosophy as it may be superficially judged. In fact, it was philosophy that first posed the questions: "What is humor?"/"What is laughter?", and it seems that this discipline has the greatest chances of grasping the mechanism of humor and arriving at its deep structure-meaning.

Humor shares with philosophy the act of withdrawal from practical aspects of life and a standard commonsensical approach. They both incessantly search for ever fresh and thus astonishing solutions, employing the power of abundant imagination, and also fantasy. They both open new perspectives, enabling experience and cognition detach from prevalent stereotypes. In this context, it is worthwhile to quote William James's interpretation of the akin characters of humor and philosophy: "Philosophy, beginning in wonder, as Plato and Aristotle said, is able to fancy everything different from what is. It sees the familiar as if it were strange, and the strange as if it were familiar".[5] John Morreal summarizes the analogy between philosophy and humor as follows: "To have cultivated a philosophical spirit or a rich sense of humor is to have a distanced, and, at least potentially, a more objective view of the world".[6]

However, this dependence between philosophy and humor is of a pretty complex character, showing in most cases its *à rebours* nature – especially while getting *more and more remote from the original source*. Thus, further in the text we would like to detect the logos-related elements in Greek philosophical concepts of humor and its accompanying laughter, and to prove – following Anna-Teresa Tymieniecka's idea – that the motto "Logos, the sense of sense, penetrates All. (...) IN LOGOS OMNIA"[7] pertinently refers also to the foundations of humor.

The question arises, how to describe this mechanism of humor, that can be related to the logos-like inventiveness and versatility? The contemporary critic, Arthur Koestler in his bisociation theory states that the bases of all original thought and discoveries, also in the sphere of humor, lie in the simultaneous perception of given phenomena jointly in traditionally separate and incompatible systems of reference, governed by conflicting rules.[8] Thinking in the categories of one system only does not allow for any original and truly creative achievements. The escape from routine is signaled in the insight which presents a familiar situation in new light and triggers a fresh response to it. The act of bisociation joins the separate matrices of experience, expanding the horizons of complex though surprisingly rich vision. It implies "living on various planes", multidimensionally. The violent collision of two matrices of perception or reasoning, a sudden bisociation of a phenomenon in two traditionally incompatible systems, evokes a particular emotional tension which finds its outlet in laughter. This very act of associating matters that commonly are considered disjunctive, results, in our opinion, from the logos source of everythingness, as it ingenuously points at the basic and sensible unity of all things, in spite of their superficial labyrinth-like or even whimsical differentiations.

It was Socrates, the model sage of antiquity, who – in his search for truth – proved to be aware of the significance of humorous elements in the solemn discourse, introducing wit and irony into the realm of philosophy. Generally speaking, the essence of irony, understood both as rhetorical device and Socratic method, lies in incongruity between the literal and deep meanings of the utterance, between its apparent foolishness and the hidden substantiality and wisdom. It can be used as a tool for the whole variety of purposes, including the cognitive-dialogical ones, as was the case with Socratic method. Irony creates a certain (often humorous) illusion, building up the opposition: serious – non-serious, god-like – clown-like.[9] The motif of game is therefore, not alien to this method. For Socrates, irony is mostly verbal, and the intended audience is expected to grasp eventually that the speaker is highlighting the literal falsity of the utterance, and not *eironeia*, which aims at deception and is predominantly malevolent.[10] Irony is thus employed to serve the enlightened and logos-based objectives of the sage's teaching, who intends his disputants to arrive at the truth, using this indirect method.

"Philosophy is when we laugh. And we laugh at stupidity"[11] – this remark by Hans Blumenberg may be pertinently referred to the concept of the classical philosopher who truly appreciated the significance of laughter and humor in the plane of the logos: Democritus, often presented in the iconography as "the laughing philosopher". Similarly to former Greek thinkers, also the author of atomic theory viewed the logos as the factor determining the proper measure and moderation, which, in turn, ensure the state of peace, balance and good mood. Joy in life is considered as highly rational standpoint, and it is only the stupid people who are unable to find any satisfaction in living.

As a philosopher, Democritus shows a detached attitude towards the follies of the world. In vivid opposition to the crying Heraclitus, he assumes the laughing mood in the face of circumstances that cannot be controlled by reason. Seneca comments upon these two opposite philosophical attitudes to the universal lack of sensibility as follows: "Whenever Heraclitus went forth from his house and saw all around him so many men who were living a wretched life – no, rather, were dying a wretched death – he would weep, and all the joyous and happy people he met stirred his pity; he was gentle-hearted, but too weak, and was himself one of those who had need of pity. Democritus, on the other hand, it is said, never appeared in public without laughing; so little did the serious pursuits of men seem serious to him".[12]

Both the Heraclitean and Democritean positions can be described, after Helmut Plessner, as two radically different reactions to the situations of crisis of expression (and such must be for a sage the absurd situation of human folly): weeping or laughing.[13] The atomist philosopher, instead of despairing of human unwisdom, was inclined to perceive human weakness and vice not as tragedy, but rather as comedy, and to manifest a mild and sympathetic approach to that which seemingly deserves contempt. To quote Seneca's evaluating indications: "We must, therefore, give our minds such a bent that all the vices of the populace may not appear hateful to us, but ridiculous, and we should imitate Democritus rather than Heraclitus. For the latter used to weep whenever he appeared in public, but the former laughed: to one everything which we do seemed to be foolishness, to the other, misery. Therefore all things must be made light of and borne with a calm mind: it is more manlike to scoff at life than to bewail it. Furthermore, he who laughs at the human race also deserves better of it than he who mourns for it. The former leaves something still to be hoped for; the latter stupidly weeps over what he despairs of being able to correct: and he shows a greater mind who, after he has contemplated all things, cannot restrain his laughter than he who cannot restrain his tears, inasmuch as he does not allow his mind to be affected in the least, and does not consider anything great, severe, or even serious".[14] Democritus's position seems to take into account more aspects, and be more promising than that of Heraclitus. His laughter is not just a simple fountain of light-hearted jesting, but rather emerges out of bitter wisdom, and manifests a benevolent acceptance of the inferiority of certain aspects of existence. For him, in the philosophically considered plane, it was the logos itself that imposes the requirement of reconciliation with the weird fortune and triviality of human endeavors. In paradoxical and absurd situations in which all ways of expression fail, humor and laugh seem to constitute the only sensible reply. The abundant and apparently irrational laughter of Democritus in the fictitious *Letters of Hippocrates* is evaluated as a proper way of reacting to life's follies. At first, though, the ceaselessly laughing Democritus is considered maniacally mad. Eventually, Hippocrates pronounces the diagnosis that the philosopher shows profound wisdom and is but "too sensible". In the face of human littleness, vice and inability to think logically, Democritus's laugh becomes a remedy preventing him from giving up in despair, and simultaneously signalizing to his contemporaries their funny irrationality. He criticized their thoughtlessness, as they seemed to have "neither eyes nor ears". The logos itself points at the adequate behavior in situations in which reason is lacking – trivializing the unwise phenomena, and indicating their insignificant and absurd character.

Even though Democritus does not believe in the possibility of improving the citizens of his native Abdera, the laughter devoid of bitterness or malice enables him to preserve balance and maintain the inner independence from the crazy human world. Such is the natural consequence of his metaphysical theory according to which the world – including man and society – is nothing more than a never-ending game of atoms ceaselessly moving in the void.

According to Michel de Montaigne, Democritus's assessment of the human condition as futile and comic is more appropriate than Heraclitus's, as it takes into consideration the truly humane values. It presents people as trivial rather than evil, and, accordingly, deserving laughter, not hatred. Likewise, the human condition should not yield to despair, but rather to an all-comprehending smile. The inhabitants of Abdera – the objects of Democritus' laugh – are treated with benevolence, as the creatures unable to see through their restricted horizon. The philosopher detects humane value in laughing at all people including himself – the value of which one-sidedly serious and highly critical approach is usually devoid.

Here we encounter a very important aspect of humor: the multi-sidedness of its approaches to given phenomena, or, to be more precise, their complex and ambivalent perception, which allows for more comprehensive outlook on things.

The main opponent of Democritus, Plato, outlined a radically different vision and assessment of the nature of humor and laughter. Contrary to Democritus, and following his master Socrates, he tends to juxtapose humor and the logos.

In this context, we should start with recalling the famous scene from Plato's *Theaetetus*, in which the Thrace servant laughs at Thales when the philosopher falls into a well while looking up the starry sky. On a different and competitive plane of reference/association, Thales's wisdom is perceived by the unwise and primitive woman as foolery, as alien to the dictate of the "sense" of common people, and consequently finds its outlet in laugh. The philosopher Thales becomes an object of the servant's amusement, who is convinced of her own commonsensical superiority. For the idealist philosopher, such amusement is contrary to reason. In the *Republic*, we become further convinced that someone who has seen the Idea of the Good, thus reaching the highest level of cognition, must appear funny both to himself and to others, as he represents the triumph over the stereotypes to which humans are so much used. The desire for laugh is presented as detrimental,[15] for the ideal sage should not give his reason up totally to emotions. In Plato's ideal state, there is practically no place for humor, as in like manner there is no place for poetry. It is only the solemn observation of the rules of logos that befits the philosopher. Its opposite – a joyful fantasy with its fictitious implications – constitutes but an obstacle on the path of pursuing the ultimate truth of eternal being. The method implied by reason and solemn love/charity turns to be more useful for such a noble purpose. Moreover, the search for amusement is not worthy of the philosopher whose legitimate task lies in seeking for the best, not the most joyful.

But even the exalted Plato felt bound to agree that humor possesses certain value in the dialectical process of cognizing the essentials. In the *Laws* he stated: "For it is impossible to learn the serious without the comic, or any one of a pair of contraries without the other, if one is to be a wise man (...)".[16] Humor helps us understand that which is actually valuable and which constitute the opposite of the comical: the solemn. Imitation of the ridiculous in comedy may be accepted only on condition that it represents a despicable contrast to the good. Thus the knowledge of the nature of the comic should serve mainly a negative and cautionary purposes: those who learn the ridicule from the position of distanced audience can avoid it in their own endeavors. All in all, the logos, alongside the virtue inseparable from it, requires that the humoristic should be given up in the face of the ultimate Good. Further in *Laws*, we read that: "(...) to put both [the serious and the comic] into practice is equally impossible, if one is to share in even a small measure of virtue; in fact, it is precisely for this reason that one should learn them, – in order to avoid ever doing or saying anything ludicrous, through ignorance, when one ought not".[17] Consequently, a comic performance is not appropriate for the liberated, i.e. the rational person. It can only suit the slave whose activities, interpreted in terms of the mimesis theory, may possess certain value as an artistic imitation of situations and features considered inferior and despicable: "(...) we will impose such mimicry on slaves and foreign hirelings, and no serious attention shall ever be paid to it, nor shall any free man or free woman be seen learning it, and there must always be some novel feature in their mimic shows".[18] The reason and laws based on it should therefore impose restrictions on amusing entertainment so that they do not turn detrimental to the logos-imbedded task of humans: "Let such, then, be the regulations for all those laughable amusements which we all call comedy, as laid down both by law and by argument".[19]

Also in the *Republic*, when setting up rules for the education of the Guardians of the ideal state, Plato presents laughter as the reaction to be avoided, because it may lead to violent (i.e. unreasonable) behavior: "Again, they must not be prone to laughter. For ordinarily when one abandons himself to violent laughter his condition provokes a violent reaction".[20] Accordingly, literature should be censored so that it does not provide bad examples of gods and heroes as overcome with laughter.

But although he despised laughter and comedy, as they were presented in shear contrast to the main objectives of his idealist theory, Plato was the first philosopher in the ancient world to undertake an attempt at clarifying the very notion of comicality. In the *Philebus*, he finds the cause of funniness in complex processes occurring within the human soul, wherein the ambivalent emotions blend up. In his interpretation, the nature of comicality or "the malice of amusement" consists in the mixture of pleasure and pain. Again, philosophically, a pleasure that results from the good is considered here, whereas a pain constitutes a feeling accompanying evil. In the case of the comical we encounter a paradoxical combination of both, in contrast to knowledge, in which pleasures are unmixed with pain.[21]

For Socrates pronouncing Plato's opinion, it is "the evil" or "the vice" that initiates the whole analysis of the comical. All vice has its source in unwillingness (or inability) to answer the Delphic inscription "Know Thyself". Consequently, a man does not actually know himself, and takes himself for somebody whom he is not. As James Wood pertinently remarks, "Comedy in the *Philebus* is condemned not because it is mimetic (as in the *Republic*), but because it is malicious, as malice in turn is condemned for its unjust co-mingling of pleasure and pain. Malicious laughter occurs in conjunction with the spectacle of the ridiculous, which is defined in opposition to the Delphic command 'Know Thyself'. Hence, malicious comedy is grounded in self-ignorance and foolishness, both that of the ridiculed and, as it turns out, of the ridiculer".[22]

In the *Philebus*, Socrates mentions three errors that can appear on the path of gaining self-knowledge: errors related to wealth (when people consider themselves richer than they are in reality), errors connected with their alleged beauty and power, and – the most widely spread one – errors consisting in considering oneself more virtuous than one really is. All these faults prevent one from truly knowing himself, and as such constitute vice. Consequently, in accordance with the theory presented in the *Philebus*, they lead to pain. In order to show the blended pleasant and painful nature of comedy, Plato offers another subdivision: though all ignorant and false opinions of oneself are disastrous, there is a qualitative difference between the vice of strong and of weak men: "Let this, then, be the principle of division; those of them who are weak and unable to revenge themselves, when they are laughed at, may be truly called ridiculous, but those who can defend themselves may be more truly described as strong and formidable; for ignorance in the powerful is hateful and horrible, because hurtful to others both in reality and in fiction, but powerless ignorance may be reckoned, and in truth is, ridiculous".[23] Plato's arguments become more clear in the light of the infamous revenge taken on his master Socrates by men in power, who were obviously mistaken in their self-estimation. On the other hand, the relevant ignorance of powerless people is not that frightening, but merely ridiculous.

Evil presented in comedy instead of evoking pain, induces laugh. Here, Plato outlines the first version of the so-called *superiority theory* in relation to laughter and humor: pleasure inherent in laugh (and humor) results from the feeling of superiority in relation to the weak, defective and inferior. However, this type of pleasure seems in most times alien to a logos-controlled philosopher that finds pleasure only in the solemn knowledge of the ultimate forms, with its highest goal: the Idea of the Good. Nevertheless, as James Wood pertinently states, "...laughter is called for precisely at the moments of greatest philosophical solemnity in order to remind oneself and others of human limitation and the need for humility in the quest for cosmic enlightenment".[24] Accordingly, it may be noticed that it is the logos itself that imposes the need for humility in the face of the primordial and the highest. Considering the earthly restrictions (the chains in the ingenuous allegory of the cave presented in the *Republic*), the highest cannot be fully comprehended, at least here, on the earth. Therefore, the critical aspect of laughter and humor may also be cognitively valuable. The most difficult to achieve and requiring maturity is the ability to ridicule oneself. In the *Philebus* we read that the philosopher laughs both at himself and others, playfully recognizing the limits of any philosophical seriousness. This is what James Wood recognizes as the philosophical *redemption of laughter*.[25]

Diogenes reversed the dialectical situation of amusing inferiority and amused superiority, mocking at all types of persons who considered themselves superior, and moving the boundaries of philosophical humor up to the absurd. In a provocative manner, he ridiculed both the common sense and laws of average people, and Plato's spiritual aloofness, for the sake of what he perceived as the primary life values. For that reason he was criticized by the rational philosopher who considered him primitive or even barbarian. He did not hide his scornful intentions, and manifested sarcasm towards the spiritual discourse. As Diogenes Laertius reports, "When Plato was discoursing about his 'ideas', and using the nouns 'tableness' and 'cupness'; 'I, O Plato!' interrupted Diogenes, 'see a table and a cup, but I see no tableness or cupness'. Plato made answer, 'That is natural enough, for you have eyes, by which a cup and a table are contemplated; but you have not intellect, by which tableness and cupness are seen'.[26] However, as we can further read in the *Lives and Opinions of Eminent Philosophers*, in his malicious satire he himself was showing a proud sense of superiority: 'Thus I trample on the pride of Plato'; and that Plato rejoined, 'With quite as much pride yourself, O Diogenes' ".[27] Paradoxically, Diogenes's own mockery seems to be incapable of discovering his own ridiculous sense of superiority, devoid of self-criticism and even ... the sense of humor.

According to Christoph Martin Wieland, the author of an apologetic book devoted to Diogenes, the controversial philosopher behaved so weirdly not because he was mocking at all people on principle, but because he looked through the absurd character of their everyday habits in which he did not want to participate.[28] Such was his way of manifesting independence from all the stereotypes imposed by society. The philosopher of Sinope in an excessive mocking manner was revealing the follies of his social milieu, thus – following the prophecy of Delphic oracle and considering it his vocation – reversing, long time before Nietzsche, the commonly accepted values. His aggressive mockery was directed against the legally regulated social order that imposes illusory and false purposes.

To those who in the name of majority ridiculed him, Diogenes had a ready answer: "When a man said to him once, 'Most people laugh at you'; 'And very likely', he replied, 'the asses laugh at them; but they do not regard the asses, neither do I regard them' ".[29] Diogenes's extravagance found its zealous defender in the person of Wieland who presented Diogenes's sarcasm – the sarcasm of the "mad Socrates" – as the humor of enlightened humanist, who applied bitter satire and mockery in order to reveal the existential and moral essences.

The theory of humor and laughter was further developed by Aristotle. It was him who reserved the very ability of laughing to human beings.[30] If we combine that characteristic of man with Aristotelian *differentia specifica* of his equally famous definition included in the *Nicomachean Ethics*, it may turn out that rationality and capability of laughing are not, after all, so alien to each other. Anyhow, considered as one of human distinctive features, ability of getting amused deserves a philosophical analysis in the eyes of the rationalist thinker.

In accordance with his sensibly moderate philosophy, Aristotle was far from abandoning the whole enterprise of amusement and comicality whatsoever. Neither was he convinced about the trivial nature of humor as Plato used to be. In his opinion, we should not totally suppress laughter that arouses from comicality, because jesting makes for the pleasant side of active life. In the *Nicomachean Ethics*, the capability of laughing in included among the social virtues, and people devoid of the sense of humor are reckoned as lacking good manners. However, as usual, it is reason that imposes restrictions on our readiness to overdo jesting and laughing, so that they remain tactful and polished. Accordingly, the Stagiryte distinguishes the tasteful and moderate type of humor from the excessive one, which cannot be accepted as a social virtue.

In the *Poetics*, looking for the essence of humor and comicality and referring to Plato's superiority theory, Aristotle states that in being amused by someone we are finding that person inferior in some way. However, "to find someone's shortcomings funny, we must find them as relatively minor",[31] not harmful or disturbing. In like manner, the Stagirate defined the essence of comedy: "Comedy, as we have said, is a representation of inferior people, not indeed in the full sense of the word bad, but the laughable is a species of the base or ugly. It consists in some blunder or ugliness that does not cause pain or disaster, an obvious example being the comic mask which is ugly and distorted but not painful".[32] In his opinion, contrary to tragedy whose characters are average or better than average, comedy presents subjects of lesser virtue than the audience. Here we despise the characters, since they are shown as in some way inferior to us. The "ludicrous", according to Aristotle, is "that is a failing or a piece of ugliness which causes no pain of destruction".[33] The distanced and minimized evil shown in comedy is unable of making any harm, and thus can be understood as fulfilling didactic purposes. It namely contains criticism of human weaknesses presented in an amusing way, and by means of entertainment teaches to avoid them. So Aristotle, like Plato, brings ethical considerations of comicality to our attention again.

However, the aspect of Aristotelian humor theory which is closest to finding the logos-source of the phenomenon, is his concept of incongruity. According to John Morreall, Aristotle is the first to suggest the incongruity theory of humor,[34] later developed by numerous thinkers, including Cicero, Descartes, Hutcheson, Kant,

Schopenhauer or Bergson, and several contemporary humor researchers who provide modified versions based on the classical idea. Incongruity theories seem to be intellectually most promising, as they focus on the formal object of amusement, attempting to outline the distinguishing features of humor. These theories expose its paradoxical and quasi-cognitive character, the aspect that – at least partially – can satisfy the philosopher. In the view of theories belonging to this group, "what amuses us is some object of perception or thought that clashes with what we would have expected in particular set of circumstances".[35] Generally speaking, incongruity theories are founded upon contrast and dialectically opposite sets of references.

In the *Rhetoric*, Aristotle states that the humorous effect is achieved by setting up a certain expectation in the audience and then surprising them with something they did not expect, delivering something "that gives a twist". "The effect is produced even by jokes depending upon changes of the letters of a word; this too is a surprise. You find this in verse as well as in prose. The word which comes is not what the hearer imagined".[36] Surprise evoked by contrast between the original line of reasoning on the side of listeners/readers with reference to the speech/text and its actual utterance induces special intellectual amusement. More universal implications of the described theory can be found in contrast between stereotypes resulting from culture, education and individual experience, and their non-conformity to the actual perceptions. By the feeling of amusement, this process leads to temporary suspension of one-track commonsensical approach to phenomena, offering an insight through the prism of dialectical relation between antithetical elements. Aristotle further explains that the surprise must somehow "fit the facts",[37] or – to use the terms of the contemporary version of the theory – the incongruity must be capable of resolution. "The more briefly and antithetically such sayings can be expressed, the more taking they are, for antithesis impresses the new idea more firmly and brevity more quickly".[38] We can compare the moment of surprise imposed by the wit to the freshness and suddenness of new discoveries in the field of any type of creativity. That is how the logos poetically operates, using brilliant antitheses and quick associations of opposed elements to evoke new perspectives of interpreting the unknown aspects of the familiar, to experience the flux between contradictory planes, to approach the dialectical fullness of being.

Let us quote Anna-Teresa Tymieniecka's words again in this context: "The world horizons that our experiences open before us appear and vanish as our focus shifts. Yet the initial spontaneity of that consciousness' emergence is not self-explanatory. It is not its own cause, neither does it carry its own 'reason' ".[39] In the view of the theories of humor based on the notion of incongruity and contrast initiated by Aristotle, also the power of humor which opens new perspectives of vision and insight may be considered as emerging from the main stream of life energy which itself arises out of the primeval rational principle that surpasses the individual. Again, applying Anna-Teresa Tymieniecka's theory, we can detect the universal metaphysical foundations of this specific absurd-like character of the individualization of *primordial positioning of life* within the logos itself. In Book I of *The Fullness of the Logos in the Key of Life* we read: "Transcendental consciousness does, indeed, posit and objective world around us but one with established or now being established forms, ways of proceeding. However, these recognized modalities and their very coming about are being existentially conditioned and have their roots not in themselves but within *the primordial positioning of life* and its individualization – positioning within an immense network of logoic forces, schemata, and routes, of which human consciousness is but a constructive knot on a larger scale".[40] Accordingly, discussing the essence of humor in the horizon of logos, we can conclude that the situations/the whole world projected or constituted by "the humorous consciousness" have their source in the powerful life differentiation of the logos itself, and symbolically point (indirectly and perversely) to its richness/fullness that seeks to find its outlet also in induviduated creative paradoxes inherent in humor.

In the ancient Greek thought we find several insights that can be employed to the interpretation of humor origins in terms of logos. Though never developed into a comprehensive theory at the times, they have inspired further reflections over the universal logoic foundations of the philosophically understood humor, with its playful form, cognitive shift and quite serious implications.

*Department of Culture Studies and Philosophy, University of Science
and Technology AGH, Kraków, Poland
e-mail: amm@agh.edu.pl*

NOTES

[1] Cf. http://www.iep.utm.edu/humor/.
[2] Sextus Empiricus, *Against the Logicians*, trans. Richard Bett (Cambridge: CUP, 2005) http://assets.cambridge.org/052182/4974/frontmatter/0521824974_frontmatter.htm, VII, 133.
[3] Anna-Teresa Tymieniecka, *The Fullness of the Logos in the Key of Life*, Book I, *The Case of God in the New Enlightenment* (Dordrecht: Springer, 2009), p. xxxii.
[4] Cf. ibidem, p. xxv: "The human quest for wisdom, for making sense of the things we believe on faith, is being pulled apart by the intellectual program of 'deconstruction', on the one hand, and by a revived religious distrust of reason, on the other".
[5] William James, *Problems of Philosophy* (Cambridge, Massachusetts: Harvard University Press, 1979), p. 11.
[6] John Morreal, "Introduction", in *The Philosophy of Laughter and Humor*, ed. John Morreal (Albany, NY: State University of New York Press, 1987), p. 2.
[7] Anna-Teresa Tymieniecka, *The Fullness of the Logos in the Key of Life*, op. cit., p. xxvi.
[8] Arthur Koestler, *The Act of Creation* (New York: The Macmillan Company, 1967), pp. 27–87.
[9] Cf. Włodzimierz Sztorc, *Ironia romantyczna* (Warszawa: Wydawnictwo Naukowe PWN, 1992), p. 6.
[10] Cf. David Wolfsdorf, "The Irony of Socrates", *The Journal of Aesthetics and Art Criticism* 65:2 (Spring 2007), p. 176.
[11] Hans Blumenberg, *Das Lachen der Thrakerin* (Frankfurth a.M.: Suhrkamp, 1987), p. 149.
[12] Seneca, *On Anger*, trans. John W. Basore, 2.10.5 (http://www.morris.umn.edu/academic/philosophy/Collier/Intro%20to%20Philosophy/Seneca.pdf).
[13] Helmut Plessner, *Laughing and Crying: A Study of the Limits of Human Behaviour*, trans. J. S. Churchill & Marjorie Grene (Evanston, IL: Northwestern University Press, 1970).
[14] Seneca, *On Tranquility of Mind*, based upon transl. William Bell Langsdorf (1900), revised and ed. Michael S. Russo http://www.molloy.edu/sophia/seneca/tranquility.htm.

[15] Plato, *Republic*, 388d–390a, Plato. *Platonis Opera*, ed. John Burnet (Oxford: Oxford University Press, 1903) www.perseus.tufts.edu/.../text.
[16] Plato, *Laws*, 816 d–e, [in:] *Plato in Twelve Volumes*, Vols. 10 and 11, trans. R.G. Bury. (Cambridge, MA: Harvard University Press; London: William Heinemann Ltd., 1967 and 1968). http://www.perseus.tufts.edu/hopper/text?doc
[17] Ibidem, 816e.
[18] Ibidem, 816e–817a.
[19] Ibidem.
[20] Plato, *Republic*, trans. Paul Shorey (Princeton, NJ: Hamilton & Cairns, 1961), 388e. pp. 633–634.
[21] Plato, *Philebus*, trans. Benjamin Jowett, 48c, http://www.greektexts.com/library/Plato/philebus/eng/786.html.
[22] James Wood, *Comedy, Laughter, and Malice in Plato's Philebus*, http://apaclassics.org/images/uploads/documents/abstracts/Wood.pdf.
[23] Plato, *Philebus*, op. cit., 49c.
[24] James Wood, op.cit.
[25] Cf. ibidem.
[26] Diogenes Laertius, *The Lives And Opinions of Eminent Philosophers*, trans. C.D. Yonge, Ch. VI. http://fxylib.znufe.edu.cn/wgfljd/pw/diogenes/dldiogenes.htm.
[27] Ibidem.
[28] Christoph Martin Wieland, *Nachlass des Diogenes von Sinope*, [in:] *Sämtliche Werke IV* (Hamburg: Hamburger Stiftung zur Förderung von Wissenschaft und Kultur, 1984), Vol. 13.
[29] Diogenes Laertius, op.cit.
[30] "... no animal but man ever laughs", Aristotle, *On the Parts of Animals*, trans. William Ogle, p. 77. http://books.google.pl/books?id.
[31] John Morreall, *The Philosophy of Laughter and Humor*, op. cit., p. 14.
[32] Aristotle, *Poetics*, trans. S. H. Butcher, ch. 5, 1449a. http://classics.mit.edu/Aristotle/poetics.html.
[33] Ibidem, sections 3 and 7.
[34] John Morreall, *Taking Laughter Seriously* (Albany, NY: State University of New York Press, 1983), p. 16.
[35] John Morreall, *The Philosophy of Laughter and Humor*, op. cit., p. 6.
[36] Aristotle, *Rhetoric*, based on trans. by W. Rhys Roberts, http://www2.iastate.edu/~honeyl/Rhetoric/rhet3-11.html#1412a.
[37] Ibidem. Cf. also http://www.iep.utm.edu/humor/.
[38] Ibidem.
[39] Anna-Teresa Tymieniecka, *The Fullness of the Logos in the Key of Life*, op. cit., p. xxix.
[40] Ibidem.

BRIAN GRASSOM

THE IDEAL AND THE REAL: BRIDGING THE GAP

The inner life most gladly, most cheerfully, most devotedly wants to be the living bridge between our present life and the ideal life, the life that we want to have.

–Sri Chinmoy Kumar Ghose[1]

ABSTRACT

The ideal and the real represent a projection of human consciousness that appears to be in a state of internal conflict. In fact, human consciousness itself might be said to inhabit a perennial interplay of the two. Whereas the result of this interplay can be by turns pleasure or sorrow, the two never appear to be completely reconciled. By tracing the identification of the real and the ideal – in the sense that they appear to the rational mind – in the art and philosophy of antiquity, this chapter attempts to clarify their meaning and proposes that they converge in art. Here art acts as a vehicle of another, transcendent and unified consciousness that is sometimes termed as 'spiritual'.

The quote above deserves careful consideration, not least as regards the phrase "the inner life", and its connotation of spirituality. If seen in a particular way, I would say that art is essentially a 'spiritual' pursuit. If we can for the moment suspend our preconceptions of these words, and allow art to fold into the same undefined area that is implied above by the "inner life", in what way and in what sense, then, does or can art bridge the gap between the ideal and the real?

First of all, we have to know what we mean by the real, and what we mean by the ideal. In human life, it seems the ideal is a vision of what reality should be – according to our deepest desires and aspirations. The real is what *actually is*. The ideal appears to be projected either in the future, as yet un-manifest, or in the remote past and therefore not immediately available to us. The real exists either in the actual present, or in the known experience of the past. The ideal is always to come, or perhaps to come again; the real is imminently now, irretrievably past, or a probability of the future.

But, in point of fact what we term the 'real' is actually also a projection – just as much as is the ideal: if we stop to consider for a moment what we mean by 'real' or 'actual' we find that it is not so easy to grasp, or even to define. Of course, we *think* we know what we mean, but that *thinking* – on closer examination – can be misleading, to say the least. This point of doubt about our relation to the 'real' world – and the consequential search for truth – is what forms the basis of rationalism, from Plato to Descartes, to Husserl – and is arguably the basis of philosophy as a whole.

But if the real is subject to doubt, then what of the ideal? Surely it must be more so, since it is already in retreat from pragmatic realism; and just as the real may prove difficult to define, then by the same token the ideal in comparison is so simple it cannot hope to represent any truth for the rational mind? Well, not quite so: if we care to look at the ancient origins of our rational philosophy, the ideal is there – standing side by side with reasoned dialectic. And it is an extremely difficult notion to shake off, or to ignore, even for the most rational of minds. For far from being diametrically opposed to rational thought, what we call the ideal is actually fundamental and integral to the way that thought constructs itself. This is something that has been explored by writers such as Derrida and Levinas,[2] and it reveals a deeper significance to the idea of the ideal, and its relationship to the real, than might be assumed superficially, or in the first instance.

My purpose here is not to delve into the psychology of the ideal – much as that would be a fascinating project in itself – for I am not well versed in psychology. Rather, I would like to approach the psychological and philosophical tension between these two – the ideal and the real – through an examination of how they interact through the medium of art, thereby not only (and not merely) grounding the dialectic in a 'case' such as art, but *a fortiori* attempting to show at the same time why art is a singularly unique and invaluable instrument of human progress and fulfilment.

A BAD HISTORY

Many people would argue that there is no need for the ideal: indeed, we would be much better off without it. For there can be no doubt that it has been the cause of much trouble. One only has to consider some of the political, social, and religious ideals of the last few centuries and their catastrophic impact upon the real world to be tempted to renounce idealism altogether. But like it or not, the ideal continues to play a large part in human life. It is in fact integral to everything we do. We pursue the ideal every day in so many ways that we are hardly aware of. We cannot help projecting our wishes and desires into our actions, from the simplest to the grandest project. For example: ideally, I will give a very interesting presentation, everything will go smoothly, there will be no technical problems, and the audience will be quietly enthralled. This is an ideal. And that pattern is repeated in many of our projected actions and situations. Then of course, we know from experience that reality has a habit of not performing in an ideal way, and so we come up with contingency plans – just in case. But there can be no escaping the fact that what we are pursuing is a form of the ideal. It is an *idea* of *perfection*, of things as we would like or wish them to be, and as we think they could or should be – if only reality would co-operate.

So we can see that the ideal is linked to *perfection*. In this sense, idealism actually is amoral – a criminal, for example, might see his ideal as the perfect crime. Nevertheless down through the ages, idealism and perfection have been linked, and perfection inextricably bound up with the idea of 'the good': unfortunately, what

Chen Yan Ning. 1972. *Chairman Mao Visits Guangdong Countryside*

we have found is that what appears to be good for some turns out to be very bad for others.

Communism and Fascism, for example, were constructed around ideals. Both considered their respective idealism to be innately 'good', although they declared themselves poles apart, and both proved disastrous for the majority of people affected by them. This can often be the result of trying to put the ideal into practice, of trying to impose it upon the real. Reality, on the other hand, might sometimes appear good, and sometimes bad. But it never appears to be perfect – at least not for very long – and so we are invariably, and ultimately, dissatisfied with it.

In speaking of the ideal and the real, we are willingly or unwillingly drawn back to Plato. In the *Republic* he set out his thoughts on the ideal society. And the spirit of this great work of philosophy has throughout history inspired many to attempt the implementation of ideals in the political and social circumstances of their time, with varying degrees of success and failure. Plato, however, was not a politician, but a writer and philosopher. Key to his notion of the ideal republic is the idea of the "philosopher-king". This idea pre-supposes wisdom and goodness in those ultimately responsible for the welfare of the republic. The reason for this is that without wisdom and goodness the republic, as envisaged by Plato, would begin to fall apart. They are the glue that keeps it together. The "King" represents the rule of law, of virtue, and the "Philosopher" wisdom and goodness. For Plato the virtues of wisdom and goodness are non-negotiable, even – it might be argued – undemocratic: nor are they, however, a 'divine right'. They are, like reason itself, 'absolutes' outside of space and time and therefore beyond human fallibility, but not beyond human understanding. Now, this very absolutism if misconstrued or misappropriated, can become the cause of unimaginable mischief – as we well know. It is important, therefore, that wisdom and goodness remain true 'ideals' in the Platonic sense: i.e. they are beyond material form. If that is the case they cannot be appropriated under

The illustration shown here is from the Leni Riefenstahl film *Olympia* (1938). Paradoxically, and unlike *Triumph of the Will*, this film is *not* notable for Nazi ideals. It *is* idealistic in its treatment of the modern Olympics as an event of international friendship and human achievement. Its stunning photography focuses on all the athletes, including beautiful sequences of the black American Jesse Owens

any circumstances in the material world: they are only visible through their 'real' and material effects. What is essential – and this is where materiality comes in – is to carefully and thoughtfully create the conditions under which they can ameliorate reality.

This is an idea that in its purest form also occurs in the history of India, and is represented in the national flag of that country: The great king Asoka was instrumental in combining temporal law with spiritual tenets, and his 'wheel' symbolises this convergence in the centre of the flag. It was evident in the rule of Charlemagne, and the English King Alfred. Similarly, it inspired Dante to dream of a blend of earthly government and divine wisdom in the Europe of his day. And it surfaced yet again in the founding principles of the republic of the United States of America, embodying a secular wisdom founded in humanism.

Now these examples might appear in themselves to be yet more manifestations of a present idealism, which is at best naively unrealistic and at worst decidedly foolhardy or even dangerous: one simply has to dig deeper into the records to find that 'reality' does not match up to our cherished historical retrospective. But to my mind, in each case they are not so much ideal historical narratives as figural prototypes of an attempted *balance* of the phenomena of the ideal and the real, which is where their true value is to be found.

A QUESTION OF BALANCE

The problem with the ideal arises when we try to appropriate it, try to bring it into being. There are two reasons why this might fail.

First of all let us consider the ideal as Plato envisaged it. According to Plato what we take for the 'real' form is not truly real. True reality is Ideal Form. Taking his "couch", (or "bed" as it is more popularly known) as an example, we can look at it from several different angles and find it is different each time. Whichever view we choose we find it cannot be the definitive form of the bed, as there are many other viewpoints showing different aspects.

Does a couch differ from itself according as you view it from the side or the front, or in any other way? Or does it differ not at all, in fact, though it appears different, and so of other things?[3]

It follows that the ideal form of the bed – the "one in nature" is not apparent. Therefore it would appear that the ideal form cannot actually be envisaged, or that it cannot be envisaged as form. He uses this analogy to illustrate the difference between ideal form and material form, and between art and truth.[4] But the point is that in his view the ideal form cannot be made manifest, for as soon as it was it would no longer be ideal, and another ideal form would of necessity arise to take its place.

Of course, we can do away altogether with the notion of an ideal form of the bed: but this would mean that it cannot be 'a' bed since there is no definitive bed to base it upon, nor can it be the definitive 'the' bed since we know there is more than one form of bed in the world. It can of course be 'this' particular bed, perhaps the one and only 'bed' known to us at the time – but, again, we have seen that *in reality* it has no definitive viewpoint and its true form escapes us. Furthermore we are throughout relying upon the general concept of 'bed'. Without the concept 'bed' the bed cannot be identified. This generality – 'bed-ness' or 'of being a bed' – is a notional and abstract one, but it is indispensable to our way of thinking. And it is fundamental to language. Even if we speak metaphorically – e.g. "she made a bed of the cold ground" – we have to admit of the concept 'bed'. And this concept is difficult – if not impossible – to define.

This raises "the question of being" as Heidegger would have it.[5] After Plato the question was not raised, but on the contrary – according to Heidegger – "forgotten". 'Being' was *assumed* to *be*. From Aristotle onwards the simple nominal category 'bed' replaced the intuitive notion of ideal form that had underpinned Plato's rational enquiry, and Plato's ideal form was subsumed within Aristotle's "form and substance", which duality eventually led to numerous representations such as material/immaterial, mind/body, subject/object, etc. It was following Heidegger's example that Derrida critiqued Plato, as the lynchpin of a way of thinking that has profoundly influenced and directed Western civilisation. In his essay *Khôra*[6] Derrida seizes upon a doubt in Plato's mind. This doubt is to whether or not everything can be divided between the "sensible" and the "intelligible", as Plato's mode of enquiry had otherwise almost invariably led him to believe. In the middle of the *Timaeus*, dividing in two a discourse itself composed of *logos* and *mythos*,[7] Plato muses on a possibility beyond both ideal and physical form that is inexplicable and irreducible in terms of pure dialectic. It is the "khôra", a place – according to Derrida – that is not a place, "an apparently empty space – even though it is no doubt not *emptiness*"[8] – which nevertheless at once both encompasses and is independent from everything else. It has to be so in order – logically – for the contingencies of rational thought to exist at all. Plato likens it to a "mother" or "receptacle", using language that is unique to this particular dialogue of the *Timaeus*. So here, at last, there is a link between two opposites such as the 'sensible' and the 'intelligible', 'truth' and 'opinion' – something that binds them into the same system, and as Derrida frequently points out, puts into question their constant prioritising. One can read this – and I think that Derrida intends it to be read so – to mean that there is another way, within and beyond the bi-polarities of thought, in which truth is revealed.

Now, this 'other' way might immediately be mistaken for a new ideal, or at the very least a different 'take' on life. There is always that danger. But the danger of misappropriation is quickly obviated if we remember that this 'other', this different 'take', occupies an "empty" space where thought has lost its autonomy in its very source, and we are empowered to renounce all bi-polarities in favour of what might be called *oneness*. It must include everything. If this indefinable space – which Plato called "khôra" – is in fact the source of everything, how can it possibly reject anything?

The second – and connected – reason why an ideal does not work in reality is that if we are looking for a peg to fit into a square hole, it has to be a square peg. Similarly, if we are looking for an ideal to apply to reality, to nature, there is only one that will do. It must conform to nature. It cannot be one of countless creations of the mind, aloof and ultimately divisive. It must come from beyond the mind, having oneness as its core. As we shall see, this oneness begins by accepting reality as it is, and ends by transforming it into infinite possibilities. In this way reality – and nature – can recognize and welcome the ideal as their true complement, present yet absent from time immemorial. Our view of reality and nature is transformed and transcended.

ART AND THE IDEAL

Now – in my opinion – the same principle holds true in and through art. We have seen many instances of the ideal in art. The ancient Greeks are usually associated with bringing the ideal into Western art. In fact, our modern concept of the ideal may well begin with Greece. For when we look at classical Greek art, we are encountering an attempt to bring rationality and the ideal into sculpture and architecture. The Greeks saw mathematics as part of the *logos* that underpinned the entire universe, the inner world as well as the outer. They used mathematics to arrive at ideal proportions in architecture, and then transcended the maths in order to enhance their designs with a beauty beyond calculation. They sought to bring proportional perfection into being, and their sculpture of the human form also reflected this. They appeared to see in the human form its divine possibilities, in terms of beauty and proportion, and this they attempted to visualise through art – as if art were a bridge between the human and the divine, between the real and the ideal. (It must be remembered that they also sought to achieve this harmony, beauty, and proportion through physical exercise, careful grooming, and graceful attire and we can only assume that they – like us – came in all shapes and sizes and degrees of beauty [or not] in the real world).

Athens, Greece. 447–431 BC. *Parthenon.* Myron. c 450 BC. *Diskobolos (Roman Copy).* Museo delle Terme, Rome

So we can see this kind of Greek art in two ways. On the one hand, it aspires to an arguably impossible ideal: but on the other it has undoubtedly glimpsed within the real the possibility of innate beauty, like the shadow or echo of something more perfect. It is as if each thing in its naturalness were endowed with a beauty and perfection that is part and parcel of its reality, if only we could see it. The universe is composed of balance and proportion. So the real in a way inspires the ideal, which in turn inspires the real. Here, beauty comes initially from truth, and then aspires to perfection – to the ideal. So that, perhaps,

Beauty is not truth, but Truth is beauty.[9]

Thus the Greeks strove for an ideal of perfection in their artistic rendition of the human form. But – if we discount for the moment possible communication with the gods – where else could they have glimpsed that perfection except in the imperfect *human* form they saw around them? Like Plato, they seemed aware of a higher perfection, of which the quotidian is a distant, dim, but nonetheless true reflection. This glimpse of perfection and the urge to realise it and embrace it may be the phenomenon we call ideal beauty. It is Pygmalion's dream, though alas we have no Aphrodite to fulfil our desire. But the fact remains that the ideal exists in the human imagination, inspired in part through contact with the real world.

From Greek art of the Classical period, then, we get the feeling that the ideal is something real; that it is possessed of a real existence – though perhaps hidden – within nature. It is in this period also that we find the stirring of *pathos*, where human feeling is realized and expressed in a subtle and refined way, as if that feeling has become important in its subjectivity, thus elevating its subject – mankind – in a spirit of self-reflective empathy and sympathy.[10]

Greek. West Pediment, Temple of Zeus at Olympia. c 460 BC. *Hippodamia Attacked by a Centaur.* Archeological Museum, Olympia

Feeling, and our compassion for it, tells us that we are capable of higher thoughts, higher emotions, and Greek sculpture tells us that the human form is somehow an expression of this latent, and inner, refinement. This greatest gift of nature enables man to achieve the highest and noblest ideal: that is, he can recognize the other's feelings, and even place them before his own. He can become selfless. This strain of self-reflection, acute observation, empathy with and sympathy for the world around us is integral to the development of art.

A similar feeling occurs when we experience landscape. From where does the pastoral image – so revered in ancient times – derive its delight? Why do we love to hear birds singing, to see the sun shining on fields and rustic lanes?

John Constable. 1821. *The Hay Wain*. National Gallery, London (Photo: Wikimedia Commons)

What distant country does it remind us of? Whence springs the ideal notion of Arcadia, even in the person untutored in the classics or ignorant of the history of painting? Is it racial memory; conditioned consciousness; cultural heritage; genetic programming; or does it remind us of something within ourselves – perhaps even an *inner* country – that we have only partially glimpsed from time to time in the outer world? If this is the case, why is it there, waiting to be discovered within?

When we look at traditional Oriental art, on the other hand, we see another side of the same coin. Here nature seems to already contain the beauty and perfection that the artist wants to communicate. The ideal, in the sense of perfection, is actually already there within the real trees and mountains. We know that nature is transient. But in Oriental art nature – or reality – is perfect not despite the fact of its apparent spontaneity and transience, but – on the contrary – *because of it*. Perfection is integral, and exists within the passing moment.

This is also true of the artistic gesture – the brushstroke, the textured line, the flow of colour – elements of oriental art that can also be traced in that of the West. They are the imprint of an otherwise un-manifest beauty: perfection in imperfection, the intelligible in the sensible, the meaningful in the meaningless, form in substance, the ideal in the real, the infinite in the finite.

Lu Yan Shao. 1980. *Mountain and Clouds*

Clyfford Still. 1948. *1948 C*. Hirshorn Museum

WAY TO GO

So art can, like "spirituality", be a bridge between the real and the ideal. Indeed, art is essentially 'spiritual' in that sense. (Let us be careful here not to ascribe to 'spirituality' the attributes of yet another misleading ideal out of touch with the real. True spirituality is anything but this, and we must use the word carefully and wisely). Art, like spirituality, cannot – or should not – be appropriated. In fact it must be marked by its absence of appropriation. Art is ultimately indeterminate in terms of its practical meaning, whilst rich in meaning of a deeper kind. This very indeterminateness contains a "more" than what is presented materially or empirically in the artwork, or its narrative. Art is not rational in the sense that it can be expressed in terms of the opposites we have already discussed – truth/opinion, sensible/intelligible etc. The ideal that art superimposes upon the real is indefinable. Unlike a static transcendent such as a fixed notion of beauty imposed from without, art's 'ideal' is somehow already integrated within its real form.

> ... the more is not simply the nexus of the elements, but an other, mediated through this nexus and yet divided from it. The artistic elements suggest through this nexus what escapes it (...) It is not through a higher perfection that artworks separate from the fallibly existent but rather by becoming actual (...) they are not the other of the empirical world: everything in them becomes other.[11]

When quoting Adorno it helps to remember that Marx saw the ideal arising from the material, and not the other way round as in Hegel. Adorno developed this theme to a very fine degree, where it is no longer identifiable even as "spirit", becoming something beyond description, even beyond language, but somehow tangible through materiality in its subtler forms of literature, music and art. But in order to be what it is, and to escape the appropriation of quotidian thought, art makes use of paradox, irony, and ambiguity, to double back upon itself as its own ineffable "other", in a conscious gesture of self-erasure.

Where does this leave us regarding the ideal and art? Derrida often speaks about a "democracy to come"[12] in his examination of the political. This essentially means that democracy as a system has – in reality – yet to dawn anywhere on earth. Government of the people, for the people, and by the people seems a mathematical impossibility. It is also, as Derrida rightly observes, the only system that opens itself to critique as part of its nature. It is this essential fragility that gives it strength – a seeming impossibility. But in fact the impossibility of true democracy means that it is "forever on the move", and the *promise* of its coming, either today or tomorrow, is actually its life and its essence. But it will never fully arrive (or will it?).

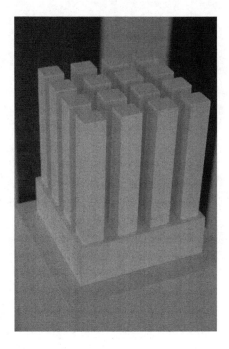

Brian Grassom. 2006. *City of Dreams*

It is the same with art. Art is forever on the move, and the artist knows this instinctively. It's 'on the move-ness' is integral to art, even when the work is static, a finished article. Art never grasps the ideal: that is always just out of reach, like Pygmalion's wish. But it awakens in us the ability to realise in its imperfection, in its movement and in its transience, the infinity, eternity and immortality that we are consciously or unconsciously seeking, and find that their creative reality lives within us – spontaneously – from moment to moment. In this sense it is a "bridge between the ideal and the real".

Gray's School of Art, The Robert Gordon University, Aberdeen, Scotland
e-mail: briangrassom@mac.com

NOTES

[1] Sri Chinmoy Kumar Ghose, from "True Spirituality and the Inner Life", a talk given at North Dakota University, October 25th 1974.

[2] Derrida's notion of *différance* is a way of thinking the replacement – without presence – of the ideal of "presence" within text. Levinas similarly critiqued the "ideal of reason" as a self-perpetuating device of reason itself. See, for example, Derrida, J., *Margins of Philosophy*, trans. A. Bass (Chicago: University of Chicago Press, 1982), pp. 25–26; and Levinas, E., *Totality and Infinity*, trans. A. Lingis (Pittsburgh, PA: Duquesne University Press, 1969), p. 33. Levinas speaks in this instance of the "desire for the invisible" or the "absolutely other".

[3] Plato, *The Republic*, trans. Paul Shorey, 2 vols. (London: William Heinemann LTD., 1935), Vol. II, Book X, p. 431. I have used this particular translation here, which differs from the better known one by Benjamin Jowett (see Note 7 below), as a point of interest. Jowett uses "reality" for "fact", and "bed" for "couch", but the meaning is the same.

[4] However, Plato does not tell us what he means by "in fact": this seemingly unassuming phrase is glossed over, not defined, an *a priori* assumption, and this is important because "fact" or "reality" thus behaves in the text in the same way as ideal form, which he insists cannot be made visible. So, the "fact" of the bed – its reality according to reason – although intuitively perceived, is, like ideal form, not available to the material senses. It is, in modern parlance, an 'abstraction' and therefore essentially an ideal form as opposed to a reality. Of course, Plato was convinced of the primacy of mathematics, and if this were a mathematical problem the "fact" of the bed might be much like the solution to an abstract equation. Indeed, a geneticist might argue that Plato was taking the first tentative steps towards a purely scientific view of life. But pure mathematics, although derived from nature, is in itself abstraction and does not answer the fundamental problem of the bed's "apparent" reality as experienced in nature by human consciousness. Neither would mathematics use such language as the "unique" bed "that God produces" (ibid. pp. 429, 427). This dichotomy in Plato's work, between the rational and the intuitive, between *logos* and *mythos* is frequently and lovingly exploited by Derrida.

[5] Heidegger, M., *Being and Time*, trans. J. Macquarrie/E. Robinson (Oxford: Blackwell, 1962), p. 2.

[6] Derrida, J., "Khora", in: *On the Name*, trans. D. Wood/J. P. Leavey/I. McLeod (Stanford: Stanford University Press, 1995), pp. 89–127.

[7] Plato, "Timaeus", in: *Dialogues*, trans. B. Jowett (Chicago: William Benton, 1952), [52], p. 457. Socrates asks Timaeus to give his account of the nature of the universe and it's creation. About halfway through this curious mix of reason and myth, "khora" is introduced.

[8] Derrida, J., op. cit., p. 103.

[9] Sri Chinmoy Kumar Ghose, *Meditations: Food for the Soul* (New York: Harper and Row, 1970), p. 53.

[10] See Janson, H. W., *History of Art* (London: Thames and Hudson, 1986), p. 131.

[11] Adorno, T. W., *Aesthetic Theory*, trans. R. Hullot-Kentor (Minneapolis, MN: University of Minnesota, 1997), pp. 79, 81.

[12] Derrida, J., *Rogues: Two Essays on Reason*, trans P. Brault/M. Nass (Stanford: Stanford University Press, 2005), pp. 78–92.

FIRAT KARADAS

HISTORICITY, NARRATIVE AND THE CONSTRUCTION OF MONSTROSITY IN JOHN GARDNER'S *GRENDEL*

ABSTRACT

The relationship between history and narrative has always been a subject of controversy among philosophers, historians and literary theorists. Is narrative the indispensable component of history? What is the function of narrative in history? How does history represent human experience with the narrative function? Is historical narrative imitation or reproduction of the past? What is the role of the historian and his constructive imagination in history writing? This article discusses these questions in the context of a literary text, Gardner's *Grendel*, which is a re-writing of the Old English epic *Beowulf*, and with reference to phenomenological and Kantian ideas of history, narrative, the self, and imagination. Relying mainly on Hayden White, Louis Mink, and Paul Ricoeur's ideas of history and narrative, the present article concludes that history is a reproduction of past actuality instead of an imitation of it. Thus, in the article the term history-making is preferred instead of history writing and history-making is regarded as bearing close resemblance to story-making. The chapter studies *Grendel* against this philosophical background in terms of how narrative plays a symbolizing, form-giving tool for consciousness in historicizing human experience and how heroism and monstrosity are historical, ideational constructs on which human experience is founded.

Historicity and its positioning of the individual subject and the society into spatio-temporal context through narrative are much debated issues in phenomenological research. The symbolic/pattern-making function of the human mind, its structuring fragments of experience into meaningful wholes and thus locating the self within history and time by "telling" it have been some of the major concerns of phenomenology. From Husserl to Heidegger and Ingarden, to Paul Ricoeur, Hayden White, Louis Mink, and David Herman, philosophers of history, theorists of narrative, and phenomenologists have discussed the indispensible relationship between narrative and history and their construction of the self as both itself and not itself. They have elaborated on the made-up and ideational world of history brought about by the narrative function and on the way the self is defamiliarized in narrative and history. Husserl thought that "the ego constitutes itself for itself, so to speak, in the unity of *Geschichte*,"[1] and saw historicity as the result of the intentional act of consciousness and its effort of creating a historical self by locating phenomenal experience into time. This idea of Husserl is a good point of departure for the argumentation of this chapter and requires further elaboration, which the present chapter aims to do. It also poses questions about the relationship between historicity and consciousness and how narrative functions as a cognitive tool for historicizing human experience. This chapter aims to study the relationship between narrative and history, and the pattern-making, metaphorizing and defamiliarizing function of narrative in history. With the supposition that authors of literature have always benefited from or been influenced by philosophy when writing their works and that literature has always played an important ground for philosophical investigations, this chapter discusses the above-mentioned philosophical issues in the context of a literary work, John Gardner's *Grendel*. The chapter studies *Grendel* in terms of how narrative plays a symbolizing, form-giving tool for consciousness in historicizing human experience and how heroism and monstrosity are historical, ideational constructs on which human experience is founded. Gardner's novel is studied not only for exemplifying the tenets of the philosophy of phenomenology concerning history-making, narrative, and the self but also to extend the discussion of these issues to the study of historical narrative construction in literature.

In the first volume of *The Philosophy of Symbolic Forms* the German neo-Kantian philosopher Ernst Cassirer argues that "[c]onsciousness is a symbolizing, 'form-giving activity'"[2] which "does not merely copy but rather embodies an original formative power. It does not express passively the mere fact that something is present but contains an independent energy of the human spirit through which the simple presence of the phenomenon assumes a definite 'meaning', a particular ideational content."[3] Thus, the existence of social phenomena depends, as Kant argues, on the purpose of the human mind conceiving it, that is, on the determining will of the subject; social phenomena is "the correlate of the 'I think' or of the unity of consciousness; it is the expression of the *cogito*."[4] The father of phenomenology, Husserl, also saw consciousness as central in the perception of phenomena, though, differently from Kant's categorizing and all-pervading mind, he studies the conscious processes of phenomenal experience and thus bridges between subject and object. In *Ideas: A General Introduction to Pure Phenomenology* (1913) Husserl studies the structures of experience as they are represented in consciousness and explores in depth how the conscious world of the perceiver acts in the physical world of objects. With his idea of "intentionality," a crucial term in his philosophy and meaning the directedness of consciousness towards its object in experience, Husserl explicates the way consciousness works in the process of the structuration of experience.

Narrative is a cognitive tool through which consciousness *symbolizes* or *structures* the human experience of time. It is "a pattern-forming cognitive system" that functions "to connect and integrate certain components of conscious content over time into a coherent ideational structure."[5] It is "a system for structuring any time-based pattern into a resource for consciousness, making it possible for cultural as well as natural objects and phenomena to assume the role of cognitive artifacts to begin with."[6] Stories tell about the actions of intelligent agents that are situated within a world together with the objects they act upon. In this respect, telling a story necessitates, in the words of David Herman, "modeling, and enabling others to model, an emergent constellation of spatially related entities."[7] Narrative thus operates as an instrument of mind in the construction of reality, and entails a cognitive process of assigning referents a spatio-temporal position in the storyworld. It provides "crucial representational tools facilitating humans' efforts to organize multiple knowledge domains, each with its attendant sets of beliefs and procedures."[8]

Taken on this ground, studying narrativity is to investigate social phenomena as a "world-spanning" network of relations taking place in the mind of the teller. The teller creates this network of relations by the cognitive activity of emplotment, which for the French philosopher Paul Ricoeur, is the essence of narrative. To delve further into this idea, we should linger more on Ricoeur because his idea of *mimesis* and metaphor carries us to the vantage point of this chapter. Ricoeur brings a new dimension to the role played by the subject in the construction of human experience relying on Aristotle's idea of *mimesis* and Kant's idea of "schematizing a synthetic operation." Ricoeur's idea is important in that it points to the subjective, cognitive and metaphorical base of not only Aristotle's idea of *mimesis* but also all literary and linguistic creations. In *The Rule of Metaphor*[9] Ricoeur states that Aristotle defines tragedy as "the imitation of human action." However, it is an imitation that elevates, magnifies and ennobles this action. In this regard, Ricoeur argues that for Aristotle *mimesis* is *poiesis*, that is, construction or creation. With *mythos* (plot) it becomes a rearrangement of human action into a more coherent form and with *leixis* (poetic language) a structuring that elevates this action. Thus, *mimesis* is something that composes and constructs the very thing it imitates. In this regard, *mimesis* is an imitation that has a double reference: a reference to reality and a self-reference, a representation of human action and a construction of that action. So, the reference of tragedy to reality is not a direct one but a "suspended" one.

Relying on Aristotle's *mimesis* and Kant's "schematizing a synthetic operation," and studying Augustine's *Confessions* and Heidegger's "within-timeness," Ricoeur concludes that human experience is characterized by discordance. The constructive imagination brings concord to this "aporia" by way of what he calls "predicative assimilation", that is, by seeing the similar in the dissimilar.

Literature, in narrative form, brings concord to this "aporia" by means of the invention of the plot. Narrative, to which Ricoeur devotes a great deal of his work, is a synthesis of heterogenous elements, a gathering-up of events and incidents "as widely divergent as circumstances encountered while unsought," a concord created out of the discord of experience, out of the divergent bits and pieces of experience. Like metaphor, narrative is a "semantic innovation" in which something new is brought into the world by means of language. Instead of describing the world, it redescribes it. Just as metaphor is the capacity of "seeing as," narrative opens us to the realm of the "as if." It attaches to the events of the world a form they do not otherwise have. Emplotment, the core feature of narrative, is thus a "grasping together," a patterning of experience, and it is one of the main functions of the imagination. It is a cognitive tool of making sense of experience and of making life plausible. In this regard, the way fragments of experience are organized in the process of emplotment depends on the plotting imagination, that is, on the story-maker.

The narrative kernel and thus the truth-claim and fictionality of history are much discussed issues by philosophers of history. There is a general agreement among philosophers of history that narrative is the core component of history. However, since narrative presents, as discussed above, a distorted picture of the events it relates, the truth-claim of history and the levels of fictionality and distortion in it are frequent objects of discussion. So, philosophers usually discuss historical narrative with an effort of distinguishing it from fictional narrative. Louis Mink, for instance, argues that historiography can be differentiated from fiction with its truth-claim and point of common sense; "historiography consists of narratives which claim to be true, while fiction consists of imaginative narratives for which belief and therefore truth-claims are suspended."[10] Differently from fictional narratives, in historical narrative the historian "does not invent but discover, or attempts to discover;" "the story of the past needs only to be communicated, not constructed."[11] However, Mink stresses that historical narrative is "a matter of fallible inference and interpretation"[12] because "narrative form in history, as in fiction, is an artifice."[13] As historical, he argues, historical narrative "claims to represent, through its form, part of the real complexity of the past, but as narrative it is a product of imaginative construction, which cannot defend its claim to truth by any accepted procedure of argument or authentication."[14]

In his analysis of historical narrative Ricoeur also begins his discussion by underlining the truth-claim factor of historical narratives. However, seeing emplotment as the characterizing feature of also historical narrative and attaching the synthesizing imagination of the historian a crucial role in history-making, he sees the story of the past in historical narrative not just as something communicated, but, to a large extent, as something constructed. Drawing on Aristotle's *mimesis*, Heidegger's and Augustine's concepts of time, and Kant's idea of synthesizing imagination, he arrives at the conclusion that history is a kind of narration in which the past, the present and the future are synthesized and our temporal experience shaped. In "The Narrative Function" he states, "to be historical, I shall say, an event must be more than a singular occurrence: it must be defined in terms of its contribution to the development of a plot."[15]

He defines history as a narration that describes a sequence of actions and experiences in two dimensions: chronological and configurational. For Ricoeur, the first may be called "the episodic" or sequential dimension. This dimension characterizes the story as made out of events. The second dimension is "the configurational one, according to which the plot construes significant wholes out of scattered events;"[16] it is to "grasp together successive events…to extract a configuration from a succession."[17] To explain the configurational dimension, Ricoeur employs Kant's idea of "reflective judgment" and states that the narrative operation in historicizing human action has the character of a judgment because to locate an event in historical time is not only to follow episodes but also " 'to reflect upon' events with the aim of encompassing them in successive totalities."[18]

As in all other symbolic forms, in history, too, the "telling" subject and its imagination play a crucial role in history-making. Ricoeur concludes that "the historicity of human experience can be brought to language only as narrativity…For historicity comes to language only so far as we tell stories or tell history."[19] This process is not a naïve one; as Richard Kearney puts it, it "involves 'the representative function of the imagination.' "[20] In history events are manipulated and given some form by the historian's productive imagination. Ricoeur states, "by telling stories and writing history we provide 'shape' to what remains chaotic, obscure, and mute...historical narrative and fictional narrative *jointly* provide not only 'models of' but 'models for' articulating in a symbolic way our ordinary experience of time."[21] The historian does this by selecting only those events that, in his estimation, should not be forgotten and structures them in narrative order. Moreover, he highlights the events that he thinks memorable and overshadows those that should be forgotten. In this regard, the act of narrating history is a "schematizing" and "synthetic" operation in which "dissimilar" events are "configured."

A philosopher of history who most openly stresses the "constructed" and fictional character of historical narratives is Hayden White. In "The Historical Text as Literary Artifact" White criticizes Northrop Frye's idea that the historical is the opposite of the mythical and argues that mythos is not the opposite of historical narrative but inherent to it. He states, "histories gain part of their explanatory effect by their success in making stories out of mere chronicles."[22] Similar to Ricoeur, he uses the term "emplotment" to explicate the way historians make stories of a past event. White defines emplotment as "the codification of the facts contained in the chronicle as components of specific *kinds* of plot structures."[23] He discusses that in their efforts to make sense of historical record, which is fragmentary and always incomplete, historians have to make use of what R. G. Collingwood calls "the constructive imagination," which is much like Kant's *priori* imagination and Ricoeur's predicative imagination. The constructive imagination makes events into a story "by the suppression or subordination of certain of them and the highlighting of others, by characterization, motific repetition, variation of tone and point of view, alternative descriptive strategies, and the like—in short, all of the techniques that we would normally expect to find in the emplotment of a novel or a play."[24] He presents that no historical event is intrinsically tragic or inherently comic or ironic. The mode of emplotment—that is, whether it is comic, tragic, romantic, ironical, and so on—depends upon "the historian's decision to configure them according to the imperatives of one plot structure or mythos rather than another."[25] The cultural heritage of the "audience" of the historian plays a crucial role in the way the historian emplots past events. White stresses that "the encodation of events in terms of various plot structures is one of the ways that a culture has of making sense of both personal and public pasts."[26] Events which appear strange, enigmatic, incomplete, and implausible are encoded in culturally provided categories by the historians. In short, the unfamiliar events take a familiar kind of configuration and events are "rendered comprehensible by being subsumed under the categories of the plot structure in which they are encoded as a story of a particular kind."[27]

Coming closer to Ricoeur's idea of imagination and metaphor, White asserts that historical narratives are metaphorical statements which suggest "a relation of similitude between such events and processes and the story-types that we conventionally use to endow the events of our lives with culturally sanctioned meanings."[28]

Dwelling further on this idea in "Historical Emplotment and the Problem of Truth," he states,

> Any figurative expression adds to the representation of the object to which it refers. Figuration produces stylization, which directs attention to the author and his or her creative talent. Next, figuration produces a "perspective" on the referent of the utterance, but in featuring one particular perspective, it necessarily closes off others.[29]

All historical narratives, as such, "presuppose figurative characterizations of the events they purport to represent and explain."[30] For this reason, histories are not only about events but also about the possible sets of relationships that those events can be demonstrated to figure. These sets of events are not immanent in the events themselves; "they exist only in the mind of the historian reflecting on them;" they are present as "the modes of relationships conceptualized in the myth, fable, and folklore, scientific knowledge, religion, and literary art, of the historian's own culture."[31] This means that historical narratives can be characterized by the mode of figurative discourse in which they encode their objects of representation.

The idea of history as a narrative construct has been an object of criticism by some philosophers of history. For instance, David Carr criticizes Ricoeur, White, and Mink's idea that historical narrative "is not imitative but creative of reality"[32] and that narrative structure is an artificial, imposed form of ordering ascribed to our actual experience. Carr argues that narrative activity is a constitutive part of action and the events of life constitute "a complex structure of temporal configurations that interlock and receive their definition and their meaning from within action itself."[33] Relying partly on Husserl's idea of protension and retention, he says we grasp a configuration extending from the past to the present even in the relatively passive experience of hearing a melody.

In spite of the criticism to their almost total disregard of the truth-claim of history and to their view of history as more fictional than factual form, this study has elaborated more on Ricoeur, Mink, and White's arguments because they present us with a theoretical framework to study historicity, narrativity, and the construction of the self in John Gardner's *Grendel*. With their ideas of emplotment, imagination, metaphorization, and the "mythic" core of narrative, Ricoeur and White's ideas are of particular importance for the purpose of this study.

John Gardner's *Grendel*[34] represents how historicity and narrativity function to configure human experience and how this configuration serves for the foundation of a civilization. *Grendel* is a re-writing of the Old English epic *Beowulf* from the perspective of Grendel—the first known monster of English literature. As already known, the old English epic *Beowulf* begins during the climax of Grendel's attacks on King Hrothgar's meadhall. It is said that before these attacks King Hrothgar enjoyed a prosperous and successful reign. He built a great mead-hall, called Heorot, where his warriors can gather to drink, receive gifts from their lord, and listen to stories sung by the scops, or bards. The kingdom enjoyed peace and prosperity until Grendel began his attacks. However, the focal point of the epic is not Grendel's attacks or the reasons behind these attacks. Attention is centered on Beowulf's heroic deeds, his rescue of Hrothgar by killing Grendel and his mother, and towards the end his slewing a dragon to save the Geats from its threats.

Relying on White and Ricoeur's ideas, we handle here the epic form as a historical narrative in which the mythic or epic side overshadows what Mink calls "past actuality." As said before, Ricoeur regarded *mimesis*, and thus literary and historical narratives, as having double-reference: self-reference and reference to external reality. This proposition is true for historical narratives with a strong claim for truth. However, in such narratives as the epic or myth, the self-reference is much more dominant that the other one. Signs of "past actuality" are almost lost in the "projected past" of such narratives. In the epic and myth the narrative world is defamiliarized and distanced from ordinary human experience by the plotting imagination to the extent that the ties connecting the two become almost totally invisible. While historical writings with a claim for truth focus on the conflicts between enemies and allies in the historical evolution of a civilization or a nation, in epical or legendary historical narratives the conflict is usually between heroes and monsters, between Apollo and Python, Perseus and the Gorgons, Siegfried and the dragon, and Beowulf and Grendel. The construction of the subject also changes from one historical narrative to the other. While the former constructs the nation and its allies by also constructing an enemy identifiable with temporal experience, the latter constructs a supernatural hero by also creating a monstrous counterpart.

In *Beowulf*, in the bard's historical narrative (because the first history-makers were usually harpers or bards) historical data are embroidered with mythic and supernatural elements; for instance, the foundation of Hrothgar's kingdom by his ancestors, the prosperity of the kingdom during Hrothgar's reign, and Beowulf's reign in the Geatland after the death of Hygelac—the ex-king of the land—are given side by side with such mythic and supernatural elements as Beowulf's heroic deeds, supernatural monsters and dragons. The historical data are even almost lost in the self-referential narrative concerning Beowulf's heroic deeds and fight with the monsters. Besides, though Beowulf is a late-comer in the history of Hrothgar's kingdom and of the war with Grendel, he is made by the harper's plotting imagination the central figure, the pros and cons of the history before him being almost totally overshadowed. Grendel, together with the dragon, is silenced throughout the narrative, and the reasons behind his fight with Hrothgar are not told. Thus, Grendel can be said to be the first outcast of English Literature as well as of the first known historical narrative of Britain.

Gardner's *Grendel* is a counter-narrative to *Beowulf*, and it retells the *Beowulf* story from Grendel's point of view, making Beowulf not appear until the end of the novel, which is the actual place he deserved in the history of Hrothgar's civilization and in the war between Hrothgar and Grendel. In *Grendel*, it is told that Grendel, a large bearlike monster and the narrator of the novel, has spent the last twelve years locked in a war against a band of humans. The main action of the novel, like the Old English epic *Beowulf*, takes place in the last year of that war, but the novel skips back in time in order to illuminate the origins of the conflict as well as Grendel's personal history. The strategy of skipping back in time gives us, as readers, the opportunity to see the reasons of his war with men against the background of his personal history, as opposed to the lacking and one-sided historical account presented in *Beowulf*. As Joseph Milosh says, unlike Grendel in *Beowulf*, Gardner's Grendel "is anything but a static character. He grows, passing through several initiations, evolving more than many a modern hero."[35] In his personal experience we learn that, as a young monster, Grendel lives with his mother in a cave on the outskirts of human civilization. A foul, wretched creature who long ago abandoned language, Grendel's mother is his only kin or companion. He is all alone in the world; he is neither an animal nor a human and thus he is excluded from both worlds. He says "I exist, nothing else;"[36] "I am a *lack. Alack*! No thread, no frailest hair between myself and the universal clutter;"[37] "I saw long ago the whole universe as not-my-mother, and I glimpsed my place in it, a hole."[38] He describes himself as "an alien, the rock broken free of the wall."[39]

One day, the young Grendel discovers a lake full of firesnakes and, swimming through it, reaches the human world on the other side. He gets fascinated with the world of men as they speak his language and are thinking beings like him. As soon as he comes face to face with human beings, Grendel becomes aware that he is dealing with no dull mechanical animals but with pattern-makers, the most dangerous things he has ever met.[40] He watches from a safe distance as mankind evolves from a nomadic, tribal culture into a feudal system with roads, governments, and militaries. He eavesdrops and observes Hrothgar's hall, philosophizes on the human world, listens to bards' songs, sometimes attacks the thanes in the meadhall to take the revenge of his exclusion, and toys with them until Beowulf comes and kills him in the last two chapters.

The tales sung by a bard named the Shaper and Grendel's relationship with them are the main focus of this chapter. The Shaper occupies the most respectful position in Heorot and displaced all the other bards after his coming to the mead-hall. Listening to the tales sung by the Shaper, Grendel gets astounded by the pattern-making and creative imagination of the human beings and fears from this monstrous imagination. The Shaper plays the most crucial role in Hrothgar's civilization because, as his name signifies, he shapes the kingdom with his tales; he metaphorically establishes for the kingdom a socio-cultural value system and a historical identity based on heroism and on the creation of the monstrous other. In other words, he functions as the history-maker of the novel and creator of a belief-system in the kingdom. The Shaper sings of "battles and marriages, of funerals and hangings, the whimperings of beaten enemies, of splendid hunts and harvests," and "of Hrothgar, hoarfrost white, magnificent of mind."[41] Emplotting human experience into history, he most of the time sings war songs. His harp mimicks "the rush of swords, clanking boldly with the noble speeches, singing behind the heroes dying words."[42] Grendel says: "If the songs were true, as I suppose at least one or two of them were, there had always been wars, and what I'd seen was merely a period of mutual exhaustion."[43] He constructs a historical narrative in his songs as if there were nothing in human life other than wars and as if the experience without war, which Grendel sees, were not lived.

Reminiscent of Ricoeur's and White's definitions of historicity as the subjective location of being into time and as the configuration of human experience through narrativity, Grendel says that the Shaper is the greatest of shapers, "harpstring scratchers;"[44] he is a shaper of the past, an "old heart-string scratcher, memory scraper,"[45] "always transforming the world with words."[46] With his songs he has changed the world, "torn up the past by its thick, gnarled roots," and "transmuted it."[47] "He reshapes the world," Grendel says, "he stares strange-eyed at the mindless world and turns dry sticks to gold."[48] His songs consist of words "stitched together out of ancient songs, the scenes interwoven out of dreary tales, made a vision without seams," and—reminding us of the phenomenological ideas mentioned above—he thus constitutes "an image of himself yet not-himself, beyond the need of any shaggy old gold-friend's pay: the projected possible."[49]

Grendel knows that Hrothgar's hall is built on bloodshed and destruction of nature, but the Shaper—"the blind selector" of historical events— tells tales as if no man in Hrothgar's hall has ever hurt a living creature or "twisted a knife in his neighbour's chest."[50] Grendel questions the fictionality and the untruthfulness of the Shaper's historical narrative throughout the novel. He is bewildered by the brutality of men and by their killing other living beings without any meaningful aim. He observes their *monstrosity* as they hack trees and build huts, kill cows, horses and men, and leave them to rot; they plunder lands, and whipped up the oxen to death while getting their piles of plunder to their land. He gets annoyed as he remembers what all men do to each other and to nature: "the ragged men fighting each other till the snow was red slush, whining in winter, the shriek of people and animals burning, the whip-slashed oxen in the mire, the scattered battle leavings: wolf-torn corpses, falcons fat with blood."[51]

The gap between humans' actual savagery and their false representation in the Shaper's narratives can be best illustrated with Chapter 4, when Grendel steps unknowingly on something fleshy as he approaches the meadhall as usual to eavesdrop the harper's songs of "elevated human action." He realizes that it is the corpse of a killed man. The clothes of the man are stolen. As if trying to show the untruthfulness of the Shaper's songs, he slung the dead body upon his shoulder and approaches the meadhall. As he approaches the meadhall, he sees the Shaper singing as usual. Though Grendel comes to the full realization that humans are monstrous beings with their way of life and their savage attitudes to each other and other living beings, the Shaper is concerned with constructing in his historical narrative the glory and the untaintedness of human beings and the brutality and monstrosity of Grendel. Telling a tale also sung in *Beowulf* on Grendel's origin, the Shaper relates that the earth was built long ago, that "the greatest of gods made the world, every wonderbright plain and the turning seas, and set out as signs of his victory the sun and moon, great lamps for land-dwellers, kingdom torches, and adorned the fields with all colors and shapes."[52] Hrothgar's civilization was the centre of this phase of constant light until Grendel comes into being. Constructing Grendel in his historical narrative as the destroyer of this edenic state and associating him with Cain in biblical mythology, he tells that Grendel gave end to this state of paradise by beginning the first feud with his brother and thus among human beings; he relates that Grendel's fight with his brother split all the world between darkness and light and identifies Grendel with the dark side, Cain. Though Grendel defeats Hrothgar's men throughout the history of Hrothgar's civilization, the Shaper establishes a heroic value system and tells how they fought Grendel, Cain, and all the other forces of evil gloriously, which, Grendel says, is a lie.

This tale that locates Grendel in historical time on the same line with Cain also takes place in *Beowulf*. In the Old English epic, it is said that the bards' songs in the meadhall angered Grendel, which is said just in passing and not elaborated on. When we observe the content of the songs in *Beowulf*, we see that they are all concerned with Grendel, posit him in historical time on the same level with Cain as the originator of all evil, and tell how he disturbed the Golden Age of humanity with his evil doings. Hence, it can be assumed that in both *Beowulf* and Gardner's novel Grendel's configuration in the bard's historical narrative as "monstrous" and as a descendant of Cain, even as Cain himself, seems to be the cause of his anger and the reason of his attacks. In a way, his attacks to Heorot in *Beowulf* and *Grendel* can be interpreted as a reaction to this religo-historical configuration.

Though Grendel is silenced in the old English epic, he voices his counter discourse to this religo-historical configuration in Gardner's novel. He states,

It was a cold-blooded lie that a god had lovingly made the world and set out the sun and moon as lights to land-dwellers, that brothers had fought, that one of the races was saved, the other cursed. Yet he, the old Shaper, might make it true, by the sweetness of his harp, his cunning trickery.[53]

The Shaper's discourse is so effective that even Grendel himself is fascinated by it and begins to believe in his own monstrosity. He intrinsically begins to like hearing his monstrosity being told in the Shaper's narrative. "Though, they, vicious animals, cunning, cracked with theories, I wanted it, yes!" he says, "even if I must be the outcast, cursed by the rules of his hideous fable."[54] Being all alone in the disordered universe and leading a meaningless life, he is intrinsically delighted to be *meaningfully* constructed in the Shaper's narrative as "monstrous."

As Judy Smith Murr puts it in "John Gardner's Order and Disorder," "Grendel, symbolically the offspring of Cain, emerges from the underbelly of the world to confront mankind...The underground world of Grendel is dark, terrifying, and chaotic, but frightening and disordered than the above-ground of man."[55] He emerges from his underground world to find himself posited against myth, the myth that the world is ordered and that fact is transformed by song. Torn apart by poetry, "Grendel must face the search for meaning and balance."[56] After the magical effect of the Shaper's narrative, he is determined to find the connection between himself and the world even though "the world is a pointless accident,"[57] and in all his efforts he is but "spinning a web of words" between himself and all he sees.[58] Torn with internal conflict regarding his existence in the universe, he visits the Dragon—another outcast in the novel "cursed by the bards' hideous fables"—to find relief for his fallible condition and clarify his mind about the human world and its entangling narratives. The Dragon forms the philosophical core of the novel and plays a critical role in Grendel's thoughts and actions in the forthcoming chapters. He makes Grendel realize that he plays a constituent role in the human world because he makes humans define themselves and shape their world. The Dragon tells Grendel,

You improve them, my boy! You stimulate them! You make them think and scheme. You drive them to poetry, science, religion, all that makes them what they are for as long as they last. You are, so to speak, the brute existent by which they learn to define themselves.[59]

Quite in accordance with the proposal in the theoretical framework of this chapter that the historian and story-teller bring order with their narratives to the chaotic human experience, the Dragon tells Grendel that the Shaper brings order with his narratives, configuring Grendel as the Absolute Enemy, the focal point on which he constructs the belief system and heroic values of his society. The reason for historical constructs is to be found in the human beings' effort to create established order and universe's refusal of "the deadening influence of complete conformity."[60]

This is what leads Grendel to think towards the end of the novel that "all order...is theoretical, unreal—a harmless, sensible, smiling mask men slide between two great, dark realities, the self and the world—two snakepits."[61] The Dragon criticizes human beings stating that they have "no total vision, total system;" they have "merely schemes with a vague family resemblance, no more identity bridges and, say, spiderwebs."[62] They have no sense of connectedness; they take facts in isolation and when they come to connect them, "ands and buts are the *sine qua non* of all their achievement."[63] Their lives consist of "crackpot theories" and absurdities, and "they build the whole world out of teeth deprived of bodies to chew and be chewed on."[64] For the Dragon, this is the place where the Shaper saves them:

He provides an illusion of reality—puts together all their facts with a gluey whine of connectedness. Mere tripe, believe me. Mere sleight-of-wits. He knows no more than they do about total reality—less, if anything: works with the same old clutter of atoms, the givens of his time and place and tongue. But he spins it all together with harp runs and hoots, and they think what they think is alive, think Heaven loves them. It keeps them going.[65]

Time, according to the Dragon, is an important tool for creating this illusion of reality based on artificial order and connectedness, but it also shows the impossibility of overcoming the absurdity of life. Though they emplot their experience by creating artificial beginnings and ends in their narratives, humans are unable to encompass all the fragments of experience in one pot, together with the beginning, the present and the end.

In this respect, causality, which is the main component of narrative, is only an imposition of order on actual human experience. In contrast to the causal time the Shaper creates in his narrative, real time is a flux because there is no time outside consciousness. Thus, the death of consciousness is the end of being and human time. The Dragon suggests, "pick an apocalypse, any apocalypse. A sea of black oil and dead things. No wind. No light. Nothing stirring. A silent universe. Such is the end of the flicker of time, the brief, hot fuse of events and ideas set off, accidentally, and snuffed out, accidentally, by man. Not a real ending of course, nor even a beginning. Mere ripple in Time's stream."[66] In this respect, as chronology in history is a human construct, there is no "Dark Ages" in history and thus no monstrous creature representing the darkness because "not that one age is darker than another."[67]

Grendel comes to a full realization of his position in the Shaper's narratives after talking with the Dragon and becomes more aware of the absurdity of life. He learns from the Dragon that human beings define themselves and make their lives meaningful by narrativizing him. Their existence depends on him, but he realizes that his own existence also depends on their narratives:

> My enemies define themselves on me. As for myself I could finish them off in a single night, pull down the great carved beams and crush them in the meadhall, along with their mice, their tankards and potatoes—yet I hold back. I am hardly blind to absurdity. Form is function. What will we call the Hrothgar-Wrecker when Hrothgar has been wrecked.[68]

He says that he existed alone before he knew human beings and the Shaper began to tell tales about him. Even his mother did not love him for himself, but for his "son-ness," his "possessedness." He asks, "If I murdered the last of the Scyldings, what would I live for."[69] Thus, it can be concluded that each side exists *in spite of* and *because of* the other.

With this knowledge in mind, he is no longer torn apart by humans' pattern-making minds and begins to rule over and make fun of their narratives. As Milosh states, "Grendel's response to their violence results in the quick retreat of his attackers and, for the monster, an increasing awareness of his power, particularly his ability to toy with them."[70] He says, "I had become something, as if born again. I hung between possibilities before, between the cold truths I knew and the heart-sucking conjuring tricks of the Shaper; now that was passed: I was Grendel, Ruiner of Meadhalls, Wrecker of Kings."[71]

His toying with Unferth, one of Hrothgar's thanes, is a good example of Grendel's ruling over and mocking humans and their narratives. When he attacks Heorot with more self-confidence and sense of absurdity about life after his talk with the Dragon, Grendel confronts Unferth's heroic—or it is better to say mock-heroic[72]—acts, which Grendel describes as "crowning absurdity."[73] In a shift from the original *Beowulf* poem, the thane Unferth—not Beowulf—represents the traditional Anglo-Saxon heroic code. Grendel says that among his fellow thanes Unferth is "like a horse in a herd of cows."[74] Unferth begins his first battle with Grendel like an epic hero, making poetic speeches that exalt his moral code and highlight his bravery in battle. He speaks, holding his sword and shaking it, "tell them in Hell that Unferth, son of Ecglaf sent you, known far and wide in these Scanian lands as a hero among the Scyldings."[75] Making fun of epics as well as historical narratives whose focus are "ideal heroes," Grendel responds Unferth's comical heroic speeches as: "I've never seen a live hero before. I thought they were only in poetry. Ah, ah, it must be a terrible burden, though, being a hero—glory reaper, harvester of monsters! Everybody always watching you, seeing if you're still heroic...Always having to stand erect, always having to find noble language."[76]

Grendel undercuts Unferth's attempt at traditional heroism by raining apples at him and turning the serious battle into a mock heroic poem, a grotesque clown show. However, though Grendel destroys the trappings of heroism, Unferth follows Grendel to his cave in the burning lake to take revenge. He shouts: "You think me deluded, tricked by my walking fairytale."[77] "Except in the life of the hero," he continues, "the whole world is meaningless. The hero sees values beyond what's possible. That's the nature of the hero. It kills him, of course, ultimately. But makes the whole struggle of humanity worthwhile."[78] Unferth encounters the same problem Grendel does: a vision of the world as essentially meaningless. But while Grendel has decided to deny the possibility of imposing his own meaning on the world, Unferth chooses to use the ideals of heroism to create meaning for himself and all of mankind, which the historian also does with his historical narrative. For Unferth, the romantic ideal of heroism is a vision, encouraged by the Shaper, that holds existentialism and nihilism at bay.

Realizing that Unferth wishes to be killed by him to be assigned the title of "a hero killed in a heroic battle with the monster," Grendel makes Unferth's heroism more and more grotesque by refusing to kill him and taking him back to Heorot as if carrying a toy. Besides he makes Unferth more and more ashamed of his situation by killing everybody except him in each attack to Hrothgar's meadhall. This part mocks heroism and represents that heroism is a historical construct through which history-makers such as the Shaper impose an ideal meaning, a totalizing view on the absurd and disordered human world.

Though Grendel mocks the ideals created by the Shaper in this part of the novel, the existence of both the Shaper and Grendel depend on each other. Toward the end of the novel, Beowulf, the central figure of the Old English epic *Beowulf*, appears to save Hrothgar from the monster. Beowulf's entrance into the novel signifies a new beginning in the history of the Scyld and the end of Grendel's "story," the story created by the Shaper. Thus, it signifies the metaphorical death of both Grendel and the Shaper. As he speaks on the death of the Shaper in Chapter 10, Grendel articulates this idea as:

> End of an epoch, I could tell the king
> We're on our own again. Abandoned.[79]

The Shaper's death leads Grendel to philosophize on his existence in the world, on his personal development, his dependence on the Shaper's historical configurations and myth-making, and on how he and his existence have ceased to exist with the Shaper's death. The below words of Grendel are of particular significance in this respect and somehow summarize the theoretical proposal of the present article about history-making and narrative:

> ...because the Shaper is dead, strange thoughts come over me. I think of the pastness of the past. How the moment I am alive in, prisoned in, moves like slowly tumbling form through darkness, the underground river. Not only ancient history—the mythical age of the brothers' feud—but my own history one second ago, has vanished utterly, dropped out of existence. King Scyld's great deeds do not exist "back there" in Time. "Back there in Time" is an illusion of language. They do not exist at all. My wickedness five years ago, or six, or twelve, has no existence except as now, mumbling, mumbling, sacrificing the slain world to the omnipotence of words, I strain my memory to regain it.[80]

The chapter ends with his mother's warning Grendel of the impending danger with the words "*Beware of the fish*"[81]—which symbolizes Beowulf's coming from the sea in the next chapter—and with Grendel's philosophical expression "*Nihilo ex nihilo,*"[82] which Grendel says pertaining to the Shaper's funeral and mean "nothing comes out of nothing." Thus, with the end of the Shaper's historical narrative Grendel's existence concurrently becomes a void and his forthcoming non-existence with Beowulf's coming is signified.

Beowulf's coming in the next chapter hints a new beginning and an end to Grendel's life. As soon as he sees Beowulf, Grendel understands that the person he is face to face this time is an extraordinary one with a huge body and otherworldly look. Differently from Beowulf in the Old English epic, he is not presented as representing the traditional Anglo-Saxon heroic code. He appears as a fantastic and supernatural, almost like a science-fiction android. "The eyes slanted downward, never blinking, unfeeling as a snake's."[83] Beowulf is not simply described as a machine; he is described as a dead man. His voice is that of a "dead thing," and his patience rivals that of a "grave-mound." These images reinforce the idea that Beowulf will be the agent of Grendel's termination. However, as a man who has risen from the dead, Beowulf also resembles the resurrected Christ. Grendel's mother tries to warn her son of his impending doom by bleating "Beware the fish"— fish being a commonly recognized symbol for the Christ figure. Indeed, Beowulf is associated with fish images several times throughout this chapter. He comes from over the sea, "has no more beard than a fish"[84] and has shoulders as "sleek as the belly of a shark."[85] Furthermore, the story of the swimming contest with Breca demonstrates Beowulf's competence in the water.

Beowulf's strange face, otherworldliness, unblinking eyes, and huge body begin to grow unsettling to Grendel after a while. He understands that "the stranger [is] no player of games."[86] He grows "more and more afraid of him and at the same time...more and more eager at the hour of [their] meeting."[87] Affirming the proposition above regarding Beowulf's Christ-like connotations and in accord with Grendel's identification with Cain in the Shaper's narrative, looking at Beowulf's features Grendel feels that Beowulf seems "from a dream" he has almost forgotten.[88] When Beowulf grips his arm with "crushing fingers...like fangs with poison,"[89] Grendel says he grotesquely shakes hand with his "long-lost brother."[90] With Beowulf's deadly grip, Grendel feels that his "long pale dream," his "history, falls away."[91] The words Beowulf whisper as he kills Grendel validate the association of Beowulf with the fish and Christ above and presents a counter discourse of hope to the Dragon's and Grendel's nihilism; Beowulf begins his lecture to Grendel by quoting the dragon, describing the present moment as a "temporary gathering of bits, a few random specks, a cloud." Actually, the writer gives the impression that Grendel in fact confronts the philosophizing Dragon instead of the Old English epic hero Beowulf. In an interview in *The Paris Review*, Gardner answers a question regarding this issue as follows:

> As a medievalist, one knows there are two dragons in medieval art. There's Christ the dragon, and there's Satan the dragon. There's always a war between those two great dragons. In modern Christian symbolism a sweeter image of Jesus with the sheep in his arms has evolved, but I like the old image of the warring dragon. That's not to say Beowulf really is Christ, but that he's Christ-like.[92]

Beowulf, the Christ-dragon, accepts the Satan-dragon's explanation of the world as a place where everything eventually dies. However, while the Satan-dragon emphasizes death and decay, Beowulf looks beyond the moment of death and emphasizes the rebirth that always follows. Calling Grendel "my brother" he says:

The world is my bone-cave, I shall not want...As you see it is, while the seeing lasts, dark nightmare-history, time-as-coffin; but where the water was rigid there will be fish, and men will survive on their flesh till spring. It's coming my brother. Believe it or not. Though you murder the world, turn plains to stone, transmogrify life into I and it, strong searching roots will crack your cave and rain will cleanse it. The world will burn green, sperm build again. My promise.[93]

He concludes his words with a phenomenological view of time and a belief in heroism and meaning in life: "*Time is the mind, the hand that makes (fingers on harpstrings, hero-swords, the acts, the eyes of queens). By that I kill you.*"[94] Beowulf's counter discourse tells Grendel that time is product of the mind; however, for a meaningful life, heroism, the Shaper's historical configurations, and everything related to human action are required. Against this background it can be said that Beowulf's killing of Grendel seems to metaphorically mean the victory of hope over nihilism, the aboveground over the underground, and perhaps more importantly, authoritative narrative discourse over the other discourse of the monster.

Returning to our philosophical framework and repeating Beowulf's words, time is the mind and history is time brought to language in the narrative form. There is a human action out there in the external world; after all it is the hand that makes. However, what the hand make, relying on the theoretical background of this chapter, are effects of such imaginative constructs as ideals, ideologies, utopian visions, heroism, freedom, and so on. Besides, human experience is implausible, fragmentary, and to some extent, absurd without the forming mind of the story- and history-maker. As the Dragon tells Grendel, the Shaper saves humanity from meaninglessness by creating an illusion of reality; creating ideals for which humans can fight to make them keep living for the future. Unlike what the Dragon thinks and Grendel later comes to think, this imposition of form on reality should not be rejected or mocked as absurd; human is a sense-making animal, and narrative and historical configurations are cognitive tools through which s/he makes sense of the world, gathers fragments of experience to configure a meaningful life vision. Thus, the Shaper, metaphorically speaking, is a basic component of all societies because, with his constructs, he makes human life organized around certain ideals and values. The configuration of a monstrous other to make these ideals of established order definable against the background of the disordered represented by the "monstrous" is ethically the main defect, but also perhaps the inevitable factor, of the Shaper's narratives.

Mustafa Kemal Universitesi, Tayfur Sokmen Kampusu, Egitim Fakultesi, Ingilizce Ogretmenligi Bolumu 31000, Alahan-Antakya Hatay, Turkey
e-mail: karadas.firat@gmail.com

NOTES

[1] *Cartesian Meditations*, 132–133.
[2] Ernst Cassirer, *The Philosophy of Symbolic Forms: Language* Vol. I. Trans. Ralph Manheim. (New Heaven, CT: Yale University Press, 1953). p. 61.
[3] Ibid., 78.
[4] Gilles Deleuze, *Kant's Critical Philosophy*. (Minneapolis, MN: University of Minnesota Press, 1984). p. 15.
[5] Leonard Talmy, "A Cognitive Framework for Cognitive Structure," in *Toward a Cognitive Semantics*, Vol. 2 417–482 (Cambridge, MA: MIT Press, 2000). p. 419.
[6] David Herman, "Stories as a Tool for Thinking," in *Narrative Theory and the Cognitive Sciences*, ed. David Herman (Stanford: CSLI Publications, 2003), p. 170.
[7] David Herman, *Story Logic: Problems and Possibilities of Narrative*. Frontiers of Narrative Series (University of Nebraska Press, 2002), p. 296.
[8] "Stories for Thinking," p. 165.
[9] Paul Riceour, *The Rule of Metaphor*, trans. R. Czerny, K. Mchaughlin and J. Costello (Toronto: University of Toronto Press, 1977).
[10] Louis Mink, "Narrative Form as a Cognitive Instrument," in The *History and the Narrative Reader*, ed. Geoffrey Roberts (London and New York. Routledge, 2001), p. 215.
[11] Ibid., 215.
[12] Ibid., 212.
[13] Ibid., 218.
[14] Ibid., 219.
[15] Paul Ricoeur, "The Narrative Function," in *Paul Ricoeur: Hermeneutics and the Human Sciences*, ed. J. B. Thompson (Paris: Cambridge University Press, 1981), p. 277.
[16] Paul Ricoeur, "The Human Experience of Time and Narrative," in *A Ricoeur Reader: Reflection and Imagination*, ed. M. J. Valdés. (Toronto: University of Toronto Press, 1991), p. 108.
[17] Ricoeur, "The Narrative Function," p. 278.
[18] Ibid., 279.
[19] Ibid., 294.
[20] Richard Kearney, "Narrative Imagination: Between Ethics and Poetics", in *Paul Ricoeur: The Hermeneutics of Action*, ed. R. Kearney (London: Sage, 1996), p. 179.
[21] Ricouer, "The Human Experience of Time and Narrative," p. 115.
[22] Hayden White, "The Historical Text as Literary Artifact," in *The History and Narrative Reader*, ed. Geoffrey Roberts (London and New York: Routledge, 2001). p. 223.
[23] Ibid.
[24] Ibid.
[25] Ibid., 224.
[26] Ibid., 225.
[27] Ibid.
[28] Ibid., 227.
[29] Hayden White, "Historical Emplotment and the Problem of Truth," in *The History and Narrative Reader*, ed. Geoffrey Roberts (London and New York: Routledge, 2001), p. 380.
[30] Hayden White, "The Historical Text as Literary Artifact", p. 231.
[31] Ibid.
[32] David Carr, "Narrative and the Real World," in *The History and Narrative Reader*, ed. Geoffrey Roberts (London and New York: Routledge, 2001), p. 146.
[33] Ibid., 147.
[34] John Gardner, *Grendel* (New York: Vintage Books, 1971).
[35] Joseph Milosh, "John Gardner's *Grendel*: Sources and Analogues." *Contemporary Literature*, pp. 48–57 (19, 1978), p. 49.
[36] *Grendel*, 28.
[37] Ibid., 29.
[38] Ibid., 158.

39 Ibid., 23.
40 Ibid., 27.
41 *Grendel*, 43.
42 Ibid., 34.
43 Ibid.
44 Ibid., 42.
45 Ibid., 46.
46 Ibid., 49.
47 Ibid., 43.
48 Ibid., 49.
49 Ibid., 49.
50 Ibid., 48.
51 Ibid., 44.
52 Ibid., 51.
53 Ibid., 55.
54 Ibid.
55 Judy Smith Murr, "John Gardner's Order and Disorder: *Grendel* and *The Sunlight Dialogues*," in *Critique: Studies in Modern Fiction* (18.2, 1976), pp. 97–108.
56 Ibid., 99.
57 *Grendel*, 28.
58 Ibid., 8.
59 Ibid., 73.
60 Ibid., 67.
61 Ibid., 157.
62 Ibid., 64.
63 Ibid.
64 Ibid.
65 Ibid., 65.
66 Ibid., 71.
67 Ibid., 67.
68 Ibid., 91.
69 Ibid., 158.
70 Joseph Milosh, "John Gardner's *Grendel*: Sources and Analogues," p. 50.
71 *Grendel*, 80.
72 Mock-heroic is a literary style that mocks classical stereotypes of heroes and heroic literature. Mock heroic works mock traditional heroes by either putting a fool in the role of the hero or by exaggerating the heroic qualities to such a point that they become absurd.
73 Ibid., 82.
74 Ibid.
75 Ibid.
76 Ibid., 84.
77 Ibid., 89.
78 Ibid.
79 Ibid., 149.
80 Ibid., 146.
81 Ibid., 149.
82 Ibid., 150.
83 Ibid., 154.
84 Ibid.
85 Ibid., 155.
86 Ibid., 163.
87 Ibid., 165.
88 Ibid., 154.

[89] Ibid., 168.
[90] Ibid., 169.
[91] Ibid.
[92] *The Paris Review*, Interviewed by Paul Ferguson, John R. Maier, Sara Matthiessen, Frank McConnell. Issue 75, Spring 1979, p. 8.
[93] *Grendel*, 170.
[94] Ibid.

ALİ ÖZTÜRK

THE POWER OF DANCE/MOVEMENT AS A MEANS OF EXPRESSION

ABSTRACT

We live in a world of movement and all of the creatures are in a gradual evolution and development. Indeed, the whole universe is in a dance/movement. The universe and all of its components are in a well-programmed movement. This movement is very effective in framing the order of the universe. The movement is the way of expression for humankind. The movement was the nucleus of the life for the primitive forms of the human; they had to be in movement in order to survive their life, to find their food and clothes, or even to protect the fire. They used movement in religious rituals and ceremonies to express their fear, happiness, passions or hope. The birth, breeding and death are all the essentials of human life and they all occur as a result of the movement. When the humankind is closer to the nature, the movement becomes more spontaneous in expression. The learning process in the early ages is performed through living and doing, therefore the movement activities are the natural method for such kinds of learning. The movement activities help individuals to develop their physical, mental, social and emotional growth. The movement activities are vital for individuals as much as art or math. By means of the movement activities, a child might recognize and be aware of his/her own physical skills and limits, as well as s/he might be able to explore and understand his/her body. The creative movement activities are also divided into two groups as personal movements and functional/physical movements. While the personal movement activities reflect the mood and character of an individual, the functional/physical movements serve for a more practical purpose such as development of motor behaviors. There is not forethought or pre-planning processes in creative movements. The individuals let themselves to the rhythms of the music and go to a totally different ball game. There is not a perfect or previously practiced movement type. Individuals learn how to perform a dance through practicing the movements and through combining those movements within the course of time. The knowledge of movement and experience in combining the movements to perform the dance lead the individuals to create a composition. All of these actions improve the movement knowledge of the children. Within the course of time, the movements become more controlled, fluent and elegant. This is also reflected in the quality of the dance (Lynch-Fraser 1991). Therefore, individuals should refresh their understanding about theirselves and about the things in their environment through the dance/movement. This restrats in learning will continue without any judgements. With reference to the above mentioned arguments, the following issues will be discussed in the present study:

- The power of dance/movement as a means of expression
- The relationship between individuals and dance/movement (The gains of dance/movement to the individuals)
- The need of dance/movement for the individuals in relation to his/her creative, artistic and aesthetical aspects.
- The relation of dance/movement with the other kinds of art

INTRODUCTION

The universe was created with a rhythm and this rhythm is still going on. From the beginning of the first movement, there were people in this rhythm. In ancient times people knew how to make a sound and they were moving like an animal with instincts. After a while they began to use movement to do something or to tell their troubles. From that, moment dance and movement began. For this reason dance is as old as human life on earth.

People were trying to explain something with the experience of sound and movement. After a while, these strange sounds and movements began to adopt each other. When the primitive people caught and brought their hunts, this awoke the curiosity of the tribe members. After this people began to turn this into a show and they were acting the moment of hunting. Therefore, they were playing in front of the people. This revealed a visual thing. This was dance/movement. Because people's using a combination of compatible or incompatible movements to express themselves was to make the sense of the universe.

We live in a world of movement and all of the creatures are in a gradual evolution and development. Indeed, the whole universe is in a dance/movement. The universe and all of its components are in a well-programmed movement. This movement is very effective in framing the order of the universe. However, it is rather complicated to understand this movement of the universe through the available senses of living creatures. The humankind, which is an important element of the universe, expresses themselves through movements that are similar to the movements of the universe. By the same time, the humankind's desire to make movement is the source of energy that is essential for the existence, survival and self-realization of humankinds. The movement was the nucleus of the life for the primitive forms of the human; they had to be in movement in order to survive their life, to find their food and clothes, or even to protect the fire. They used movement in religious rituals and ceremonies to express their fear, happiness, passions or hope. The birth, breeding and death are all the essentials of human life and they all occur as a result of the movement. The more the humankind is closer to the nature; the movement becomes more spontaneous in expression. In other words, the movement is an essential and a natural need for humankinds for self- realization, self-improvement and expression. The movement could be categorized under two main areas, which are the substantial movements that were the genetic component of the humankind and the movements that come to scene with reference to the demands and desires of individuals. The babies, for instance, react with a variety of movements and they cannot stand still. These movements can be defined as the physical movements that meet the basic movement needs of the organisms. That is, it can be claimed that these instinctively encoded movements are the main functions of the living organism and have vital roles for humankinds. For instance, the physiological system of humankinds such as nutrition, excretion and circulation are reflex movements, which function within their own program and continuum. If there is any malfunction in these reflex movements, it might cause serious diseases or even the death of the person. The reflex movements are programmed as specific life cycles, which are repeated throughout the life of humankind within an identical rhythm. This rhythm or movement is also called as biological rhythm. All of the creatures in the world function within their own biological rhythmic cycles.

The movement capacity of humankind can be found in the biological rhythm of every people. Thus, humankind is in a search of other movements that might meet their expectations and such a deliberate search leads to new pursuits. During the search for the new movements, individuals use their imagination and creativity within their own potentials and capacities. As long as the experience of individuals is increased in performing movements, they become more creative in performing new movements and their desire to perform new conscious movements also increase along with their experiences. The movement might be a natural reaction to the music or any sound and it is physiological oriented. The dance, on the other hand, is one of the movements that come to scene in relation to the conscious demands and desires of individuals and it is an art, which consisted of creative movements. The movement could be turned into dance when it includes creativity and imagination. Although every kinds of dance include movement, it does not mean that every movement is a dance. Dance includes features of the movement, which are created by the individuals consciously. Dance keeps the play-think methods, physical activities, heuristic and fictional powers, as well as emotional responds together. The instrument of the dance is the body of humankind and the human needs to practice a variety of movements and actions in order to use this instrument functionally. Therefore, the human needs opportunities to develop these movements.

The movement activities help individuals to develop their physical, mental, social and emotional growth. By means of the movement activities, a child might recognize and be aware of his/her own physical skills and limits, as well as s/he might be able to explore and understand his/her body. According to Dewey (1900), the learning individuals should be active. Since, mentally and physical activeness of individuals is the fundamental requirement of the learning process. For instance, play, inquiry, fieldwork and self-expression were used by Dewey as teaching techniques in his laboratory school in 1896. This is a key indicator of using movement activities for teaching where the learners learn through doing and experiencing. In this respect, it can be claimed that as much as learning of art or math, the movement activities are also vital for the learning process of individuals.

The creative movement activities are also divided into two groups as personal movements and functional/physical movements. While the personal movement activities reflect the mood and character of an individual, the functional/physical movements serve for a more practical purpose such as development of motor behaviors. There is not forethought or pre-planning processes in creative movements. The individuals let themselves to the rhythms of the music and go to a totally different ball game. There is not a perfect or previously practiced movement type. Individuals learn how to perform a dance through practicing the movements and through combining those movements within the course of time. The knowledge of movement and experience in combining the movements to perform the dance lead the individuals to create a composition. All of these actions improve the movement knowledge of the children. Within the course of time, the movements become more controlled, fluent and elegant. This is also reflected in the quality of the dance (Lynch-Fraser 1991).

DANCE/MOVEMENT AND INDIVIDUALS

The most basic learning strategy of an individual is his/her endeavor to distinguish the world through using his/her own physical skills. Along with other activities, the movement activities provide great opportunities for individuals for their self-developments. By means of dance, which is a creative movement, individuals detach their self from the outside world, forget everything related to their self and let their self to the rhythm of the music and they trance to the imaginary worlds. They continue to search for new movements freely by means of the senses that the music and rhythm evoked in their minds.

As Husserl (1950) claims, like phenomenology, which suggest to restart learning in order to a better understanding of the world, the individuals should refresh their understanding about their selves and about the things in their environment through the dance/movement. This restarts in learning will continue without any judgments. An outstanding example of using dance and movement in such restarts in learning and search can be seen in "semah" (Dervish's whirl) which was introduced by a famous Turkish philosopher Mevlana Celaleddin-i Rumi. It is believed that purifying from his/her physical and mental ties with the world, the individual, who perform movement through semah, becomes closer to "Allah" (God)–the creator of the universe. The individuals, who integrated with the universe by means of the rhythm and movement that was provided by semah, carry on learn and relearn process on his/her own.

Individuals use their body language for the first time through the physical movements. That is, they try to establish a nonverbal communication with their environments by means of physical movements. For instance, a newborn clearly expresses his/her needs through blinking, crying or contractions. S/he gets his/her first impression related to the world and his/her self through physical movements. The relationship with the environment creates a social identity for the individuals in the society. Since, socialization improves the sense of satisfaction as well as creativity.

The individuals are free to reflect their personalities through performing their own styles in their creative movements. If the movement is limited by external factors, the freedom and creativity in the movement are also restricted. Although a special rhythm and a movement figure was considered as the base of some of the dance and movements, there is not a very specific movement type that individuals should imitate in their own movements. The creative movements might come out at any time when the individuals feel free or wanted to perform a movement and this movement can be associated with a music, poem, and rhythm or even with silence. Through hearing their heartbeats and rhythm of their pulse or feeling their sensations and thoughts, children might use their body as an instrument. They might be everything that they imagine and it depends on their imaginations. The movements they have performed are the expression of their existence (Edwards and Nabors 1993).

The freedom, which forms the baseline of the creative movements, enables the individuals to use their imaginations freely and lets them become flexible, fluent and authentic. The functional movement activities include various actions that facilitate the development of small and gross motor skills. The functional movements also provide opportunities for individuals to act out their different feelings as well as to discover new skills and movements. The functional and personal movement activities also have an effect on the body development of individuals. Since, small and gross motor skills include various exercises that support the development of the children healthily. The individuals might enrich their internal experiences if they notice what they wanted to performed and what they expressed through their movements.

People dance for pleasure, satisfaction, exchange of opinions and fellowship, expression or for mutual advantages. The dance might be both the means of message and the message itself at the same time. Just think about the children, who waves, sings, forms a circle, falls down, chuckles and stands up to repeat all of these movements. They do all of them not for the sake of launching a communication with others but just for fun, learning or for new searches. In this respect, dance can be a self-serving activity and even it may have a unique language.

MUSIC, POETRY, NARRATION AND DANCE/MOVEMENT

Music is a natural way of bringing out the creative movements. Solely use of the body might be enough to produce music or rhythm for some of the creative movements. The start of a strong beat or rhythm of a clear-cut music is also the start of a movement. Murmuring to oneself or thinking as "do the same action slowly now", "do it in another direction or in different pace now", "try to do the same but be a bit more lively now" help bringing out the creativity in the movement (Jordan 1972).

Such a condition helps individuals become aware of new opportunities. Hence, an ordinary exploration in the world of music and movement might be transformed into an unexpected happiness. Listening to music is a natural way for discovering the creative movements. Different types of music and rhythm should be selected for the first practices. For instance, teacher might initiate a teaching and learning process through playing music. The selected music should have a strong beat and rhythm and as well as a straightforward structure. The children should not be informed about what they will listen to. The teacher should let students listen to the music and ask their feelings related to the music. While the children listening to the music, the volume can be turned down and they might told to make a circle. The teacher should talk about the music one by one in the circle. Most probably, some of the children might already be fascinated by the rhythm of the music. The teacher might also join to the circle. The children can freely move in the room as they wish and do whatever the music tells them. The children might perform some movements such as clapping their hands, stomping to the floor, dancing and yielding during the activity. Such activities might serve the purpose. Playing a slightly soft and favorable music can be used to give children the feeling of conflict or might used to relax the children after the activity. This approach can be performed by toys, puppets or parts of the body such as hands, foots, or fingertips in different occasions with different groups.

The imagination of the individuals is the only limitation for performing rhythmic movements through using one's own body, objects or someone else's body. The children solve the conflict while they are struggling with movement activities and music. They use their logic to define which instrument sounds like a thunderclap, or how a scarf wings in the air. They create a schema by means of the musical instruments, their body motions or the words in the song. They learn the concepts related to the numbers while they are singing a song related to the numbers or while they are clapping their hands or stomping their feet on the ground with a pace. They learn thinking symbolically while they are imitating the elephant walk or while they are jumping like a rabbit.

Likewise, various fields of art also support the expressionist aspect of the dance/movement. For instance, like the rhythm or using an effective language, the poetry has also importance in the movement activities. It is not always necessary to use rhymed verses or prose. Every single word that was heard might be connotation of a character. What is important here is to listen to the sounds and express the connotations by means of the body. The discrimination of the quality and character will follow it. The facilities of the text and potentials of the body gain meaning through the practice.

The use of types and characters in movement help to understand the relationship between the endeavor and mood. The ratio of the tension in the body, the movement rate and the range, reflects the person's instant character, age and mood. Hence, a character in a stretched and bending body, with intertwined hands and rickety motions with a squint look might give the impression of an angry old man. This is the expression of the movement as a whole picture. Slight gestures and facial expressions might imply great expressions (Bruce 1965).

THE NARRATION POWER OF DANCE/MOVEMENT

Dance/movement is the origin of every human activity and used in all of the expressive arts, since, the movement is a means of transmission of the inner world and energy of human to outwards. Literarily, movement indicates more than a physical motion. Movement is an activity, which includes the whole of a person. That is, not only physical motion but also intellectual, emotional and intuitional aspects of the person are included in the movement activities. Therefore, it can be claimed that the dance is an important movement activity and all people should be encouraged to dance.

Naturally, all kinds of expressional activities include a communication aspect and movement is a way of communication. Sometimes people start a communication unconsciously and sometimes the communication is carried out with a conscious endeavor. The power of communicating through motions or actions is usually connected with the endeavor of the babies to express their needs in their pre-language use period. Although in some cases it is believed that the habitual behaviors related to the motions are not characteristic, thus, they are used to mask the mood and personality, in fact, every motion or actions reflect the mood and personality of people throughout their lifecycle. However, a real communication exists only when the expression is interpreted by the receiver of the message. For instance, the communication in the children's play, where they create spontaneous movements, generally performed and developed within a form of ritual.

Children seem more compatible with the body and as they do in everything, they use their whole body to communicate. Thus, the creative process begins. In order to complete this process, the combination between skills and kinesthetic sense is required. Sight, touch and balance sense that is in kinesthetic contributes to awareness of motion. In order to develop these feelings, to communicate effectively and to get the creative movement, there is a need to offer opportunities to individuals. This situation will help them to have more confidence about their creative talent and provide them to have more positive relationships with their physical and emotional environment.

The ability to communicate through motions has a natural importance and it has a key role in the education process. Step by step, this ability leads up to the scene and dance. What is more, it has a central power in itself that is independent of the language power. Thus, the individuals become more social and more collaborative, and by means of developing their personal expression powers, individuals can create a balanced communication with their partners.

Another purpose of the dance/movement is to guide the individuals be aware of the connection between the emotions in the physical perceptions and the mental awareness. However, all of these skills should come together in order to complete the creativity process. What is more, people should be familiar with their body and identify their emotions in order to increase the sensitivity of the movement. Identification of these emotions and blending them into the movements might prevent the movements to become mechanical actions. The familiarity of the body, which supplies our physical presence and positioning in the space, will provide a confidence in performing our daily actions. Such features help to use the body in creative movements with multi purposes (Russell 1987).

While exploring the features of the movement, people also experience the unfamiliar features as well. What is more, while exploring the unfamiliar features, people can allow their less salient features to exist, thus in turn, this situation help them to be ready for the conflicts that they might face. The creative movement is dependent to the body more than other creative actions. Providing opportunities during the creative movement activities enable the individuals to trust themselves in terms of their creativity and establish a close relationship with their physical and emotional environments.

CONCLUSION

Movement is a means for survival and search for new pursuits. By means of the movement, people can communicate with various levels spontaneously and the dance/movement transmits the consciousness of human being to upper levels. The communication that is established by the help of movement also satisfies the natural needs of people concerning the movement and rhythm. Such a movement might also increase the motivation of the people in the works they performed.

The power of the dance also includes a metamorphosis skill and a skill to run over the extraordinary things. While dancing, people change their emotions and thoughts through inspiring, enjoying, relaxing, cultivating and pondering their minds.

The learning in the real life is a whole process of activities, which lead people to struggle and develop skills to express their selves. One of the ways of supporting individuals to search for new opportunities is to lead their instant interests towards the actions, like transforming the instant emotions of people to the dance. Dance is a way for recognition and communication. Almost all of the societies used the dance as a means of personal or cultural communication and as a supplier of physical and emotional needs. No matter how adults opine the dance, everyone should watch the dances of children in order to see their inner enthusiasm.

While dancing, the children will be dismissed with the outside world, forget everything about themselves, they give themselves to the rhythm of music, and they dive into another world. In children's dance, there is not any perfect, pre-made or practiced form of motion. These movements can be associated with the effort of the child to know the world. Child's using his/her physical skills to recognize the world is the most basic learning strategy of the child. Compared to other activities, activities including motions are great opportunities for the whole development of children. In these creative movements, any previously thought or planned thing to put into practice cannot be seen.

Movement is the fundamental baseline of universe and it forms the structure and background of all substances in the universe. For instance, unless the particles of a metal perform a movement, the structure of it cannot be shaped or its color cannot be seen. That is, all of the features that form the metal become a unique thing through the movements of the particles. Likewise, the human beings become creative, lively, sensitive to others and acct in a rhythm by the help of the art of movement and the experience that emerged as an outcome of movement. People, who perform movements creatively and actively, can easily associate their physical, emotional and intellectual skills as a whole and express themselves without the help of the words. The expression power of movement seems a more confidential way for the people, who have difficulty in expressing their thoughts verbally.

The performance of the movement should not depend on any form and shape and should be far from any judgment or evaluation. Individuals can be aspirant for explorations in such cases and might perform unique movements, as exploring the individual experiences in the phenomenology. People should relinquish their minds and bodies and should be ready for new explorations and experiences. Working with our bodies is a valuable gift for us and it is worth to work through.

Anadolu University, Eskisehir, Turkey
e-mail: alio@anadolu.edu.tr

REFERENCES

Bruce, V. 1965. *Dance and dance drama in education.* Oxford, New York: Pergamon Press.
Dewey, J. 1900. *School and society.* Chicago: The University of Chicago Press.
Edwards, L.C., and M.L. Nabors. 1993. The creative arts process: What it is and what it is not. *Young Children* 48(3):77–81.
Husserl, E. 1950. *Die Idee der Phänomenologie.* Fünf Vorlesungen. [The idea of phenomenology. Five lectures], ed. Walter Biemel. The Hague, Netherlands: Martinus Nijhoff, 1973.
Jordan, D. 1972. *Childhood and movement.* Oxford: Basil Blackwell.
Lynch-Fraser, D. 1991. *Playdancing [electronic resource]: Discovering and developing creativity in young children.* Pennington, NJ: Princeton Book Co.
Russell, J. 1987. *Creative dance in the primary school.* Tavistock, Devon: Northcote House Pub. Ltd. (3. Edition).

SECTION IX
INTERSUBJECTIVITY, FREEDOM, JUSTICE

JOHANNES SERVAN

MAKING HISTORY OUR OWN – APPROPRIATION AND TRANSGRESSION OF THE INTENTIONAL HISTORY OF HUMAN RIGHTS

"Only ethical justice is ultimate justice" (ca. p. 203–4, H&B)

ABSTRACT

In this chapter I will sketch out a framework of an intentional history of human rights. This framework is based on the idea that intentional history – as the retracing of the essential argumentative steps that justifies our current situation – is a narrative mode of both appropriation *and* transgression of history as our own.[1] The traditional way of understanding intentional history, as we find it in Edmund Husserl's latest work *The Crisis* (1936), emphasized the appropriative dimension which is meant to give a critical and rational account for the arguments that justify our current and globally embraced perspective on philosophy and natural science. In the case of human rights the ambition will be to justify its global hegemony and to give us a satisfying reason to be loyal to this tradition. This gives us the motive for retracing the steps of ethical-political thoughts backwards from the present through the UN declaration after the end of the Second World War, to the Classical democracy in Greece. In addition to this historical reduction as a questioning back, I suggest that we will have to include the transgressive dimension of intentional history. The appropriation of a history as our own involves both the generation of a normal lifeworld, a homeworld, and a liminal experience of a co-generational outside, an alienworld. The transgression of narratives implies in this sense the liminal experience of someone living in a history which generates an abnormal, i.e. alien homeworld; a mythical alien. The account of transgression as a generative dimension becomes ethical imperative due to the recognition of the irreducibility of the homeworld/alienworld structure in experience. Ultimately we will never be at home, with our own history, without the alienworld as the experience of an inaccessible generative depth that calls our ability to appropriate into question and sets limit-claims for the universal synthesis of humanity. The radical consequence of this account is a shift from the genetic idea of a universal ethical humanity[2] that will realize itself in the global political state founded on a constitution of human rights, to the generative idea of a unity of home and alien that expresses itself through their difference and does not depend upon a resolution in a higher synthesis. (Steinbock 1995, p. 246).

INTRODUCTION

When Edmund Husserl wrote *The Crisis of the European Sciences and the Transcendental Phenomenology* (1936), one of his central questions was: How did the current norms of science and philosophy constitute historically? More specific, he asked for the deep motives for a positivistic restriction of the ideal of science. Husserl found the answer to his question in the critical reflection on the intentional history from Galileo, through Descartes and the empiricists, most importantly Hume, to Kant. This historical reflection is not carried out within a phenomenological attitude, but is rather a teleological-historical answer informed by a phenomenological insight of the generational process. This insight in its turn was motivated by the transcendental significance of birth and death.

On one level this teleological-historical reflection in *The Crisis* can be read as a successful and exemplary appropriation of a traditional narrative by an insider: The development of a genuine humanity through the birth of Greek philosophy is put into question by the fall of the traditional rationalism, the following global dominance of positivism and its parallel skepticism for metaphysics (problems of reason: knowledge, valuation, ethics, history, freedom). By employing the narrative tools such as a "myth of origin" and a "crisis" (turning point, something is at stake), Husserl provides a competing narrative to the modern ideal of science and its foundation in the mathematical abstraction of nature, challenging the universe of mere facts by calling it a positivistic decapitating philosophy of its genuine metaphysical questions. In the end this could be a loss of something unique in the Greek idea of human life for the whole of human kind:

> To be human at all is essentially to be a human being in a social and generatively united civilization; and if man is a rational being (animal rationale), it is only insofar as his whole civilization, that is, one with a latent orientation toward reason or one openly oriented toward the entelechy which has come to itself, become manifest to itself, and which now of necessity consciously directs human becoming. (p. 15, Husserl 1936)

On another level Husserl's new definition of an historical humanity implicitly presuppose an ideal of a universal synthesis of the lifeworld. The genuine character to the European homeworld, as a generatively united civilization, seems to rely on whether or not it bears within itself an absolute idea which can become manifest as a universal philosophy. Apparently, there wouldn't be a *real* crisis if it was confirmed that "Europe" was merely an empirical anthropological type, like "China" or "India". It would only lead to the conclusion that Europe and the West was governing the world on the basis of historical non-sense, i.e. by brute force. Is there nothing in between these two extreme points of existence? Could we not recognize the European as our own, without having to condemn the Chinese or Indian normalities as to be mere empirical types?

It is important to notice that the generational process of becoming normal, like a child growing up, involves the appropriation of cultural and traditional ways, existing long before this child was born. Growing up the child appropriates a homeworld, described in generative phenomenology as "the pregiven lifeworld horizon as a mode of delimiting styles of interaction, of life and of sense" (Steinbock 1995,

p. 109). New possibilities are opened up as familiar and normal by a simultaneous limiting off. The favoring of a universalistic reasoning with its entelechy of unity, even if endlessly distant, seeks to transcend this delimiting and therefore ignores the irreducibility of the liminal structure of homeworld/alienworld. In other words, a universalistic project seeks to eliminate that which makes the teleological-historical self-reflection, a Selbstbesinnung, exactly an appropriation of a *homestory*: a making of the history as our own in its co-constitutional relation to an alien story.

In this chapter I will argue that the shortcomings identified at this second level can be met by an account of transgression as a second, co-generative dimension [appropriation being the first] of the liminal structure of homeworld/alienworld. I will also argue that this concern becomes even more acute when considering an example more directly related to the generation of our ethical context, like for instance the global community of human rights and the struggle for their recognition among the national states. In this case we can no longer restrict the discussion of generational ruptures to an epistemological problem of historical continuity, like with Thomas Kuhn's analysis of scientific revolutions or Michel Foucault's archeology of knowledge. Rather we must consider it as an ethical imperative to be able to respond to the liminal structure of homeworld/alienworld that is generated out of these historical breaks, as irreducible and not as something we can or should overcome.[3]

HISTORY FROM WITHIN – HISTORY AS TELEOLOGY

But first, let's look closer at narratives as the organizing principles of history. Generally we mean by a "narrative" a sequence of events represented in the mode of literature, theatre, cinema, musical, etc. It's usually restricted by a beginning and an end, it got roles of agents and it is obviously told by someone, the narrator, which can also be in plural as We. In *Time, Narrative and History* (1986) David Carr makes a point regarding this plural form of the subject arguing that the group, as a unity of a temporally extended multiplicity of experiences and actions, is constituted by a narrative prospective-retrospective grasp of what has been and of what is projected to come. Defending his point against the skepticism of narratives as mere social constructs projected on a in itself chaotic causally determined reality, Carr claims that: "the temporal structure or organization of experience and of action is not different from a story that is told about it; rather, the experience or action is embodied in and constituted by the story told about it." (p. 149, Carr 1986).

In the context of writing an intentional history, we hold this responsibility in a mode of self-reflection. We are aware that the history we appropriate as our own will embody and constitute our experiences and actions. An important moment in this story is the generational process of birth and death as reoccurring breaks in the progress of the developing realization of the idea of philosophy. The significance of these breaks seems to concretize differently according to various institutions of activity. Projects in the form of a deductive system, like mathematics and formal logic, seems virtually untouched by the transition from one generation to the next. The coherence of the system disguises the implicit necessity of succeeding generations of loyal colleagues. An historical argument is more explicitly dependent on the generational succession, but is too often represented in the form of a complete dialogue with an overarching telos manifested in the resent conclusion or predicted soon to come due to a new attitude towards it. We've already located this structure in Husserl's intentional history, but there's an interesting openness in the crisis (in the sense of being an important turning point), that may serve as a leading clue for a more nuanced way of writing an intentional history.

The "crisis", is qualified by David Carr as the "quintessential element of narrative" and at the same time, related to the plural form of the agent, what he calls "the stuff of communal life" (Carr 1986, p. 159). The crisis is an important element for communal life because it identifies external and internal threats to the survival of the group. Related to human rights are of course the stories of oppression or exploitation and the survival through liberation, triumph over adversary, etc. (Carr 1986). In Husserl's concern it was the survival of the European homeworld as a philosophical tradition guided by the idea of reason. The concern was real – there was really something at stake. Contrary to Hegel, Husserl sees the crisis as a genuinely open and indetermined situation where the continuous fate of philosophy and humanity can be broken or at least take an unfortunate direction guided by "lazy reason". To the extent that we can identify the tradition of human rights with the tradition of philosophy, this crisis is also affecting the survival of a rationally founded community of human rights.

By questioning the violent and totalitarian tendencies of the history of progress in the Western intellectual tradition the "postmoderns"[4] have certainly also challenged the phenomenological project, especially concerning the central role that phenomenology has given to the transcendental subject and the problem of foundation. Generally holding that these philosophers are engaged in a generative project similar to the generative phenomenology,[5] I will focus on some problematic aspects of the teleological-historical reflection entailed by these post-structural or postmodern concerns, and then try to defend it by showing how a phenomenologically informed historical reflection may contribute in an original way of solving the problems of a generative philosophy. The post-structural method of Michel Foucault and Jean-Francois Lyotard's suspicion towards meta-narratives certainly puts into question the expectancy to find a continuous, step-by-step developed intentional history with the integrity of a deductive system, it also – especially related to principles of a global/universal humanism such as the human rights – calls into question the ideal of a foundation for a universal community.

Still, this French rupture in the modern tradition is too easily presented as a clean break with its tradition.[6] Rather it seems more reasonable to claim that the phenomenologist and existentialist anticipated the postmodern critique of the Cartesian essentialism and fundationalism, and that this intentional, metanarrative forgetfulness left the movement with a problem of interiority or subjectivity, where attributes of the subject such as intentionality and volition are curiously substituted by a personification of discipline, language games, etc. This seems to imply, if not an overarching telos, then at least the tendencies of an historical determinism driven by the "big subject" of discursive fields which leaves the individual subject with no choice and no genuine crisis. For instance, there seems to be no distinction between the motives (constituted by traditional narratives, of for instance human rights) and the acts of a social movement (carried out by the civil rights movement who made these narratives their own). (Kruks 2001) Nothing is at stake which calls to be acted upon by self-aware and responsible subjects.

Carr points to a similar problem in the interpretation of Hegel (Carr 1986). He claims that one should not read Hegel's philosophy of history as a theoretical project, but rather as a practical one. This implies that the project of a universal community should be judged not as a more or less accurate description of an exterior historical truth, but as a more or less convincing (likely, persuasive, coherent, moral-politically appealing) story of our historical past and future. We might object to Hegel's claim on the French revolution as a common European or even Human experience, on the premise of being one among several competing[7] narratives which might be appropriated or negated by us from within a historical community. The narrative organizes the historical past and projects a future where the individual is given a more or less concrete part to play. In this way Carr is able avoid the obliteration of the individual while accounting for the constitutive function of narratives for the unity of a group. The "We" is sustained from within by its aware and loyal members. (Carr 1986) The question then is: Why do we accept one narrative instead of another? Why does each of us come to recognize the same narrative as coherent and perceive its calling as worth answering?

THE APPROPRIATION OF NARRATIVES – GENERATION OF A NORMAL LIFEWORLD AS HOME

"Appropriation is a form of sense constitution that takes up pregiven sense as stemming from a homeworld and its unique tradition." (p. 180, Steinbock 1995) The uniqueness of the homeworld signifies both that it can neither be egologically founded (because it as a "Stamm" precedes it), nor can it be the all encompassing foundation for the world (which would eliminate the radical other). In the becoming of a group, such as a people or a state, the members appropriate a degree of awareness through different modes of communication (from the reversible intropathy of intercorporeal intimacy, to the irreversible and distant character of virtual communication) corresponding to various levels of community. The narrative modes are decisive for the constitution of an historical depth in experience, something small children don't have.

When members of a community appropriate a sense of their history through a narrative grasp of their past and their future, it is no longer just an objective grouping of sensitive and emphatic subjects: It becomes a community of reflexive self-awareness, a "we". Carr (1986) claims that "[...] the group is posited by its members as subjects of experience and actions in virtue of a narrative account which ties distinct phases and elements together into a coherent history." (p. 155) Contrary to an abstract, objective perspective, the historical horizon of a living community will always be situated from within, from the middle of where it comes from and where it's going.

Through accepting a narrative we come to perceive experiences and actions as common to us rather as belonging to me isolated. Understandably not everyone will be the author of the defining narrative which organizes this common, historical horizon. Most members will merely receive, evaluate more or less critically and eventually accept the stories told by the spokesmen and leaders. In most groups, from the smallest families to the large publics of modern society, there will be rival accounts of the identity and teleology of the community. And, on many occasions we'll get the chance of reconsidering our childhood beliefs or previous judgments about the coherence of traditional and new narratives. In large communities such as modern nation-states it is even a condition of democracy that the members are able to hold a complex of competing and conflicting narratives into play, integrating experiences and actions of many different groups. (See narrative imagination, Nussbaum). This all seems to correspond in general with the Hegel's concept of mutual recognition and Mill's idea of tolerance. In the terms of generative phenomenology: "Communication would be normal when it constitutes a concordant community of understanding that is optimal, integrating a rich diversity of perspectives in a shared unity." (p. 212, Steinbock 1995).

Even if this is a mere formal description of the optimal givenness of the coherent story, it gives us a clue of the best and most persuasive story as organizing a rich unity of perspectives in analogue to the kinesthetic-spatial system of intercorporeality. According to this principle, the appeal and persuasiveness of the homogenic narratives of the aftermath of a war, or of what might be as much a forced result: the winner of a scientific debate (See Hobbes and Boyle), will be seen as degenerational because the opponent version is no longer spoken of, though it might be revitalized later on. This degeneration is not just an abstract character of a narrative, it is something that will concretely materialize in the density of a culture as a failure to generate the best possible ethical context of which we are responsible in our ethical task as "functionaries of humanity".

In Anthony Steinbock's reading of *The Kaizo* and *The Crisis* he identifies a redefinition of the ethical task in Husserl's work from an ethical self-regulation (the givenness of individual genesis; the virtues of good life/the categorical imperative) to the more concrete generative dimension of the ethical context as communal and historical, of which follows that self-responsibility is at the same time before the other. The new definition of the ethical task:

[...] realizing the optimal in the ethical life means renewing the cultural community in its historical self-transformation, its institutions, organizations, and cultural goods of every kind: In short, realizing the best possible of the homeworld is the renewal of its generative force. (p. 205, Steinbock 1995)

This renewal of generative force requires a persistent attentiveness. The teleological-historical reflection is therefore called upon as a mode of historical inquiry, as a responsible and critical appropriation of history through an active examination and renewal of our tradition from within. So far so good. The problem is that the generative definition of the optimal is merely formal and cannot be specified in advance. What is most likely certain though is that Husserl's intentions of expanding an immanent ethical reform – as a devoted continuation of the Greek ideal – from our homeworld to an all encompassing world, as a "universal ethical humanity", would actually destroy the generative force inherent in the liminal structure of homeworld/alienworld. In order to free ourselves from prejudices and open up new possibilities in the continuous becoming of an always indetermined/renewable content of the optimal narrative, we should welcome competing optima/normalities. In order to be able to respond to this without taking over the responsibility for the alienworld (which again would reduce it to the same), we will not only have to be critical in our appropriative encounters, but also responsive in our transgressive encounters. Ending this short chapter I will suggest the further development of this work by pointing at some decisive aspects of the liminal structure of homeworld/alienworld regarding transgression.

TRANSGRESSION OF NARRATIVES – RESPONSIVITY TO LIMIT-CLAIMS

As we saw in the beginning of this chapter, appropriation is simultaneously a liming-off that manifest itself in a horizon of pregiven, familiar styles of communication. By growing up in a culture, becoming member of a homeworld, we will at the same time become strangers to other styles of communications such as languages and narratives. The encounter with the alien disrupts our expectation of the typical, the other as behaving normally and according to the same coherent narratives as me. At first, this break might seem to be easily overcome by a simple appropriation and the eventual universal synthesis seems to be within reach. But this is soon proved more difficult than at first, and in the end impossible if the break is heavy enough.

First, it's important to notice that the different levels of communities (from families to national states) are not organized as concentric circles. It is not as if the national states are the end product of the synthesis of all lower levels, rather it is a level of fragile integrity surviving in a field on conflicting, criss-crossing and intertwining communities (family-profession, class-country, religion – civic duty, See David Carr).

Second, Husserl draws and important distinction between recognizing someone as normal, as member of my community, and recognizing cultural difference not only as abnormality, but as having the integrity of a normal tradition that is not normal for us. (p. 242, Steinbock 1995) This implies an asymmetry and irreversibility which cannot easily be overcome, but that still has an openness as accessible in the mode of inaccessibility.

If we want to renew our intentional history of human rights, we will have to take these accounts into considerations. It will open up the future as unpredictable and indeterminable, resisting our narrative typification, but at the same time give us possibilities that are new and unique and not resting on the symmetrical inversion of our normality.

University of Bergen, Bergen, Norway
e-mail: johannes.servan@fof.uib.no

NOTES

[1] I'm of course inspired by the way narratives is used in the existential analysis of Dilthey and Heidegger, see Carr's *Time, Narrative and History* (1986). Another interesting concept I would like to work on later is "narrative imagination" and its significance for the experience of the other and the alien. The general though is that narratives is complex categorical structures that is dependent on communication in order to be constituted/generated.

[2] Quoted from the Kaizo-articles: "To a true human world-people over all particular peoples and to an over-people encompassing them, to a unitary culture, to a world state over all single systems of states." (H XXVII 58.f.) from Steinbock 1995, p. 238.

[3] This ethical imperative is of course related to the problem of recognition as we find it in Kant (Groundworks, kingdom of ends) and Hegel (Phenomenology of Spirit, the cunning of reason). It is a problem exactly because they both presuppose that recognition must be based on sameness, and not on radical otherness. We find this point even more explicit in Levinas critique of Heidegger's concept of Mitsein. Anthony Steinbock has shown that Husserl's generative attention might give Levinas original description of revelation a new dimension of generative depth.

[4] Due to the influential works of especially the French post-structuralists, narratives have become an important subject, not only for the study of literature, but for the reflection of method and representation in sciences such as history and social anthropology. It has given these fields of study a new sensitivity to the use of the available tools of representation. Unfortunately, I will not be able to discuss in depth the challenges toward writing an intentional history put forward by for instance Jacques Derrida's deconstruction and critique of the metaphysics of the present, Michel Foucault's archeology of knowledge through discursive formations and Jean-Francois Lyotard's suspicion towards metanarratives. See Jacques Derrida *Speech and Phenomenon* (1973), Michel Foucault *The Order of Things* (1966) and *The Archeology of Knowlegde* (1969), and Jean-Francois Lyotard *The Postmodern Condition* (1984).

[5] See Steinbock's article on generative philosophy and teleology in Alterity... (ed. Dan Zahavi)

[6] See Jürgen Habermas *The Philosophical Discourse of Modernity* (1991) and Sonia Kruks *Retrieving Experience* (2001)

[7] In a crisis we may also see a conflict of competing narratives, both in order to define the situation as a crisis and in order to find a convincing way to deal with it.[7] First one must make a persuasive story of how the community is threatened by something, like for instance economic recession or mass extinction of species (justice as prison/punishment vs. recovery/restitution). Then, if one accepts this story as defining a crisis, one must find ways of dealing with it in order to ensure the survival of the economic or ecological system.

JAN SZMYD

VITALITY AND WOBBLINESS OF UNIVERSAL MORAL VALUES IN THE POST-MODERN WORLD: CREATIVITY AND REGULATIVE FUNCTION OF THE LOGOS OF LIFE

ABSTRACT

Transformations in the sphere of ethical values in the modern times are complex and dynamic. Their impact on the quality of collective and individual life of societies in the globalising world is increasing. The process of describing this phenomenon may be facilitated by the category of "wobbliness of moral values." This term refers both to the existence, the functioning of values and their relation to human agents: intellectual and emotional reception, internal acceptance, impact on the recipients' personality, their stances, behaviour, etc. In the first case, the category of "wobbliness" denotes the extension or the narrowing (in current social, cultural and civilisational contexts) of practically recognised and realistically complied with basic moral values and principles, as well as obligations resulting from them. This is the "yielding" under the impact of external factors (economic, political, demographic, ecological, etc.) and at the same time the "straightening" – in a specific time and place – of the backbone of elementary values and principles of conduct, the brightening or the darkening of the significant meaning of objective criteria for basic goods and moral choices, the integration or the disintegration of historically and practically verified systems of values and rules of conduct. In the second case, this category denotes the increase or the decrease in readiness and the need of personal acceptance and practical implementation of a circle of basic values and ethical principles, an increased or a lowered level of understanding and appreciation of the significance of moral culture in the life of people, the awakening or the muffling of the personal and collective moral sensitivity, the opening onto or the isolation from the other people, the revival or the decay of empathic, altruistic and humanistic references in inter-human relations. A characteristic feature of moral wobbliness – in both its meanings – is primarily the fact that in spite of differentiation between internal or external moral qualities, the moral core embedded in the human nature remains untouched in its foundations. This relatively durable and solid core of morality is inherent in the human nature, excluding pathological or deformed cases and particularly extreme external circumstances. The ethical element, known as the conscience or the moral feeling, constitutes the so-called natural morality, confirmed both in the contemporary philosophy (phenomenology, existentialism, personalism, eco-philosophy, recentivism, etc.), as well as in certain social sciences (developmental and personality psychology, neuro-psychology, social anthropology, recent morality theories and other theoretical and empirical concepts).

INTRODUCTION

Modern times, with their characteristic phenomena such as globalisation, technological and IT processes, as well as the growing influence of mass media, commercialism and consumerism, stimulate and dynamise the "wobbliness" of the existing order in the sphere of basic ethical values and principles and in the area of personal experiences, evaluations and moral choices with a power that is greater than in any previous age. However – let us emphasize this – the modern age does not interfere strongly and deeply (as will be shown) with the moral factor (no matter how it is called and explained in theory) that is rooted in the human nature; the modern age does not cause a caving in of the basic core of elementary ethical values and principles, which obviously has fundamental significance for the correct shaping of social life, including family life and the individual and collective existence of people in the unfavourable durability and universal nature of moral and axiological orders in the era of "liquid modernity."

Therefore, theories that herald the collapse of moral foundations in the present, their decay or irreversible regression, seem philosophically and scientifically unauthorised; in a social aspect, they are destructive and deceptive. A similar evaluation can also refer to radical axiological relativism and the so-called destructivism which are influential nowadays [post-modernism, neo-pragmatism, chaos philosophies, etc. Cf. R. Rorty, J. Derrida et al.].[1]

However, the situation in contemporary moral reality is very complex. This is indicated by the fact that the indisputable and, unfortunately, increasing process of relativisation of values in the consciousness and stances of the "post-modern" man is accompanied by a contradictory process which is becoming more and more visible and is gaining increased recognition. This is the so-called process of creating various anti-relativist, constructivist and holistic concepts. These concepts are present mainly in various types of environmentalist and ecological philosophies (T. Rolston, H. Clinbell, M.E. Zimmerman, A. Leopold, H. Skolimowski et al.), in more recent life philosophies (A.T. Tymieniecka, P. Singer, K. Lorenc et al.) and in dynamically developing trends of global ethics and eco-ethics (H. Jonas, P. Singer, Z. Bauman et al.).[2]

For example, in ecological philosophies (as well as in other trends of global philosophy), the issues that are worthy of attention are the new universal values, such as acknowledgement that people from all cultures, civilisations, traditions and regions of the world constitute equivalent autotelic values, thence moral obligations towards all people should be equivalent; that life in all its manifestations and forms (not only human, but also non-human), understood holistically, is also a value in itself ... an autotelic value. Moreover, it is necessary to emphasise that new elementary principles and moral requirements are being formulated, such as, e.g., an empathic attitude to the entire eco-sphere and acceptance of the role of a responsible and careful guardian of this eco-sphere by every human being, acceptance of the equivalent rank of its every component, not only human, anthropological, but any other, etc.[3]

Obviously, everything in a man who is a human being in evolution and in the world created by him (including the world of values and moral principles) is changeable and relative, "liquid" and interdependent. Nevertheless, the "liquidity" and relativity of anything in this sphere of being and existence, including the sphere of human existence (the individual and collective existence of people), its nature, spirituality and humanity, species and civilisational development, should not be absolutised as it is done by some authors who are ideologically tied to post-modernism or other intellectual orientations that are philosophically close to it.

Expressing the idea metaphorically, the "liquid" world and "liquid" situations contain "substances" that have been, so to speak, more or less "temporarily hardened" and systems and states of affairs that are temporarily solid; in other words, there are "crystals" which, flowing in the general stream, preserve their un-softened and relatively dense structure.

This group includes certain features of human nature, certain ingredients of subjectivity and the individual and collective identity of people, people's individual and communal "I", e.g. personal, ethical, national and human "I." Here we also have some basic universal values and norms, whose evaluation, selection, compliance with or relinquishment of has to be differentiated from the issue of their "existence", expiry or revival of their natural sources; their actual withering or demise has to be differentiated from their stronger or weaker wobbliness or shakiness. At the same time, it is necessary to assume that this wobbliness or shakiness does not determine their sources, roots or their existence in general, but only shows their greater or smaller "trembling", "waving", changing of functionality and dynamics and manifold dependency on subjective factors (psychical and personal) and external ones (situational, social, cultural, civilisational, etc.).

Therefore, it can be said that our civilisation, the Western civilisation at its current crisis stage of functioning (up to a certain degree crisis), does not cause any clear moral regress in the majority of societies, social groups and professional milieus. There is no clear step back in the development of moral culture of the majority of people. What is more, this civilisation does not create any major moral stupor, any excessive weakening of moral sensitivity and reactions of human consciousness, especially in situations and circumstances which definitely require such stances, e.g. in the case of natural disasters. Quite the opposite: this civilisation constantly creates certain possibilities for moderate moral progress, and in any case it does not create insurmountable barriers.

On the other hand, the Western civilisation greatly intensifies the "wobbliness" of basic values, including moral values and norms and their "liquidity"; in principle, it does not threaten the so-called "natural morality", i.e. the main source and mainstay of these values and norms. Therefore, in relation to the comprehensively understood moral reality, its attitude is ambivalent. That is why determination of this civilisation as "liquid" does not seem to be entirely justified.[4]

In general, it can be said that moral reality (ethical theories and systems, culture and moral practice) is not, in its nature, identical with civilisational reality (technology, material infrastructure, social and political institutions, current life standards and styles). Moreover, it can be said that these two realities, separate and yet interdependent, do not develop in parallel: civilisational progress is not accompanied by moral progress, which is exemplified by the fact that certain communities on lower levels of civilisational development obtain a higher quality of moral life than highly developed societies and that in principle their moral culture – as well as their entire culture – is mostly incomparable with moral culture and other cultural ingredients of highly developed societies.

On the other hand, one cannot deny the fact that there is a certain interdependency of specific types of civilisations and moralities which has been shown by social and cultural anthropology and the history of culture and civilisation.

Generally speaking, certain types of civilisations and stages of their development have been conducive to morality: they revived it and stimulated it in various manners – and they are still performing this role.

Other types of civilisations have been fulfilling – as it turns out – an ambivalent function with respect to morality. On the one hand, they have been inspiring and cultivating morality, providing it with strong developmental impulses, good conditions for its successful shaping, yet on the other they have been slowing down its progressive changes or even deforming and destroying it (obviously only these aspects that can be deformed or destroyed, as not everything in morality, as we tried to show above, can be completely deformed or destroyed).

The technical civilisation of the West at its current post-industrial and post-informative, "liquid" and "post-modern" stage of its development exemplifies this type of ambivalent attitude to morality.[5]

Speaking pictorially, economic, political and informative realities, as well as globalisation processes that overcome this civilisation definitely "beset" the moral reality and subjugate it brutally and uncompromisingly; these processes often make the moral reality a kind of "ghetto, where basic ethical values and principles are pushed in mechanically and involuntarily, yet quite effectively."

The globalising economic system, which not only causes and increases exploitation of people by people, breaks inter-human solidarity on a local and international scale, generates and aggravates social injustice, opens the gap between poverty and affluence even wider, has ominous power in this area. It is also responsible for something that is socially and morally worse, i.e. it creates situations of exclusion from this system of growing groups of humans and transforms people into "disposable creatures" who are unwanted and ready to be thrown away and who are, at the same time, helpless with respect to this system.

This heralds the birth of new social and economic alienation which is very dangerous for people. It deprives more and more people not only of decent conditions to live, but also strips them of their basic rights and dignity, and intensifies not only the "wobbliness" of basic moral values, but also threatens (using the metaphoric terminology adopted in this text) the roots of morality as such (morality can be "uprooted" for some time in certain areas and in certain conditions).[6]

In general, we live in such a civilisation and at such a stage of its development when there is great intensification of various threats to morality and when the thesis that human nature is ambivalent with respect to morality and has not been disparaged (i.e. human nature is at the same time moral and immoral and both immorality and morality are the integral ingredients of human existence, or immorality is, in a certain sense, an inseparable part of morality, just as "unnaturalness" is, in a certain sense, an inseparable element of "naturalness").

As a conclusion to these divagations, it can be stated definitely that a present-day man remains, in spite of numerous difficulties created by modern civilisation with respect to his spirituality and in spite of his ethical ambivalence and increasing "moral wobbliness", a "homo ethicus" or *"homo moralis"* and that this human status is supported by the peculiar feature of his nature, which is called the "natural morality."

THE CONCEPT OF "MORAL WOBBLINESS"

Transformations in the sphere of ethical values in the present-day world are – as we tried to show above – very complex and dynamic. Their impact on the quality of collective and individual life of many people from almost all societies of the globalising world is constantly increasing. However, the terms used to describe this very important and characteristic process of axiological, ethical and moral changes (intensifying on a daily basis and increasing their speed), such as "crisis", "relativisation", "destruction", "collapse", etc. are not sufficiently precise and adequate; they are not satisfactory with respect to their informative content. It is easy to demonstrate their heuristic and cognitive weakness at some examples (which will be presented in a further section of this article). However, it does not mean that they are completely useless and redundant with respect to describing the features and properties of the process that is of interest to us. The category of "moral wobbliness" is going to be helpful in the description and cognitive explanation of this process. It is adopted here as the leading category.

The author of this article will try – on the one hand – to provide this metaphoric term, derived from colloquial language, with an accurate connotation which, at the same time, will be wide enough to explain its character and the core of the process; on the other, attempts will be made to show its complexity and multi-dimensional nature, which is one of the purposes of such investigation. Therefore, the main course of changes in ethical and related values (cultural, aesthetic, customary, etc.) in the contemporary world will be called "wobbliness."

In order to avoid a potential misunderstanding and possibly even a surprise, it is necessary to emphasise at the very beginning that this term will be used to determine one of the most characteristic and, at the same time, crucial type of changes in the contemporary world of values, especially moral values and moral life in general. It is not the only change and probably not the most important one, yet it is very significant and extensive. It is introduced here in order to avoid numerous misunderstandings and ambiguities which commonly appear in present-day debates on changes in ethical and other values.

Therefore, the attempt at more precise determination of the term "moral wobbliness" will commence with a statement that it does not refer to the existence and the functioning of basic (universal) values; these values are deemed relatively durable; the term "wobbliness" refers only to their changeable relations with the human agent: their intellectual and emotional reception, internal acceptance, impact on personality of their recipients, the recipients' life stances, life style, behaviour and individual and communal existence of people in contemporary times.

Speaking more precisely, the category of "wobbliness" with the connotations adopted here determines, first of all: extension or narrowing, increase or decrease (in current social, cultural and civilisational contexts) of domains of basic moral values (practically accepted and factually recognised), as well principles and obligations resulting from them; secondly, it denotes the "yielding" under the weight of external factors (economic, political, demographic, ecological, etc.) and simultaneously the "straightening up" (in analogy to Pascal's reed) of the substantially vital – in a given place and time – "backbone" of order of elementary values and standards of behaviour; thirdly, it refers to the brightening or the darkening of the importance and significance of objective criteria of basic goods and moral choices; it denotes the integration or the disintegration of existing cohesions (verified historically and practically), systems of values and rules of conduct; fourthly – and this is the most important meaning of the category of "wobbliness": it denotes (it can denote) an increase or a decrease in readiness for personal acceptance and practical implementation of a specific group of basic ethical values and principles; it is the growing or the lowering of the level of proper understanding and appreciation of the role and the meaning of moral culture in human life; it is the awakening or the "freezing" of personal and collective moral sensitivity; it is the opening onto or the isolation from other people; it is the revival or the withering of empathic, altruistic and humanist reactions in human interactions; it is the revival or the drying up of humanitarian and caring tendencies and motivations of individual persons or entire human communities.

Obviously, "moral wobbliness" does not denote all changes in morality; it does not encompass – because it cannot – all types of its changes and transformations. It is restricted to – as emphasised above – a limited, yet very characteristic and important (especially in the context of contemporary civilisation), manifestations of changes. Speaking strictly, it is limited to violations of the general condition and stability, as well as periodical weakening of vitality, strength and functionality of the moral factor in the individual and collective life of the contemporary man, i.e. it denotes a temporary impairment of its role in people's lives.

Primarily, "moral wobbliness" denotes periodical shaking and weakening, narrowing and dilution or excessive singling out of subjective references to the existing world of ethical values and temporary extinguishment or suppression of the dynamics and vitality of elementary tendencies and ingredients of the "natural morality", i.e. the predisposition and moral inclination that have been shaped and became rooted in the human culture through a complex process of human development, in the course of biological, cultural, social and civilisational evolution.

A general premise (as it will be explained in detail in a further part of this article) is a thesis in line with which the aspects that are relatively durable and common for a given species (i.e. the system of common and universal values and principles and natural human reactions, feelings and moral stances or – using different words – moral predispositions and inclinations that are rooted deeply in the constitution of every human being) are subject to staggering, impairment and even collapse in our "natural morality."

This type of changes which is called the "moral wobbliness" here is primarily conditioned by specific cultural and social changes related to the current stage of development of the technical and information civilisation with its leading processes, such as globalisation and medialisation of life, its excessive pragmatisation and instrumentalisation, and such characteristic phenomena as consumerism and functional reification of inter-human relations.

Therefore, it is clearly visible that the issue of durability and moral wobbliness discussed here is not solely a philosophical and theoretical issue. It is also a scientific and empirical phenomenon and, up to a certain degree, a commonsensical and sensory/intuitive one. It is not only a clearly cognitive issue, but also an important current practical problem. In relation to this, this text devoted to it does not have a strictly philosophical nature, but is, in a certain sense, a multi-disciplinary cognitive and practical (mainly ethical) presentation. On account of tight ties of this subject with the individual and collective life of a modern man, the issue is up-to-date, legitimate and potentially even indispensable.

THE CONCEPT OF "NATURAL MORALITY"

A characteristic trait of "moral wobbliness" – in its broad (as can be seen) meaning – is primarily the fact that in spite of greater or smaller wobbliness of internal or external moral qualities (values and principles), the "root of morality" embedded in the human nature remains practically untouched in its natural substratum. In any case, it cannot be "uprooted" easily; therefore, the "tree of morality" remains intact, disregarding the intensity and the frequency of "wobbliness" and the time and the social or existential situation in which such "wobbliness" takes place.

This primeval and relatively durable bud or, as it is called here, the "root of morality" is inherent in the human nature (excluding pathological or deformed varieties of this nature or exceptionally extreme external conditions). This ethical element, known as conscience or moral feeling, constitutes – as the author of this text discussed in a more detailed manner in his other texts – the so-called "natural morality."[7] This morality has been confirmed in numerous directions of contemporary philosophy (e.g. in phenomenology and neo-phenomenology, neo-psychoanalysis and humanist psychology, in Christian personalism and the so-called philosophy of dialogue, eco-philosophy and recentivism, etc.) and in certain social sciences (developmental psychology and personality, neuro-psychology, social anthropology, recent theories of morality and other theoretical and empirical concepts).[8]

In this article, attention will only be drawn to certain theoretical depictions of the phenomenon of human nature and the confirmation of – using the terminology adopted here – "natural morality" and the "moral wobbliness" that accompanies it in specific social and civilisational concepts in the works of selected philosophers: Roman Ingarden, Anna Teresa Tymieniecka, Peter Singer and Richard M. Hare.

Selection of such varied philosophical and ethical stances during a discussion on the issues of interest to us is not accidental. It results from a premise that various roads of theoretical and empirical cognition can lead to determination of a specific state of affairs or contribute to the solving of a given problem in numerous complex, cognitive, philosophical, scientific, commonsensical or common (natural) processes. These attempts derive from various intellectual options, points of view and manners of perceiving reality. Completely diverse theories and philosophical systems can lead to a common cognitive goal. Numerous meaningful and frequently astonishing examples of this characteristic and slightly surprising epistemological "affliction" are provided by modern physics,[9] most recent philosophy[10] and, in particular, historical and modern ethic as a cognitive domain.[11] It turns out quite frequently that various scientific theories, differing philosophical concepts and varied ideas and ethical tendencies have a given "common denominator" and "similar points of access", which will be exemplified in a further part of this text.

For example, the issues of "natural morality" and "moral wobbliness" are the meeting point for R. Ingarden, A.T. Tymieniecka, P. Singer and R.M. Hare. Therefore, it seems worthwhile to take a look at this meeting of diversity and to follow not only the obvious differences in the presented standpoints, but also the puzzling, and, slightly astonishing, close (though unintentional) relations and similarities among them.

"NATURAL MORALITY" AND "MORAL WOBBLINESS" IN A PHILOSOPHICAL APPROACH: SELECTED STANDPOINTS

Roman Ingarden's Concept of Man and "Natural Ethics".

Roman Ingarden, the co-creator of a well-known classical variety of phenomenology and author of famous philosophical works in his ontological, anthropological and ethical investigations (relying on a specific method of philosophical and experimental cognition – the so-called internal and external experiences) examines the specific, real and, at the same time, primeval "being-ness of man", his unique human nature, which is the core and the centre of the human "I", the personality and subjectivity and the source and the foundation of an extensive sphere of his creative acts and deeds, including moral deeds. The analysis of the character of this type of examination of subjective and objective reality is omitted here, along with its special ingredient, i.e. a man understood as a bodily, psychical and spiritual being.

The above-mentioned acts and deeds cannot always be assigned to a specific human agent; nevertheless, they are perceived by the philosopher as durable and unchangeable natural "equipment" of a human being as such, his durable and continuously up-dated dispositions and skills. In individual periods and circumstances of human life, they are subject to various changes and modifications; they undergo a process of intensification or expiry, yet they are always the real and the natural property of every person. They are the real feature of every human "I"; they are people's inseparable element, an inalienable feature and "disposition", similarly to the immanent liberty of human beings and responsibility that is integrally related to this liberty.

Man, a specific man (the phenomenological approach to cognition offers only the idea of a specific man) is a one-time bodily-psychical-mental creature with specific and temporaneously limited "continuity of existence" in which there is a constant exchange of matter between the body and its environment. From birth to death there are new physical and psychical processes constantly taking place; new cells are created and destroyed and the "characteristic and regulated developmental process of growth, maturity and aging until decay or disintegration of parts [...]"[12] is taking place. This relative "continuity of existence" is the fundamental condition of the entire spirituality of human beings, their internal "I", of all creative deeds and moral stances, liberty and responsibility, intellect and thought, will and action, sensitivity and emotionality.[13]

The primeval nature and the spiritual subjectivity of human beings dependent upon it also release various interpersonal references in the "continuity of existence": social, moral, emotional, etc. and create various references to values, including ethical values (virtues) such as justice, bravery, love, fairness, faithfulness, faith, hope, mercy, impartiality, nobleness, responsibility, etc.

It is worth emphasising that R. Ingarden accentuated the specific nature of ethical values. They cannot be reduced to utilitarian values, vital values or to hedonistic or moral values. They belong to the reality which is created by man by his conscious creative acts; man transforms this reality or provides it with new meanings. The subject of these values is always the human personality; they result from human deeds, acts of will, decisions, etc. Decisions of people with respect to a specific manner of conduct or approval or disapproval of own deeds are decisive for the creation of values. They are fully encompassed by the sphere of activities of a human being. They are, however – as has been mentioned before – real, i.e. people fight for them, they seek them or even die for them, yet these values exist independently from subjective determination of specific persons.[14]

Ingarden's concept of a human agent and the world of values assumes the possibility and, at the same time, the inevitability of "wobbliness" of references of a specific human agent with respect to the world of values (ethical, social, aesthetic, etc.), which is ontically independent from people, yet also co-created by them. At the same time, this concept confirms the values' durable "rooting" and one-off dependency upon the primeval and "reviving" human nature. The "reviving" of human nature takes place in individual persons and their respective lives. The elements of "natural morality" are positioned within the realm of primeval nature.

ANNA TERESA TYMIENIECKA'S UNIVERSALIST ETHICS AND THE ISSUE OF "NATURAL MORALITY" AND "ETHICAL WOBBLINESS"

A similar stance, yet explained and justified in a completely different manner (metaphysically- anthropologically and, in a certain sense, politically) in philosophical anthropology and ethic is represented by Ingarden's pupil, Anna

Teresa Tymieniecka.[15] In majority of her philosophical ideas, Tymieniecka is independent from her master; she is the leading representative of the contemporary neo-phenomenological movement and the author of numerous fundamental works in almost all areas of philosophy.

Similarly to Ingarden and many other outstanding modern philosophers (who are mentioned in a further part of this text), Tymieniecka provides serious and suitable explanations and justifications for the thesis adopted and developed here, in line with which the characteristic shakiness and wobbliness of moral stances and ethical order which is almost common in the modern times, but does not yet entail (and in principle cannot entail) a decline of their fertile soil, the source and relatively durable back-up for vitality and further development, in spite of the shakiness and wobbliness of elementary feelings, sensitivity, reactions and moral behaviour, as well as basic, socially and historically shaped, ethical values. Moral culture, which is nowadays subjected to strenuous tests (thence the discussed wobbliness and shakiness), is not only an issue of more or less successful socialisation and proper, but not sufficiently good, education or programmed learning (e.g. via ethical and cultural education), but it is also an issue of more or less skilful and spontaneous derivation of indispensable elements and ethical impulses from the source of primeval natural morality, which is still rich and vital (as the author of this text is trying to show in this article). This morality is potentially and, in general, strongly embedded in evolutionary and historically shaped human nature.

A. T. Tymieniecka also provides direct or indirect support for the thesis of the author of this article that in contemporary times we are dealing not so much with "a crisis of morality" but – temporarily – with "a crisis in morality"; a crisis of various manifestations of its social and individual functioning.[16]

A brief discussion of Tymieniecka's standpoint with respect to this issue is presented below.

A human being is, by its nature, a moral agent. At the same time, human beings are creative agents and cognising subjects. An integral ingredient of a human being's individual morality is the so-called "moral sense of human condition", which is realised in the so-called "source experience." It is tied actively and in manifold ways with the intelligible sense, the aesthetic sense and the sacral sense. All of them are the manifestations of the Logos of life. The above-mentioned senses are the basis of primeval experiences (moral, cognitive, aesthetic and religious) of their agent, which bear fruit in subjective morality, science, philosophy, art and artistic creativity and in religious beliefs. The "Logos of life", in which all potential primeval sources of all morality are embedded, is the manifestation of the cosmic Logos; on the other hand, the cosmic Logos is the manifestation of the eternal Logos (nature). Therefore, morality, as well as other virtualities of the human soul (cognitive, aesthetic and religious), appear at a specific stage of evolution of the Universe, in the process of continuous beingness, i.e. in the course of "ontopoiesis." A human entity is pervaded with the rights of the Cosmos. In its being, existence, actions and behaviour – also in moral acts and behaviour – it is a microcosm.

Morality embedded in the moral sense has a social nature. It is rooted in the social Logos and appears in relation towards others, e.g. in the relation "I" – "you" and "I" – "they", implementing various modalities of life, such as solidarity, intimacy, affiliation, guilt and others. In essence, it does not need any external principles and rules of conduct or specific standards and ethical codes. They are necessary mainly on account of their insufficient attractiveness, stiffness and a tendency to unify human choices, decisions and stances. The moral sense is autonomous with respect to any normative ethic and social rules; it is constantly developing and has not been finally shaped; it is subconscious and spontaneous and in its activity it is supported by the intelligible and aesthetic sense, thereby gaining certain rationality, sensibility and "beauty", as well as purposefulness.[17]

The moral sense may influence the sense and the quality of the individual and collective life; it may open the human agent onto another people, recognise inter-human relations and introduce the feeling of kindness and justness into human stances. First of all, the moral sense may show a sense of moral conduct, stimulate ethical evaluations, i.e. open the platform for the functioning of the moral conscience which, according to Tymieniecka, is a "deliberating and justifying factor" that expresses care for another human being and considers what we owe each other, taking into account both the welfare of individual persons and the collective life (the individual and the general welfare). This last attitude encompasses motivation of individuals to live socially and to establish social assistance institutions and to practice inter-human justness and solidarity.

In this place, it is necessary to mention one more thesis of A. T. Tymieniecka's moral philosophy, i.e. the statement that moral sense is closely related to the issue of natural human rights; these are rights vested in every person on account of their ontic status.[18]

In the context of such versatile and unfinished "interpretation" of the world and the man by A.T. Tymieniecka, which is performed in constant tension and cognitive effort, the concepts of the "moral sense" and the natural morality are being deepened and enriched. This is how the fundamentals of modern subjectivity and conscience ethic are being created. This ethic is individualised, subjective, personal and autonomous; it is not related to codes; it is strictly personal and interpersonal; it is subjective and yet, at the same time, pro-social and, in a certain sense, ecological and global. This is how the project of a new humanist constructive ethic (yet not normative) is being created. This ethic not only "reflects" the character and the transformations of the contemporary moral reality in a certain degree; it also discloses a significant ability for becoming a part of the main trends and tendencies of such transformation. Because of these aspects, this is the ethic of "here and now" and is functionally vital. This is the ethic that is able to build a strong support for ethical optimism, which is so needed these days. It sustains the belief about constant relation of ethics to the human nature and to the currently threatened humanity.[19]

PHENOMENOLOGY OF LIFE OF A.T. TYMIENIECKA AND BASIC ETHICAL PROBLEMS

A question is raised about the name that should be given to the ethic contained in the above-presented complex of metaphysical and anthropological thoughts of A.T.

Tymieniecka, called the "phenomenology of life."[20] How should it be positioned in the wide spectrum of contemporary ethical systems? And first of all, how should its main goals and tasks be interpreted and how to answer the question about its relation to the subject matter of this article?

Let us try to answer these questions.

First of all, this ethic has neo-classical character with clear references to ancient ethics, in particular Aristotle, and to ethical concepts of outstanding representatives of numerous directions representing later European ethical tradition: Kant, Schopenhauer, Bergson, M. Scheller, R. Ingarden, E. Levinas, E. Mounier, T. de Chardin, A. Schweitzer and other contemporary morality philosophers.

On the other hand, this ethic is radically modernist, with bold use of modern ethical ideas of leading representatives of ecological and globalist ethics, *inter alia* H. Rolston, T. Regan, P.W. Taylor, P. Singer et al. These diverse and rich sources of the discussed ethic, especially the classical and modern ones, influence the circle of its basic principles, values and the main goals and tasks set before it.

As far as principles and tasks of the discussed ethics are concerned, attention should be drawn to the fact that the most important ones (the classical tendency is revealed here) are these which are meant to shape the internal harmony of people, their versatile spiritual development, in three basic spheres: intellectual, moral and aesthetic. They are used to protect and stimulate the creative activity of people and people's self-creation, as well humanisation of life and inter-human relations.

These are primarily such principles and values as life and self-individualisation of life, intellect as the main signpost for human conduct (including moral conduct), measuring and creative wisdom, courage and moderation, prudence and tolerance, restraint and deliberation, order and harmony, justice and spirit of joint activity, existential solidarity, the greatest happiness principle, the golden rule, the highest middle way and the right measure, the human moral excellence, and the requirement to counteract evil and brutality – in defiance of the destructive instinct embedded in the human nature.[21]

To these, so to speak, classical principles and ethical values derived – as can be seen – from greatest and most universal systems from the past, A.T. Tymieniecka's ambitious or even heroic ethics (ethics focused mainly on vital values) adds several new, yet, in a certain sense, also universal principles and ethical and pro-ethical values from the contemporary trends of ethical thought, mainly ecological, environmentalist and globalist. These are:

- life in harmony with "Nature";
- acceptance of responsibility for Nature, in particular for the forms and manifestations of life developing in it: starting from the life of minerals, plants, animals and ending with human life;
- discontinuation of thoughtless and catastrophic "hurting" and devastation of the eco-system; irrational wasting and destruction of the soil, water, air, mineral, plants and animals;
- acceptance of the fact that we are not vested with – in reference to Nature – special rights and claims, and that we are encumbered with liabilities and obligations

towards its non-human "settlers", which means that in the area of Nature, we cannot do things that are only pleasant and convenient for us, and solely take care of our interests and specific human goals; we also have to take into account the "rights" and interests of all other creatures, i.e. we have to take care of the common welfare of the Planet;
- acceptance of the role of a submissive carer of Nature, wise and prudent, with broad imagination and possibly accurate predictability, its intelligent and rational manager; a good and wise farmer with respect to its resources; an effective "defender" of its riches;
- acceptance of the role of somebody who can secure its stability and balance, in particular the role of a "guard" or a "custodian of everything that is alive";
- acceptance of the role of a "guardian" assigned to ensure balance and harmony of the entire biosphere, welfare of its individual elements, and therefore welfare of the entirety; a "guardian" assigned to remove, as far as possible, disharmony and destructiveness, disorder and chaos; a "custodian of life equilibrium".

In other words, man in relation to the ecosystem and to other people should primarily be the "moral creature" (ethical man, *homo ethicus*).

The term "moral man" has a double meaning here: normative and empirical. In each of these meanings, man is a creature with a constructive reference to the phenomena of moral shakiness and wobbliness, moral disorder and decay. In the first case, by means of consistent acceptance and fulfilment in the ontopoiesis of life of a wide spectrum of the above-listed principles and moral obligations, man effectively becomes the virtuous man. It is easy to notice that we are dealing with elements of virtue ethics. In the second case, the man who determines himself morally and obtains his ethical identity – in situations of moral shakiness and wobbliness – acquires an ability to have a complete moral life thanks to the specific moral qualities vested in his nature, in the form of feelings, emotions, inclinations, interests and moral motivations. Man acquires human moral excellence, which a peculiar synthesis of advantages and virtues of man as a moral being.[22]

At the end of this brief summary of A.T. Tymieniecka's philosophy of life and ethic, let us try to interpret her main goals and show them in the perspective of desires that are involved in the practice of human life in other contemporary ethics.

If we interpret the intentional and teleological side of the discussed ethic correctly, the main and, at the same time, specific goals of A.T. Tymieniecka's ethics are:

- demonstration of groundlessness of currently fashionable and influential (especially in certain intellectual circles) ideas about clear and constantly deepening disorder and chaos of almost everything in the contemporary world, including disorder and chaos in the world of moral and other values and the pessimistic (even catastrophic) mood that accompanies them.[23]

Going against the tide in relation to numerous contemporary ideas and the general "climate", A.T. Tymieniecka justifies her different and moderately optimistic standpoint on the basis of metaphysical concept of the Logos; her concept is developed in the major work entitled "Logos and Life" [2000],[24] where she formulates a thesis in line with which "listening" to the voice of the universal Reason, omnipresent in the entire "living Cosmos" (including the human life), and being guided by its intuitive and experimentally given guidelines leads to desired and sensible choices, provides rational criteria for choices and moral evaluations and proper life orientation. It also facilitates departure from temporary hesitation and moral wobbliness, rationalises the ontopoiesis of life and leads to moral progress.[25]

The above thesis, as well as all other statements with metaphysical character, can be deemed disputable and empirically unverifiable and maybe due to this A.T. Tymieniecka refers to empirically confirmed facts whilst justifying further specific goals of her ethics. She refers to the elements of natural and anthropogenic morality, which are described in various manners, yet are unanimously confirmed in developmental psychology and personality psychology; these elements take the form of specific predispositions, tendencies and moral "skills." According to the author of the "Logos of Life" these are – as emphasised above – moral sense, i.e. human ability to have moral reactions and inclination to do good things, "reasonable measure" and "moral measure" and the "human moral excellence", etc.

All these dispositions and moral inclinations allow man, in majority of life-time situations, dilemmas and moral choices, etc. to "be moral", in spite of difficulties, hesitations and wobbliness with respect to values, temporary collapses and regressions in the ethical stance. They allow people to practice the principle of the "happy medium", harmony and equilibrium; this principle is particularly important in the epoch that is constantly balancing between drastic extremities in numerous areas of human life, e.g. between a radical relativism and moral nihilism and various types of extreme absolutism and ethical fundamentalism.

Both the "obedience" of the cosmic vital Logos and "identification" with its tendencies, as well as references to natural moral potential (this potential, according to the author, is evidently and durably present in the human nature) allow for reaching for certain specific and sceptically perceived ethical goals, such as looking for the lost "compass of life", i.e. a sensible goal for the human existence, the humanising process of life and, in spite of increasing difficulties, obtaining "moral progress" (in an individual and social dimension) in the individual and collective life, developing deeper spiritual life, cultivating practically applied moral virtues, or – in general – setting the civilisation on an ascendant course, i.e. saturating it with authentically humanist values of life, with better and wiser moderation in the sphere of decisions and choices, with a permanent desire for knowledge and satisfaction of this truly human desire, with appreciation of the values of higher culture and ongoing aspiration to the objective truth and to the life's equilibrium.[26]

ETHICS OF PETER SINGER AND RICHARD HARE VS. "MORAL WOBBLINESS"

Theoretical concepts and ideas of two outstanding ethicists from a California university (educated in Oxford), i.e. Peter Singer[27] and Richard M. Hare,[28] are similar to the ethics presented above.

The former draws attention to the modern, cultural and social concept of "constancy" and relative "invariability" of human nature and morality that is related to it, and which, in the course of its natural evolutionary development (in psychological, social, civilisational and cultural context), is gradually becoming anthropologically "consolidated" and integrated, showing more and more distinctly these aspects that are common, whilst weakening and pushing aside these elements that are "various", "different" and "other." In this morality, its characteristic conflicts, tensions, hesitations and dilemmas are becoming more and more assimilated. This evolutionary process clearly shows that in all types of ethics and moralities which have been created by people throughout their history and have been shaped in their subsequent communities and cultures, there is something indisputably common. There are various confirmations for the fact that people, in the course of time, are getting closer together mentally and morally; their morality and its main base – nature – have certain repetitive features, transpiring in almost all societies (small, large, developed and under-developed, ethnically and culturally diversified and uniform). This is not contradictory to the ongoing development (progress) of morality practiced in life and in theoretical ethical thought. What is more, there is a certain similarity between selected features of human nature and the nature of "long-living", "intelligent" and "social" mammals.

In this respect, one of the statements of P. Singer gains particular significance. Singer claims that "ethics is not [...] a senseless collection of fragments assigned to various people at various times. In spite of historical and cultural differences in beliefs with respect to moral obligations, our beliefs are drawing closer together. Nature has its fixed features and there are only few manners of co-existence of human beings and their development."

"In fact," continues P. Singer "certain features of human nature are repeated in all societies and they are common to all long-living, intelligent and social mammals. These features are revealed both in our conduct and in the conduct of all primates."[29]

Societies that are healthy in the sphere of collective psyche, mentality and social character and which do not succumb to significant disintegration and decay trends are becoming more and more alike with respect to their moral culture, in spite of their sizes and ethnical and other differences. This process also encompasses the area of values recognised by them, the hesitations that they experience and their moral dilemmas.

"[...] the aspects which, in a given society or in a given religious tradition," says Singer "are considered virtues, are probably virtues in other societies; moreover, a group of virtues esteemed in the great moral culture will never be a basic part of a collection of moral vices in another culture." Exceptions from this rule are short-lived and they refer to societies at the stage of collapse or final decay. On the other hand, within the scope of every tradition, we observe the same fluctuations[30] and the same manifestations of "moral wobbliness."

Moral similarity growing in the historical and cultural process between various societies is also related to increasing similarity in the content of various ethical concepts and systems. This is the case with respect to Western culture, e.g.:

"The history of Western philosophical ethics shows that [...] starting from most ancient thought – the Greek thought – until the modern times we can find the same old beliefs and the same old disputes once in a while."[31] In relation to this, "[...] we are able to reach an agreement with respect to the basic sense of good and evil, just as we reached an agreement in other areas of intellectual life."[32] It turns out that the so-called golden rule is [...] a central category in several great ethical systems.[33]

The thesis about similarities and closeness of major ethical directions became one of the main investigation threads in the so-called "universal prescriptivism" created by the famous contemporary ethicist, Richard M. Hare.[34]

According to this thesis, adopted as a leading investigation premise within the scope of ethical issues, attempts are made at identifying and determining the common features of most important ethical directions and working out a certain constructive synthesis (a general "ethical theory") confirming and justifying in it, logically and empirically, ethical universalism.[35]

"Universal prescriptivism," explains R.M. Hare "is an attempt at determining which errors and which accurate intuitions are hidden in other common ethical theories. This allows for avoiding mistakes in every such theory, at the same time preserving their accurate intuitions and allowing for their synthesis."[36] On the other hand, the term "ethical theory" denotes an attempt at making the content of questions about morality more precise. What is the meaning of sentences used in a moral discourse? What is the nature of moral terms or morality itself? If these attempts have a successful outcome, we will obtain epistemological data that is important for the ethical theory: aspects that we could we rely on when giving rational answers to our moral questions. Maybe there are no such answers; maybe there is only our moral feeling or customs imposed on us. On the other hand, "if we can consider moral problems in a rational manner, it would mean that there has to be a certain moral truth or facts that can be discovered."[37]

This relative ethical universalism, i.e. the thesis about similarity and repetitiousness of certain ideas and ethical concepts in various systems of ethical thought, reflects and confirms relative moral universalism, i.e. similarity and repetitiousness in various societies and standpoints of majority of people (however, not all) of basic feelings and ethical intuitions, reactions and moral behaviour, or identical or similar ethical doubts and dilemmas.

In R.M. Hare's "universal prescriptivism", special attention should be drawn to the emphasis on certain foundations which do not lose their significance in the currently undertaken evaluations, choices and types of moral behaviour, i.e. elements of every type of "living" morality that is practiced in life and not only expressed intentionally and verbally. The emphasis is on the great, and even decisive, role of practical wisdom and common sense in moral life (similar to Aristotle's *phronesis* or the practical reason of Kant and similar concepts of other thinkers) and the so-called "wisdom of the ages", i.e. an accumulation of verified intuitions, thoughts and individual knowledge and evaluations of moral situations which have been shaped in life experiences of people (both the individual and collective).

Historically shaped practical wisdom of life and the "wisdom of the ages", only seemingly aged and old-fashioned, cannot be omitted in modern evaluations, choices and moral stances, if such acts are to be relatively accurate, just and decent. What is more, they are vested with a so-called ethical authority. This derives from the fact that the "wisdom of the ages [...] is a result of reflections of many people in various situations."[38]

In his "universal prescriptivism" R.M. Hare also formulates certain valuable, even though disputable and difficult to apply in practice, praxological recommendations for moral conduct and, at the same time, rules of moral self-determination in various situations and inter-human relations or, using the category of "moral wobbliness" applied here, extrication from the condition of wobbliness and shakiness of moral stances.

These are some of these recommendations:

- it is necessary to work out such dispositions and features that are conducive to the conduct recommended by an "impartial thinker, capable of a perfect critical examination of moral issues";
- it is necessary to develop such dispositions and features which – in case there is such need – would give us a skill of practicing spontaneous, intuitive or even involuntary moral acts or conduct, especially when lack of time makes intellectual examination impossible;
- critical deliberation of a moral deed should be undertaken only when moral attitudes, worked out previously by general intellectual dispositions, are in conflict, "even though we will doubt our potential even then";
- we should "develop the same [moral – J. Sz.] intuitions to which intuitionists were making references throughout the history of ethics and morality and strong inclinations in order to pursue such intuitions and other moral feelings (for example love) which will strengthen these intuitions";
- it is necessary to assume that "moral convictions common to thinking people are those which should be nurtured";
- it is necessary to comply with the "golden rule" (of Kant) according to which "it is necessary to act towards others in a manner that we would like them to act towards us and to love thy brethren as thyself";
- in the end, it is necessary to always treat a human being as a goal and never as a means.[39]

It is easy to notice that in the above "recommendations" regarding the circumstances and the manner of making evaluations, choices, decisions and moral deeds, an important role is assigned both to the intuition and the reason, even though the former has usually the main role. In general, they combine the imperatives of Kant's ethics with ethical requirements of utilitarianism, not omitting the basic indications and categories of ancient ethical thought, e.g. the current utility of the "just desire" of Plato in the shaping of moral attitudes or the "practical wisdom" of Aristotle. It is also possible to perceive a close relation of this ethical standpoint to the main ideas of the philosophy of life and ethics of A.T. Tymieniecka. This confirms the hypothesis contained in this article that the main trends of the modern ethical thought are significantly integrated in explaining and solving of the main moral problems of modern times, including the problem of the "natural morality" and the so-called "moral wobbliness."

University of Krakow
e-mail: janszmyd@interia.pl

NOTES

[1] Cf. L. Gawron (ed.), *Filozofia wobec XX wieku* (Wydawnictwo Uniwersytetu Marii Curie-Skłodowskiej: Lublin, 2004); Zygmunt Bauman, *Liquid Modernity* (Polity Press: Cambridge 2000), Polish edition: Płynna nowoczesność, trans. Tomasz Kunz (Wydawnictwo Literackie: Cracow 2006).

[2] Cf. Witold Mackiewicz, *Filozofia współczesna w zarysie* (Wydawnictwo Uniwersytetu Warszawskiego: Warsaw 2008); Józef Bańka, Wiesław Sztumski, *Ekorecentywizm jako idea ochrony środowiska człowieka współczesnego* (Wydawnictwo Naukowe "Śląsk": Katowice 2007).

[3] Cf. Henryk Skolimowski, *Wizje nowego millenium* (EJB Wydawnictwo: Cracow 1999); Anna Teresa Tymieniecka, "The Fullness of the Logos in the Key of Life, Book I, The Case of God in the New Enlightenment," *Analecta Husserliana. The Yearbook of Philosophical Research. Volume C* (Springer: Dordrecht 2009).

[4] Cf. Jan Szmyd, " 'Odczytywanie' współczesności – możliwości, ograniczenia, funkcje społeczne i życiowe," in *Państwo i Społeczeństwo VII, No. 3* (Cracow: 2007) pp. 7–16.

[5] J.F. Collange, C. Mengus, *Communication et communion: perspectives theologoques et ethics* (Medias et charite: Paris 1987) pp. 95–97.

[6] Halina Promieńska (ed.), *Etyka wobec problemów współczesnego świata* (Wydawnictwo Uniwersytetu Śląskiego: Katowice 2003).

[7] Cf. Jan Szmyd, "Kryzys moralności w świecie ponowoczesnym" in *Edukacja Filozoficzna* (Uniwersytet Warszawski, Instytut Filozofii 46: Warsaw 2008) pp. 57–76; *idem*, "Pogoda dla etyk sumienia – nie dla kodeksów i nihilizmu moralnego" in *IDO – Ruch dla Kultury*, Vol. 10, pp. 11–16 and other works of this author.

[8] Cf. *inter alia* Tadeusz Kotarbiński, *Sprawy sumienia* (Warsaw, 1956); Włodzimierz Szewczuk, *Sumienie, Studium Psychologiczne* (Książka i Wiedza: Warsaw 1988); Józef Bańka, "Sumienie jako poręczenie moralne wyboru najlepszego" in *Etyka wobec problemów współczesnego świata*, Halina Promieńska (ed.) op. cit.; Peter Singer, *One World. The Ethics of Globalization* (Yale University Press, 2002), Polish translation *Jeden świat. Etyka globalizacji* (Książka i Wiedza: Warsaw 2006).

[9] Cf. R. Penrose, *The Basic of Quantum Mechanics* (Cambridge University Press: Cambridge 2000); P. Williams, *Uncertain Journey* (Publishing House: New York 2001); Janusz Czerny, *Czy prawo Moore'a detronizuje osobę ludzką* (Wydawnictwo KOS: Katowice 2005).

[10] Cf. A. Bronk (ed.), *Filozofować dziś. Z badań nad filozofią najnowszą* (TN KUL, Lublin 1995); Zygmunt Bauman, Keith Tester, *Conversations with Zygmunt Bauman* (Polity Press in association with Blackwell Publishers Ltd. 2001), Polish edition *O pożytakch z wątpliwości. Rozmowy z Zygmuntem Baumanem*, trans. E. Krasińska (Wydawnictwo Sic!: Warsaw 2003); *Człowiek i Świat. Współczesne dylematy. Rozmowy Zdzisława Słowika* (Biblioteka "Res Humana": Warsaw 2007).

[11] Peter Singer, *A Companion to Ethics* (Basil Blackwell Ltd. 1991), Polish translation: P. Singer (ed.) *Przewodnik po etyce* (scientific editor of the Polish edition: Joanna Górnicka) (Książka i Wiedza: Warsaw 2000); H. Promieńska (ed.), "Etyka wobec problemów współczesnego świata" op. cit.

[12] Roman Ingarden, *Książeczka o człowieku* (Cracow, 1987), p. 114.

[13] Ibid., p. 121.

[14] Cf. Roman Ingarden, *Wykłady z etyki* (Warsaw, 1989), p. 274.

[15] The main philosophical work of A.T. Tymieniecka is a four-volume treatise entitled "The Logos of Life" published between 1988 and 2000. On the other hand, one of her last works is entitled "The Fullness of the Logos in the Key of Life". Book I. The Case of God in the New Enlightenment (Springer: Dordrecht, The Netherlands 2009).

[16] Cf. Jan Szmyd, "Post-Modernism and the Ethics of Conscience: Various 'Interpretations' of the Morality of Post-Modern World. Role of A.T. Tymieniecka's Phenomenology of Life" in *Analecta Husserliana, The Yearbook of Phenomenological Research, Volume CV. Phenomenology and Existentialism in the Twentieth Century. Book 3. Heralding the New Enlightenment* edited by A.T. Tymieniecka (Springer: Dordrecht, Heidelberg, London, New York 2000), pp. 111–122.

[17] Cf. A.T. Tymieniecka, "The Moral Sense. A Discourse on the Phenomenologiae Foundation of the Social World," in *Analecta Husserliana* (1983), XV, pp. 3–78; *eadem* "The Moral Sense and the Human Person within the Fabric of Communal Life. The Human Condition of the Intersection of Philosophy,

Social Practice and Psychiatric Therapeutics. A Monographic Study" in *Analecta Husserliana* (1986) XX, pp. 3–100.

[18] Cf. A.T. Tymieniecka, *The New Enlightenment. A Review of Philosophical Ideas and Trends*, Vol. 32 (Hannover, NH 2008), pp. 3–4.

[19] Cf. A.T. Tymieniecka, "Czy istnieje świat? Nowe spojrzenie na podstawy sporu Husserl – Ingarden – rozważania ontopojetyczne," in *Roman Ingarden i dążenia fenomenologów w 110 rocznicę urodzin Profesora*. Post-conference materials prepared by Czesław Głombik (ed.) (Wydawnictwo Gnobe: Katowice 2006), pp. 36–46.

[20] The full names of this concept of thought: "Phenomenology of Logos and Life and Human Condition."

[21] Cf. A.T. Tymieniecka, *The New Enlightenment*, op. cit.

[22] Ibid.

[23] Ibid.

[24] Ibid.

[25] Ibid.

[26] Cf. Carmen Cozma, "Anna Teresa Tymieniecka's ethics: an inspiration for the contemporary world," in *Phenomenological Inquiry. A Review of Philosophical Ideas and Trends*, edited by Patricia Trutt-Coohil, (October 2009), XXXIII, pp. 23–34.

[27] Peter Singer, a philosopher, ethicist, one of the most outstanding and influential contemporary intellectuals; professor of philosophy and director of the Centre for Human Bioethics in Monach University in Melbourne, professor of bioethics at the University Center for Human Values at the University of Princeton, co-editor of international magazine "Bioethics," author of numerous works on ethics, bioethics, global ethics and social philosophy, including: *Democracy and Disobedience* (Clarendon Press: Oxford 1973); *Animal Liberation: A New Ethics for Our Treatment of Animals* (Aron Books: New York 1975); "Animal and the Value of Life" in *Matters of Life and Death* (Random House: New York 1980); *Practical Ethics* (Cambridge University Press, 1979), *The Expanding Circle* (Ferrar, Straus and Giroux: New York 1981); *A Companion to Ethics* (Basil Blackwell Ltd. 1991) Polish edition *Przewodnik po etyce. Pod redakcją Petera Singera*. Scientific editing of the Polish edition: Joanna Górnicka (Książka i Wiedza: Warsaw 1998); *One World. The Ethics of Globalization* (Yale University Press, 2002), Polish edition: *Jeden świat. Etyka globalizacji*, trans. Cezary Cieśliński (Książka i Wiedza: Warsaw 2006).

[28] Richard M. Hare, one of the most outstanding modern morality philosophers, professor at the University of Florida, retired professor at the University of Oxford, author of famous works in the area of philosophy and morality theory, including: *The Language of Morals* (Clarendon Press: Oxford 1952); "Waiting – Some Pitfalls" in *Agent, Action and Reason*, R. Binkley (eds.), (Toronto University Press, Basil Blackwell: Oxford 1971); "Nothing Matters" in *Applications of Moral Philosophy* (University of California Press: Berkley 1972); *Moral Thinking: Its Levels and Points* (Oxford University Press, 1981); *Essays in Ethical Theory* (Oxford University Press, 1989).

[29] P. Singer (ed.), *Przewodnik po etyce*. Scientific editing of the Polish edition – Joanna Górnicka, op. cit., p. 589.

[30] Ibid., p. 590.

[31] Ibid., p. 589.

[32] Ibid., p. 590.

[33] Ibid., p. 591.

[34] R.M. Hare, "Uniwersalny preskryptywizm" in *Przewodnik po etyce*, P. Singer (ed.), op. cit., pp. 499–511.

[35] Cf. Ibid.

[36] Ibid., p. 499.

[37] Ibid.

[38] Ibid., p. 510

[39] Ibid., p. 509.

MICHAEL GUBSER

A TRUE AND BETTER "I": HUSSERL'S CALL FOR WORLDLY RENEWAL

ABSTRACT

My article argues that Husserl's late phenomenology centered on an ethics of worldly responsibility. This revision marked a considerable departure from the Brentanian axiology of his earlier seminars, and it introduced a new *ēthos*, never fully developed, of ethical engagement through philosophy. In Husserl's twilight years, the world, not ego, received primary accent. This worldliness – outlined in the *Kaizo* essays of 1923–24, then buried under the egology of the late 1920s – reemerged in the 1930s *Crisis* work and surrounding manuscripts. Ostracized from Nazi society, a beleaguered Husserl raised worldly ethical concerns to new philosophical distinction, although he never wholly extricated them from either egological subjectivism or the well-known Cartesian mechanics of intersubjectivity. As a result, Husserl's late ethics, like the *Crisis* text itself, is a potent but incomplete harbinger of new phenomenological lines. My essay also suggests how Husserl's intellectual scion, Jan Patočka, appropriated and radicalized his mentor's ethics in his own phenomenological activity, elevating worldly responsibility to the pinnacle of philosophical life.

As incarnate beings, wrote the Czech philosopher Jan Patočka, citing his mentor Husserl, humans transcend their individual world and its material limits through freedom; we are, in his words, "beings of the far reaches [*bytomnosti dálky*]" who make higher commitments and bear ethical responsibility as part of our essential being.[1] "We live turned away from ourselves," explained Patočka in university lectures delivered in the late 1960s. "[W]e have always already transcended ourselves in the direction of the world, of its ever more remote regions."[2] The world and its objects manifest themselves as possibilities, and our freedom, with its concomitant responsibilities, opens a life that reaches beyond enclosed self-concern, a life that can be lived into the distances of the earth and the depths of other beings.[3] These ethical impulses, which informed Patočka's heroic dissidence in the 1970s, grew in great part from his encounter with Husserl's late philosophy.[4]

An elliptical phrase, Patočka's "beings of the far reaches" evoked Husserl's phenomenology of the lived body situated in the surrounding world. As Anthony Steinbock notes, Husserl's *Ideas II*, drafted in 1913, had already introduced the body [*Leib*] as the "zero-point" of orientation, in which all concepts of distance and direction, near and far, took root.[5] The physical, kinaesthetic sense of farness outlined there and at other points in Husserl's oeuvre hints at the wider significance of Patočka's summation, at the way that distant aspirations and endeavors redound upon subjective encounters. The "there" far away is a relative term that pivots on my near surroundings. Though close distances may be bridged and each *Dort* made *Hier*, farness ultimately presents itself as an infinite horizon of effort stretching before me. For Patočka, who met Husserl in 1929 and studied with him in 1933, this corporeality betokened a much wider ethical *ek-stasis*, as human beings were called to live beyond themselves into the world:

> [O]ur doing transcends itself in the direction of that totality to which our *ou heneka* [final cause] is the key and which, in virtue of that, merits the title of "world," the "natural" world of our life. Or, philosophically speaking, it is not this concrete context, structured by our active life, that merits the name, as much as that about the very foundations of our actual life that makes such a structuring possible – the worldhood of the world toward which the human *Dasein* transcends himself.[6]

The world is the place of human activity, and Husserl's Czech disciple saw an ethical commitment to the world beyond self as a central demand of his mentor's thought, a call to live responsibly in the far horizons of nature and society, beyond the egological self in the shared environs that co-constituted our experience. The corporeal human lived within concentric rings of worldliness: The carnal body inserted a spiritual soul and ego into a social and cultural context, which in turn presupposed the wide horizon of an infinite world.[7]

The provisionality of Husserl's final ethics is undeniable even for Patočka, but equally apparent is his firm conviction, reiterated in print and conversation, that phenomenology must tackle moral and social crises.[8] Perhaps Patočka's unavoidable entanglement in the agonies of the Czechoslovak mid-century pushed his phenomenology further toward engagement, but the common dismissal of Husserl as purely epistemological overlooks the worldly concerns and ethical agenda of his late career.[9] Admixing Brentanian terminology with a Kantian sense of imperative and the spirit of Fichte's absolute ought, the postwar Husserl modified his early axiological project in favor of a personal and social ethics of duty and intersubjective worldliness.[10] As an early indication of the new direction, his ethics courses of 1920/1924 ended with a discussion of the best possible life as one subordinated to a personal calling [*Beruf*] and an overarching norm. Instead of the value taxonomies of his youthful career, the key question became a personal and social one: "Was soll ich tun?"[11] In answer, he urged a lifelong striving, always insufficient, for a self-regulating and norm-governed vocation, an effort applied not only to the "hypothetical" (i.e. instrumental) desiderata of adopting appropriate means for chosen ends, but also to the categorical imperative of weighing final goals.[12] In what James G. Hart has called the "ethical reduction," Husserl defined a purposeful existence as one in which actions were viewed from the holistic perspective of a unified life in its worldly situation, a task whose horizons were infinite and historical.[13] This phenomenological attitude allowed one to identify a personal norm as the guidepost of a moral life; with regular practice, fidelity to this norm could be habitualized.[14] Husserl's manuscripts from the mid-1930s, coupled with the *Crisis* work, took this worldly ethics still further by demonstrating the new centrality of intercommunal worldliness in his thought.

One can, in fact, identify three distinct formulations of post-axiological ethics in Husserl's later oeuvre.[15] The most well-known, exemplified in the Fifth Meditation, but present as early as the *Ideas II*, focused on the *egological* establishment of otherness and intersubjectivity through empathy. This formulation exercised great influence, notably in French philosophy, because of its systematic and public exposition, but it has also invited criticism for an incomplete disentanglement of the other from the solipsistic ego. Empathy for an other, moreover, fell short of Husserl's aspiration for a communal and worldly outlook, and thus a second formulation of Husserlian ethics can be called *personalist*. As opposed to egology, which stressed the awareness of the other through the rational and corporeal self, a personalist ethics embraced a life motivated by individual but outwardly-directed determinations, rather than the lonely ego and (perhaps) her body. The motivated life had both rational and pre-rational dimensions, but its focus was the temporal and cultural individual regulated by self-given norms. The personalist ethic, echoing Brentanian and Kantian imperatives, took as its goal a purposive, self-regulated life. With seeds in Husserl's prewar lectures, this vision took shape in the seminars of the early 1920s, and we find it applied to communities – to "personalities of a higher order" – in the *Kaizo* essays on renewal. As Hart has shown, numerous manuscripts argue that the "I" already contains the other within it, that individuals within a community "penetrate one another" in forging a communal personality. "[W]e do not only live next to each other but in one another," Husserl contended. "We determine one another personally."[16] These claims, of course, challenge the adequacy of the philosophical individualism found in *both* the egological and personalist visions, and they point to the worldly concerns that increasingly preoccupied Husserl from the 1920s onward. As opposed to the thrust of his younger phenomenology, a third, *worldly* ethic, evident in desultory sketches, took the intersubjective realm, not the unified ego, as man's founding experience and basic ethical situation. In earlier writings, the mention of social personalities already indicated the possibility of a trans-individual subjectivity, and the postwar introduction of pre-rational drives, little-known beyond Husserl scholars, suggested the primordiality of protective, nurturing communities that preceded egological awareness. That Husserl never settled on a final vision or reconciled his views makes for a perplexing but bountiful foison.

WAR AND RENEWAL

Husserl's worldly ethics marked a new postwar emphasis in his thought. In the first half of his philosophical career, especially during the years between the *Logical Investigations* and World War I, he strove to elaborate an ethics based on Brentanian premises. Franz Brentano's promise of ethical certainty won a coterie of followers in its time, most notably his star pupil.[17] If by 1900 Husserl came to reject his mentor's psychologistic assumptions, he always upheld the call to philosophical clarity and universality. He also retained his teacher's fundamental ethical tenets – that moral insights based on feeling could be universalized through cognition; that these insights, like logical judgments, enjoyed the corroborating evidence of pure perception; and that the chief practical imperative was to choose the best among possible options.[18] In this early phase, Husserl's main criticism of Brentano's ethics was that it proffered a theory only in outline – failing, for example, to distinguish noetic (mental) judgment from noematic (object) value. He would cultivate Brentano's "fruitful seeds" by expounding a scientific apparatus for ethics to parallel the rational underpinnings of logic.[19] This endeavor, notably advanced in the Göttingen seminars, led to the elaboration of new subfields and coinages encompassing the theoretical and practical technicalities of moral experience: a noetic theory of ethical acts, an axiology of values, an apophantics linking ethical acts with their objects, and a formal moral praxis. More zealously even than his professor, Husserl espoused an ethics that was analogous to scientific logic.

The prewar seminars, however, already exhibited aporia that pointed toward later ideas. For one, Husserl's ethics revealed a tension between the description of moral phenomena and the prescription of proper conduct. Many early notes were taken up in detailing the subfields of ethical theory, describing the regions and logics appertaining to moral acts and values. At the same time, however, Husserl embraced Brentano's categorical imperative to do the best that was possible in each situation. Yet the shift from abstract description to the declaration of an imperative "ought" was not smooth, for it lacked a full conceptualization of the contexts within which ethical directives operated.[20] A second tension emerged in the contradictory drives to universalize and localize. While Husserl's empirical descriptions and categorical imperative were meant to ground a universal science, his guiding moral principle was not formal in the Kantian sense. Instead, he insisted that a concrete imperative – the content of the formal call to do the best that is possible – could only be specified in a particular time and place of action. It took the form of a local universality, of an "anyone in my circumstances should do as I do." In this form, Husserl's early ethics already implicated context in moral acts, an insight that would progressively deepen until he arrived at the notion of the lifeworld.

World War I introduced a new moral urgency to Husserl's project. The war was a personal tragedy for him – taking one son and injuring another – and the postwar years brought economic hardship and mounting dismay over Germany's social collapse. "The war," he wrote, in an article for the Japanese journal *Kaizo*, "revealed the falsehood and senselessness of this culture," prompting him to seek anew the purpose of his philosophical lifework.[21] This intellectual demarche was not wholly unprecedented: Husserl's posthumously published *Ideas II*, drafted "in one stroke" in 1912 and emended over the subsequent decade, served in part as a précis for future ethical concerns and promised far more than the logical formalism of Husserl's earlier work.[22] But in the postwar years, these aperçus blossomed into a new vision of a philosophy that would spearhead a cultural renewal by helping men to transcend political and material differences and nurture transnational ideals.[23] As early as 1917, in three lectures on Fichtean idealism delivered at Freiburg, Husserl declared the wartime crisis "a time of renewal [*Erneuerung*]."[24] Husserl's new ethics prized the recovery of human ideals as a domain of life experience, one that allowed men to dedicate themselves to the project of moral rejuvenation by envisioning a world that was not yet. "[I]t is his essence, that he must form an ideal for himself as a personal I and for his whole life, indeed a double, both absolute and relative, and strive toward its possible realization." For both individuals and societies, this ideal stood as a "'true' and 'better I'", an "absolute conception" that encouraged personal and social endeavor.[25] More than simply a goal-setting mechanism, idealization laid the groundwork for a pure and universal ethic and an individual absolute ought [*absolute Gesollte*].[26] Indeed, the assertion of ideals in the face of the empty facticity of modern science was nothing less, in his view, than the recovery of true humanity.

Although only three of five *Kaizo* articles appeared in print – and none were published in Europe – the opuscule helped to consecrate Husserl's ethical turn by offering his most sustained analysis of social and cultural life prior to the *Crisis* text.[27] His decision to publish on the theme of renewal, a topic prompted by the journal's title (*Kaizo* means renewal in Japanese), was driven partly by the need to bolster family finances. But the invitation from a former student also afforded the chance to reflect on the social collapse Husserl perceived around him and to outline a program of reform led by a rational philosophy determined to recapture its theoretical-*cum*-practical position as an existential guide. The first article introduced the theme of individual and social renewal and set as a goal the establishment of ethical norms for the modern world. "Renewal," declared its opening sentence, "is the general call in our present age of suffering and is heard throughout European culture."[28] Husserl's address urged the move from an ill-defined "natural feeling" of community and desire for reform to a rational individual and social renewal led by philosopher-functionaries.[29] As he had already lamented in his 1910-11 manifesto "Philosophy as a Rigorous Science," the human sciences lacked a rationally grounded *Wissenschaft* that could function as mathematics did for natural science. Humanistic methodologies embraced either a purely factual empiricism or a relativistic *Weltanschauungsphilosophie*; neither approach was grounded in true experience.[30] The missing science he invoked was distinct from naturalism because it did not seek the theoretical explanation of facts or elaboration of laws; instead, it sought to outline an *a priori* study of norms based on the open possibilities of a rational humanity, norms that could guide prudent action and lead a disillusioned mankind toward greater insight and humaneness. If laws established universal causal links, a norm defined a human possibility, an ideal role contained within each of us though never fully realized. Only a science embedded in the human world, he insisted, could annul the shameless "political sophistry" and Spenglerian "pessimism" of his age.[31] The motivation that drove Husserl's career from the start – the desire for a foundational rationalism – worked here as well, animating the call for a scientific grounding of ethical impulses. The difference, however, was a stunning move from the logical and epistemological concerns of his prewar phenomenology to an outright call for an individual and social ethics based on the elaboration of human possibilities, a practical "mathesis of the spirit and of humanity" that could guide human betterment.[32] While he retained a terminology, the enterprise changed.

Yet Husserl's new ethical concerns drew on an earlier methodological vision. Since the *Logical Investigations*, mathematics had exemplified the missing science of the *a priori*; every experiential reality, whether factual or imagined, contained its own mathesis, its own essential grammar that could be gleaned through the intuition of essences, or what Husserl called eidetic intuition. This core phenomenological technique relied on the imagination to vary perceptual objects so as to identify the core *invariants* or essences that defined them. In the *Kaizo* series, Husserl talked of applying this method to human phenomena in a way that had more direct implications for the social world than mathematics did for natural systems. Calculation, according to Husserl, remained "reinen Phantasiedenkens," a purely ideal reflection of real nature. Every human reality, by contrast, contained within itself a "pure possibility" that could mold and change it; man was distinct from animals because he stood under the norm of possible experience, not simply fact.[33] In this regard, the methodology of *Wesensschau* had distinct social and anthropological implications. Only mankind consisted of selves characterized by an "Innerlichkeit" that allowed men to re-imagine and remake themselves. This interiority and mutability meant that humans could not be explained solely by the causal laws of naturalistic psychology. Echoing Kant's renowned distinction between man and nature, the first *Kaizo* essay rehearsed another side of Husserl's familiar critique of psychologism: Not only did the fallacy conflate logic with psychology; it also voided humanity's distinctive interiority, which formed an essential condition of the ability to differ from oneself, to change, to regenerate, to renew. The opening of the third *Kaizo* article reveals that the three themes introduced in the first two essays – the call to renewal, the systematization of ethics, and the essential openness of man – were, in fact, one: "The renewal of humanity [*Menschheit*] – both individuals and men in society [*vergemeinschafteten Menschheit*]," it said, "is the highest theme of all ethics."[34]

Under the dispensation of renewal, man was both subject and object of ethics, a free individual capable of judgment and exertion, not bound by the circumstances of the moment or locked into biological reflex. Humans could take a perspective that encompassed past and future, overseeing their lives and forging commitments based on a survey of prospects. In order to promote reform, Husserl called on people to view their actions under the rubric of their own best possible life. This aspect allowed one to choose a purposive norm as a guide to right and wrong. Indeed, the call to subordinate one's life to higher ethical goal became for Husserl the expression of a new imperative: "To be truly human, lead a life that you can justify with thorough insight, a life of practical reason."[35]

The focus on an "echt humane Leben" was more than simply a methodological position, adopted or relinquished for theoretical aims. It was, for Husserl, a multi-stage transformation whose process formed the renewal he advocated. The first stage required individuals to commit to a calling [*Beruf*] based on the self-conscious review of personal circumstances and possibilities. This effort lifted an individual life from vague yearning to a conscious and guided commitment. But a life regulated by the demands of a calling was not yet fully ethical. While one could judge disparate activities against an overall goal and thus forge a kind of direction, it was important to evaluate the moral significance of that goal. A life regulated by a calling remained pre-ethical [*vorethische*] as long as it stayed within the framework of a particular profession and did not compare absolute aims against each other or against wider social needs. A calling, Husserl insisted, entailed merely relative value, whereas ethics concerned the absolute. Although the bridge between the pre-ethical and the ethical remained murky in this adumbration, he did suggest one motivation for moving from individual dedication toward a wider striving for ethical purpose: personal dissatisfaction.[36] If the postwar collapse left people longing for moral direction, individual efforts at reform soon yielded to malaise when they were unable to justify choices according to wider criteria. Yet this discontent spurred further social and ethical development among some individuals, who recognized the imperative to move beyond the dreary "infinity" of dissatisfactions toward a life focused around freely chosen paragons, a life defined by "the consciousness of rational responsibility, or the ethical conscience."[37]

Yet Husserl's essay did not stop with vague overtures. Although the ideal ethical life – variously called "das absolute Gesollte" or "die absoluter personaler Vollkommenheit" – was not fully achievable, Husserl offered concrete techniques for identifying and approaching it.[38] While it was not feasible to judge one's actions against an exemplar as a matter of constant daily practice, renewal could be engendered through a process of habituation, launched by conscious striving that gradually settled into more passive routine. Habituation required a commitment to regular procedures of thought and action, to a steady method of reform. "[T]he truly human life, a life of never-ending self-development," he wrote, "is, so to say, a life of 'method,' the method for the ideal humanity [*Humanität*]."[39] Tireless self-training and eventual habituation could secure an "ethical personality," first as outward expression and then as an inner will that fueled the ongoing process of individual renewal.[40]

Achieving individual ethical renewal, of course, was but a half-victory, for it ignored our duties to others. A superlative individual remained error-prone in an unreformed society, and thus a new "Menschenform" required social regeneration – extending beyond communities to nations and international humanity – and a shared ethical will.[41] Husserl used a host of metaphors likening societies to individuals: a community was a distinct "personality of a high order," a "many-headed [*vielköpfig*] and yet unified subjectivity"[42] that manifested its own distinct style or cultural character, an *ethōs* determining its moral and cultural outlook.[43] Ultimately, individual and social renewal were mutually implicated. If communities could not simply be reduced to a sum of individuals, it was also the case that "true human societies" could only exist when they had as their members "true individuals."[44]

He also proposed a pragmatic strategy for transmitting reform to wider sectors. Achieving an integrated culture ultimately fell to dedicated activists who could explain new possibilities and offer a unified vision to their confreres. Husserl described this activism as a "spiritual Huygens principle," with each reformed individual a node of wider moral renewal. Through writing and speaking, education and persuasion, these advocates would gradually transform society, person by person, from a collection of individuals into a *Willensgemeinschaft* rooted in a common tradition and shared vision.[45] These ethical envoys took the role of spiritual authorities, whom Husserl likened to mathematicians – rational instigators rather than political leaders – though he noted that cultural and religious dignitaries could also spearhead a *Willenszentralization*.[46] Presaging his famous characterization in the *Crisis*, Husserl designated philosophers the supreme functionaries of ethical renewal, "the appointed [*berufenen*] representatives of the spirit of reason." Philosophy, in turn, would become a universal science dedicated to both the theory and practice of cultural revitalization.[47] This drama, at once utopian and technical, would lead to societies dedicated to the progressive *Technik* of reform, culminating in a rational *Übernation* and ultimately a world Imperium governed by ethical ideals.[48] This perfectionist vision, which combined aspects of Aristotelian society with Platonic philosophical leadership, was not authoritarian, he maintained, because submission to rational authority would be given freely. He did not, of course, consider the goal fully attainable. Renewal was a historical process of living toward a regulative ideal; it was not a final achievement. The ethical increase of mankind was gradual and asymptotical, originating in the individual habitus and social *ethōs* of communities and growing steadily toward fruition. In Husserl's own language, an understanding of the dynamic process of ethical renewal required the tools of a genetic rather than a static phenomenology, procedures that could grasp time and constitution in a temporal understanding rather than as structures or functions of experience.[49] As Hart found in other social and ethical manuscripts, Husserl perceived renewal as an ever ongoing "movement," an infinite entelechy that approached but never met its aim.[50]

The posthumously published fifth *Kaizo* essay closed the prospectus with a macrohistorical survey of Western striving toward a rational ethical culture. The lengthiest of the sections, it highlighted the interplay between a religious worldview that presumed normative communities and a scientific outlook celebrating individual freedom and rational endeavor. According to Husserl, European thought had launched two essential movements. First, in the person of Plato, he found the cultural forefather who defined philosophy not merely as theory but as "vernunftigen Lebenspraxis," a union of theory and practice in one life-transforming science that foreshadowed the aims of phenomenology.[51] The second, modern movement, epitomized by Galileo and consummated in the Enlightenment, advanced human striving through mathematization and the universalization of scientific reason. Cultural renewal would reconnect these two strands – faith and reason, norm and law, theory and practice – which had been severed in modern life.

That a Husserlian renewal would have profound political implications is undeniable, though Husserl's political remarks are sparse and perfunctory, at times suggesting statist leanings, elsewhere a more egalitarian communalism.[52] One of the fullest sketches of the social import of his call came in the well-known 1935 Vienna lecture, one of his last public exposés. The community of philosophical reformers, he averred, was not meant to impose a Platonic hierarchy, but to serve as a model for others to emulate.

Philosophical knowledge of the world creates ... a human posture which immediately intervenes in the whole remainder of practical life with all its demands and ends ... A new and intimate community – we could call it a community of purely ideal interests – develops among men, men who live for philosophy, bound together in their devotion to ideas, which not only are useful to all but belong to all identically. Necessarily there develops a communal activity of a particular sort, that of working with one another and for one another, offering one another helpful criticism, through which there arises a pure and unconditioned truth-validity as common property. In addition this interest has a natural tendency to propagate itself through the sympathetic understanding of what is sought and accomplished in it; there is a tendency, then, for more and more still nonphilosophical persons to be drawn into the community of philosophers. ... The spread ... occurs as a movement of education, far beyond the vocational sphere.

Of course, conservative state leaders, fearful of losing authority, would react to this cultural efflorescence through persecution, but Husserl insisted that truthful ideas would outlast "empirical powers."[53] It is not hard to see why such a program, originally expressed in the midst of the Nazi consolidation, would inspire Patočka and fellow philosopher-dissidents.[54]

For those familiar with the 1931 *Cartesian Meditations*, the *locus classicus* of Husserlian ethics, a surprising feature of the earlier *Kaizo* series is its lack of concern for the phenomenological problem of intersubjectivity – or more properly, the elision of individual and social ethics.[55] Five years before the *Meditations*, the *Kaizo* articles, as Donn Welton remarks, took it for granted that the individual existed in social relations rather than presuming the need to ground intersubjectivity in the mechanics of empathy.[56] Societies only become truly human, Husserl remarked without further explanation five years before the Meditations, "when they have as their bearers true individuals [*echte Einzelmenschen.*]"[57] Five years after the *Meditations*, empathy and otherness again took a subordinate position in the *Crisis*, leaving the impression that Husserl viewed intersubjectivity as a technical facet within the wider problem of world apprehension and moral renewal, rather than as the crux of an ethics. In this light, the call for *Erneuerung* was not simply a cul-de-sac of Husserlian moral theory, but a new and central commitment to the human social world. Like Brentano, he saw ethics as a key to philosophy's practical relevance, especially in a time of crisis.

It must be said that Husserl's approach to social ethics in these essays failed to offer a convincing framework for ethical duty or practical social action. Indeed, Husserl himself seemed to be aware of its inadequacies in several passages that point to later developments in his thought. One of the main tensions of the analysis is between the overt social dimensions of the argument and the persistent Cartesianism of his ethical life reduction. Is the individual or society primary? At this stage, despite metaphorical equations between individuals and personalities of a higher order, Husserl had not yet developed his theories of empathy or horizons to such a degree that the project of renewal could be seen as anything but individually driven. Yet if a society was more than the sum of individual egos, as he insisted, then the greater whole was not yet explained, and there was no clear intersubjective juncture among reformed individuals that lifted them to a higher social plane. Nor did the universality of reason square with the particularity of individual and cultural norms. Husserl's commitment to a rational subjective experience would remain throughout his life, and it became one of the more fraught elements of his inheritance. All the same, his later work grappled with the nexus between a primordial subject and an *equally* primordial affective sociability, between intersubjective reason and a pre-rational *Triebsystem*. In a manuscript from 1921, Husserl introduced the notion of a community of joint striving [*Strebensgemeinschaft*] and a community of love [*Liebensgemeinschaften*] in which mutual contact and communication led to a shared motivation that ethically elevated the whole. In these *Willensgemeinschaften*, "every awakened person (ethically awakened) deliberately sets before himself his ideal I as an 'infinite task.'" The origin of personality lay in empathy and in social acts, in a social world that pre-existed and grounded the ego.[58] But so too did social ethics, which grew from a communal love and mutual friendship whose perfect embodiment was Christ.[59] But these incipient notions made little impression in the *Kaizo* articles. The fact that Husserl felt the need to define a philosophical and ethical basis for activism betokened a shift in his phenomenological mandate, the embrace of a world of action that needed philosophical foundation. And the collaboration between phenomenology and social activism promised mutual benefits: If a world without philosophy was ethically directionless, a philosophy divorced from human societies remained nugatory and dry. To be relevant for a troubled age, phenomenology had to become a social philosophy. But it would require the subsequent decade for Husserl to recognize the superficiality of his earlier pronouncements and devote greater care to the phenomenology of intersubjectivity and the dynamics of cultural renewal.[60]

THE CRISIS

The *Kaizo* essays stand as a crucial prehistory to Husserl's final opus, *The Crisis of European Sciences and Transcendental Phenomenology*. Though less socially and ethically explicit than its 1920s precursor, this incomplete final work was suffused with the despairs of its time. Like the postwar years, the German depression and Nazi rise brought Husserl professional and personal hardship. Facing straitened family finances, barred from university as a non-Aryan, rejected by his protégé Heidegger, and forced to publish outside Germany, the septuagenarian briefly considered abandoning his homeland for posts in California and Prague.[61] As his workload increased and political life darkened, Husserl tried to remove distractions by avoiding newspapers and narrowing his practical engagements to pleas on behalf of self and family.[62] Yet somehow, in this atmosphere of threat and isolation, Husserl achieved one of his most feverish bouts of philosophical labor, writing for six or seven hours daily as he elaborated a new account of the social and cultural lifeworld. The main argument of the resulting work is well known. Modern European humanity was experiencing cultural crisis because of a loss of meaning. The natural sciences, despite their commanding authority and technical efficiencies, had failed to provide humanity with a higher life purpose. Beneath the sheen of Western life, men struggled to glean significance from fractured and competing worldviews, and science was unable to explain the ultimate ends of their futile striving. Husserl's was, to paraphrase Eliot, a hollow age of hollow men. This crisis, he argued, took centuries to manifest, inherent even at the Greek inception of philosophy, but a crucial watershed came with the Galilean Renaissance, when natural philosophers disseminated a universalist mathematical science that dismissed subjective experience as mere *doxa*, unworthy of scientific concern. The fateful loss of original experience became especially acute in the nineteenth century, when the cult of positivist fact reached its apex and industrial advance lost a connection with deeper human urges. The disciplines fragmented; scientists became "unphilosophical experts;" and scientific rationality, while supplying life's technical accoutrements, quenched none of the thirst for greater meaning.[63] The result, said Husserl, was rampant skepticism and a turn toward mystical sources of meaning.[64] The tendency of naturalistic science to sever itself from experience by valorizing a narrow objectivism produced a cultural void that invited extremism of all sorts. If a purely calculating reason lay at the root of modern crises, however, the abandonment of reason was not a proper response. Modern cynicism and irrationalism, both painfully prevalent in Husserl's old age, were symptoms of the scientific abdication of duty, not solutions to it. To counter their appeal, he hoped to reground science in an original subjectivity that bound together reason and meaningful human experience. The posthumously published third part of the *Crisis* famously elaborated a new concept of the lifeworld as the intersubjective ground of experience, from which all rational pursuits emerge.[65]

Despite the obvious social and ethical motivations of the argument, Husserl's retention of familiar epistemological and scientific trappings mask the essay's moral agenda. But this gloss should not deceive us. Husserl's thought had shifted over forty years, and social ethics came to occupy a central position in his initially logical enterprise. Yet even where commentators grant this ethical cynosure its due, analyses focus primarily on the theory of empathy articulated in the Fifth Cartesian Meditation. This emphasis, while important for understanding Husserl's legacy in the ensuing decades, enables historians to pigeonhole him as a latter-day Cartesian and ignore broader, if less systematic socio-ethical commitments that also shaped his influence.[66]

The desultory concern for empathy in the *Crisis* underscores this point. Husserl devoted far more space to the lifeworld concept, which placed intersubjectivity on a more primordial and communal basis, than on the empathetic affirmation of a dyadic Other.[67] Perhaps acceding to some of the philosophical novelties of the renegade Heidegger, Husserl *emeritus* increasingly saw intersubjectivity as a primordial characteristic of the lifeworld experience, available to intuition without the need for empathetic verification.[68] Indeed, the world did not require assurances of Others by lonely egos because it was already multitudinous in its experiential constitution. Object perception itself, noted Eugen Fink, took the intersubjective world as a fundamental assumption.[69] Though we can only see object profiles, we always perceive the whole, an impossible perception except through the admixture of an infinite number of compossible views. To perceive a thing was to presume an all-sided world of co-viewers emerging from sense-constitution without the empathetic verification of corporeal analogues. In this account, the singular ego of Husserl's earlier works came to appear as an abstraction from primal intersubjectivity. Instead of the binary relationship of a solo subject and its duetted other in worldless *a capella*, the late Husserl ventured the phenomenological primordiality of a multitudinous, symphonic whole.[70]

Husserl's parerga from the 1930s elaborated the primordiality of worldly situation still further. According to one sketch, pre-intellectual drives, rooted in parent-child and sexual relations, sustained a primal worldliness that pre-existed egological rationality.[71] Another described the earth as a body and suggested a trans-egological apperception that blurred the self into a transcendental worldhood.[72] At once a body [*Körper*] and the "ground" of all bodies, the earth, encountered by individuals as a constant call to activity, formed a sphere of belonging-together, an "entire system of perspectives" whose "style" privileged nearness but also acknowledged the distance of other views.[73]

> What is to be emphasized here is that I can always go farther on my earth-ground and ... always experience its 'corporeal' being more fully. Its horizon consists of the fact that I walk about on the earth-ground, and going from it and from everything that is found there I can always experience more of it.[74]

This claim, of course, has epistemological and ethical implications, and it is easy to see how a nascent moral responsibility grew from the experience of world apperception. Indeed, Husserl described worldly and thingly awareness with a concept he had initially used to explain intersubjectivity: empathy.

> The fixed system of sites of all perspectively accessible external things for me is obviously already constituted through self-propelled walking, and also, that I can carnally bring everything and every object closer (at first directly on the 'face of the earth,' but also indirectly, by means of empathizing with birds I understand flight, and then by idealizing I have before my eyes the ideal possibility of an ability.) ... I can approach every site and be there, and thus my flesh is also thing, a res extensa, etc., that is mobile.[75]

Empathy not only assured the existence of sentient others; it also opened the world for experiential insight. The ethical valence is hard to miss, whether one interprets it as a responsibility to know and wonder or as a duty to acknowledge the being of others and respect the world as such. To be sure, the emphasis on a core sphere of nearness, on a world "for me," introduced a tension between a privileged self and the far, transcendent reaches. To read the writings after 1929, in fact, is to enter unexplored territory: Does phenomenology dismiss worldliness as a naïve assumption in favor of the primacy of ego, whence we restitute the world as a necessary presupposition of our experience? Or is the world pre-given as a constitutive horizon, even the fundamental ground, of our being – not a precondition derived from egological insight, as per the Cartesian approach, but a direct encounter prior to the constitution of the unified self? On this question, of course, rest some of the great debates of twentieth-century philosophy, and Husserl foreshadowed their direction by leaning toward the latter in his final years, though without ever abandoning his subjective commitments.

Indeed, Husserl took as his final phenomenological task the project of "initiat[ing] a new age" by reconstituting the social and communal homeworld for its individual members. "To be human at all," he argued, "is essentially to be a human being in a socially and generatively united civilization; and if man is a rational being ... it is only insofar as his whole civilization is a rational civilization." This whole, he emphasized, was not simply static and fixed; it evolved historically, generatively, through a kind of social and cultural entelechy.[76] And while Husserl's vision of renewal, his lifeworld of home and abode, of tradition and culture, grew from specifically European cultural premises, the rebirth of reason, he believed, promised to ramify across the modern globe.[77] Echoing the Brentano of yesteryear, Husserl's new philosophy would ground a new humanity.[78]

We must be careful not to narrow Husserl's reformation solely to a cultural or spiritual process. A renovated phenomenological reason in a "new age" would provide a firm grounding not only in unified historical traditions but also in the "normative relatedness" of things. Not only the intersubjective human world, but the physical world as well had a distinct character that unified subject and object in a temporal and spatial whole, while at the same time preserving them as distinct experiential poles.[79] Physical things, Husserl insisted in the second part of the *Crisis*, possessed an "empirical overall style [*empirischen Gesamtstil*]" of worldly "belonging together" rooted in the "invariant general style" of experience.[80] "[U]niversally," he remarked, "things and then occurrences ... are bound a priori by this style, by the invariant form of the intuitable world."[81] The argument for a stable intersubjective *ēthos* discernible from intuitable experience allowed Husserl to avoid the relativism of extreme subjectivity, the unknowability of Kantian noumenalism, and the cult of empirical fact that prompted his phenomenology in the first place. It also ensured that Husserl's world was not simply a collection of disparate sensations or hyletic data, but rather a "whole" – a "unity" and not "mere totality."[82]

This experiential world-style formed a forgotten bedrock for the scientific systems of mathematization, mechanism, and technology. For the natural scientist since the Renaissance, geometry and mathematics had become a language of nature, a *mathesis universalis* whose clarification was the infinite labor of modernity. The subjective stylizations of nature in myth, religion, and personal experience were forced to give way to the higher truths of precise measurement, and residues of belief and faith were simply the unconquered terrain of future science. For Husserl, however, this mathematization, while a powerful and positive movement, threatened to void the human experience from which it originated. He made an example of geometry, whose limit shapes he traced back to the practice of surveying and measuring designed to accommodate human needs. A precise geometry divorced from this ground could achieve technical mastery, but its feats were increasingly disconnected from experience. Indeed, the tools of modern science were so potent that it was quite easy to ignore, as Galileo did in his astronomical revolution, the practical and historical traditions from which they emerged. The lifeworld was not only overlooked by modern science; it was degraded and replaced by an idealized calculus deemed more real than the subjective confusions of daily acquaintance. Only mathematics, and not experience, was epistemologically valid – and with this monopoly, the human world was lost.[83] It must be stressed that Husserl did not reject modern science or dispute its achievements; his was no traditionalist backlash. In calling for a return to "the naiveté of life" in order to transcend the "philosophical naiveté" of science, he did not mean to deny the latter its insights.[84] Indeed, his project had the sense of a Kantian critique, validating scientific reason by delimiting its sphere of expertise. And yet, the emphasis was different from Kant's, for the crucial concern of his late thought was to recuperate the domain of experience for a phenomenological science whose methods were not natural scientific. The mistake of modern psychology was that it tried to annex the subjective field for the causal world of genetic science rather than recognizing the primacy of experiential intuition.

Thus, Husserlian worldliness encompassed not only an experiential intersubjectivity that was broader than the mechanics of empathy, but also a non-naturalistic, pre-scientific, and norm-governed engagement with things.[85] Objects themselves, we might say, *pace* Brentano, harbored a kind of intentionality, and tending toward subjective and intersubjective relationship that helped to establish an environing world. And a relational intentionality came to designate the fundamental style of the earth as a ground of subject and object *Zusammengehörigkeit*, not simply an object of mental acts. Husserl himself may have shied away from this implication: After all, he clung to the pure transcendental self of the epoché, the "distancing abstention," until his death.[86] Yet the transcendental Husserl and the worldly Husserl are not wholly irreconcilable. For it is the preservation of the subjective moment in worldly experience, of the "Welt-All" there *for me*, on which his distinctive argument about philosophical responsibility rests.[87] From its origins in ancient Greece, the "'philosophical' form of existence" attempted to supplant mythology with a new form of theoretical and practical self-mastery: "freely giving oneself, one's whole life, its rule through pure reason or through philosophy." This "superior survey of the world," according to Husserl – a "universal knowledge, absolutely free from prejudice, of the world and man ... frees not only the theorist but any philosophically educated person."[88] And a philosophical reason, as we have seen above, encouraged practical autonomy as well. Thus, modern philosophy had as its foremost task the exercise of ethical responsibility and the promotion of social renewal in a driftless world. It sought to recapture for the modern life what the ancients had introduced in Athens.[89]

According to the guiding ideal of the Renaissance, ancient man forms himself with insight through free reason. For this renewed 'Platonism' this means not only that man should be changed ethically [but that] the whole surrounding world, the political and social existence of mankind, must be fashioned anew through free reason, through the insights of a universal philosophy.[90]

But by reducing thought to mere problem solving, modern positivist science forgot its ethical mandate and "decapitate[d] philosophy," substituting faddish *philosophies* (in plural) for the quest after theoretical and practical responsibility, the cult of fact for the search for meaning.[91] Through the renewal of philosophy, Husserl sought nothing less than a renascent Europe – or, better, a European culture that could "renew itself radically" by reinvigorating a hollowed philosophical tradition. In this avant-gardist spirit, Husserl declared philosophers the "functionaries of mankind" who bear "responsibility for the true being" of humanity.[92] This formulation marked a crucial shift from his earlier program, a reconception of theoretical responsibility as world-responsibility, situated in the human community. In other words, Husserl came to see phenomenology not primarily as a philosophy of mind, an epistemology, a logic, or even an ontology, but as a philosophy of our embeddedness in and engagement with the world. He came to see it as an ethics.

Department of History, James Madison University, Harrisonburg, VA 22807, USA
e-mail: gubsermd@jmu.edu

NOTES

[1] Patočka, *An Introduction to Husserl's Phenomenology*, trans. Erazim Kohák (Chicago: Open Court, 1996), 135.

[2] Patočka, *Body, Community, Language, World*, trans Erazim Kohák (Chicago: Open Court, 1998), 36. These lectures were recorded by students from memory, not directly transcribed.

3   Ibid., 36, 103.
4   In fact, their joint involvement in a 1934 Prague colloquium launched Husserl on the *Crisis* project that would shape Patočka's own phenomenology as well as his later dissidence. Husserl could not ultimately attend the conference, and he asked Patočka to read his letter to the participants. For an account of this colloquium, see the chapter "Achter internationaler Kongreß für Philosophie in Prag (1934)" in Ludger Hagedorn and Hans Rainer Sepp, eds., *Jan Patočka: Texte, Dokumente, Bibliographie* (Freiburg: Karl Alber, 1999), 176–87. In late 1935, despite travel restrictions, he finally presented his *Crisis* argument in Bohemia before the *Cercle philosophique de Prague pour les recherches sur l'entendement humaine*. Beyond this, an early Patočka essay appeared in the same first issue of the Belgrade journal *Philosophia* that published Part I of Husserl's Crisis, along with essays from other members of the newly formed Prague Philosophical Circle. On this, see Petr Rezek, "La 'phénoménologie de l'esprit' de Patočka dans le contexte du Cercle philosophique de Prague," *Les Cahiers de Philosophie* 11/12 (Winter 1990/1991), 103–115. To be sure, Patočka's thought in the heady days of the Prague Spring and the bleak sequel of early normalization exhibited equal parts Husserl and Heidegger. He preferred Heidegger's sense of worldly context as the *Zuhandenheit* of active use rather than Husserl's *Vorhandenheit* of theoretical contemplation and phenomenological constitution. Indeed, his apocalyptic tenor in early 1970s mirrored Heidegger's in the age of "Only a God can save us," although Patočka's seems more comprehensible in the dark days of normalization. Yet his themes and terminology reflect Husserl's ethical program, the sense of Europe's cultural crisis, and the need for a renewed scientific-*cum*-social purpose. And Patočka's most extended Husserl analyses and encomium, *An Introduction to Husserl's Phenomenology* (Chicago: Open Court, 1999) came from this era.
5   Anthony J. Steinbock, *Home and Beyond: Generative Phenomenology after Husserl* (Evanston, IL: Northwestern, 1995), 115; Husserl, *Ideen zu einer reinen Phänomenologie und phänomenologischen Philosophie*, Zweites Buch, Husserliana IV (The Hague: Nijhoff, 1952), 158–59. The discovery might be dated even earlier to the Winter lectures of 1910/11 (published in English as *The Basic Problems of Phenomenology: From the Lectures, Winter Semester, 1910–1911* (Dordrecht: Springer, 2006)), where Husserl first introduced the 'natural world concept,' adapted from the positivist philosopher Richard Avenarius. The English-edition editors note that Husserl frequently referred to these lectures in his later career. Some scholars trace Husserl's insights into corporeality and worldliness to his 1907 introduction of the phenomenological *epoché*, the reduction from the natural assumption of circumstantial reality to an immediate experience that preserves worldly attributes without affirming their external reality. See, for example, Emmanuel Housset, *Husserl et l'énigme du monde* (Paris: Edition de Seuil, 2000).
6   Patočka, "Edmund Husserl's Philosophy of the Crisis of the Sciences and his Conception of a Phenomenology of the 'Life-world'" in *Jan Patočka: Philosophy and Selected Writings*, ed. Erazim Kohák (Chicago: Chicago, 1989), 235. In fact, during his months in Freiburg, Patočka worked more closely with Husserl's assistant Eugen Fink than with the master phenomenologist himself.
7   The situational priority remained ambiguous for both Husserl and Patočka, with individual corporeality at times receiving stress while elsewhere open-ended horizonality seemed to win phenomenological precedence.
8   As Dorion Cairns reports, Husserl insisted on the importance and attainability of a phenomenological ethics. See Cairns, *Conversations with Husserl and Fink* (The Hague: Nijhoff, 1976), 50–60. Recent scholarship has inaugurated a new understanding of Husserlian thought that departs from the narrowly epistemological view of an earlier era and highlights his interest in worldliness. See Ulrich Melle, "Husserl's Personalist Ethics," *Husserl Studies* 23 (2007), 1–15; R. Philip Buckley, *Husserl, Heidegger, and the Crisis of Philosophical Responsibility* (Dordrecht: Kluwer, 1992); Dan Zahavi, *Husserl and Transcendental Intersubjectivity* (Columbus: Ohio State, 2001); Janet Donohoe, *Husserl on Ethics and Intersubjectivity: From Static to Genetic Phenomenology* (Amherst: Humanity Books, 2004); James G. Hart, *The Person and the Common Life*; Steven Galt Crowell, *Husserl, Heidegger, and the Space of Meaning: Paths toward transcendental Phenomenology* (Evanston, IL: Northwestern, 2001); Donn Welton, *The Other Husserl: The Horizons of Transcendental Phenomenology* (Bloomington, IL: Indiana, 2000); Welton, *The New Husserl: A Critical Reader*, ed. Donn Welton (Bloomington, IL: Indiana, 2003); Steinbock, *Home and Beyond*.

[9] Erazim Kohák, in *Jan Patočka*, argues that the tenor of Patočka's active philosophy differed markedly from Husserl's 'contemplative' as a result of the pressure to confront national tragedy, a claim that underestimates Husserl's own social and ethical despair.

[10] The Brentanian imperative of living best possible life appeared in Husserl's prewar Göttingen lectures on ethics as well, although it stood within a framework of taxonomic value hierarchies. See Husserl, *Vorlesungen über Ethik und Wertlehre, 1908–1914*, Husserliana XXVIII (Dordrecht: Kluwer, 1988). Its postwar reappearance in the Freiburg ethics seminars brings out the Kantian dimension of lifelong self-legislation and regulation: an ethical life is one governed by clear intentional norms. Yet Husserl modified Kant's formulation, partly due to the influence of his wartime study of Fichte's *System of Ethics*, by emphasizing its substantive rather than formalist quality. Husserl gave three speeches on "Fichte's Menschheitsideal" in 1917–18 at the University of Freiburg, published in Husserl, *Aufsätze und Vorträge (1911–1921)*, Husserliana XXV (Dordrecht: Kluwer, 1987), 267–93. For an analysis of Fichte's influence on Husserl, see James G. Hart, *The Person and the Common Life: Studies in a Husserlian Social Ethics* (Dordrecht: Kluwer, 1992). For an examination of the relation between Husserl's and Brentano's ethics, see Michael Gubser, "An Image of a Higher World: Franz Brentano and Edmund Husserl on Ethics and Renewal," *Santalka* 17: 3 (2009): 39–49.

[11] Husserl, *Einleitung in die Ethik: Vorlesungen Sommersemester 1920/1924*, Husserliana XXXVIII (Kluwer: Dordrecht, 2004), 245.

[12] Unlike Kant, Husserl conceded the impossibility of following this procedure with constancy.

[13] The term comes from Hart not Husserl. Hart's *The Person and the Common Life* provides an extended analysis of Husserlian social ethics. For a shorter account, see Hart, "The Absolute Ought and the Unique Individual," *Husserl Studies* 22 (2006), 223–240. The classic Husserlian term "reduction" denotes the taking of a particular perspective in order to remove assumptions and reveal new insights. His phenomenological method hinges on key reductions that open up new domains of insight: the transcendental reduction (which suspended, or *bracketed*, the assumption of natural reality, thereby opening the world of pure experience); the eidetic reduction (in which imagination varies a perceptual object mentally in order to identify the core *invariants* or essences that came to define particular things); and Hart's ethical reduction from fragmentary experience to one's own whole life.

[14] Other documents from the early 1920s also attest to this new phenomenological sociality. See, for example, Husserl, *Erste Philosophie (1923/24): Zweiter Teil: Theorie der phänomenologischen Reduktion*, Husserliana VIII (The Hague: Nijhoff, 1959), 296–97.

[15] These were, I emphasize, different but overlapping positions, not discrete stages. There is a longstanding debate among Husserl scholars as to whether his work is best understood in terms of successive phases – static, genetic, generative; epistemological, Cartesian, worldly – that superseded one another, or as a career-long continuity in which with different themes were progressively elaborated. I do not engage this debate deeply. While I do see World War I as an important watershed in Husserl's thinking, many of the themes he developed in his postwar oeuvre were adumbrated in prewar writings, and some of the terminology from his early career lasted into the later. My concern is to highlight the increased attention devoted to social and ethical concerns in the 1920s and 1930s, a trend most commentators acknowledge whether they see it as marking a break with his earlier phenomenology or developing prospects already contained within it.

[16] Quoted in James G. Hart, "I, We, and God: Ingredients of Husserl's Theory of Community," in *Husserl-Ausgabe und Husserl-Forschung*, ed. Samuel IJsseling (Dordrecht: Kluwer, 1990), 136. Cf. Husserl, "Gemeingeist I. – Person, Personale Ganze, Personale Wirkungsgemeinschaften. Gemeinschaft – Gesellschaft," in *Zur Phänomenologie der Intersubjektivität: Zweiter Teil: 1921–1928* (The Hague: Nijhoff, 1973), 172, 174.

[17] On Brentano's ethics, see Michael Gubser, "Franz Brentano's Ethics of Social Renewal." *Philosophical Forum* 40: 3 (Fall 2009): 339–366.

[18] Husserl, *Vorlesungen über Ethik und Wertlehre, 1908–1914*, Husserliana XXVIII (Dordrecht: Kluwer, 1988), 90–101. See also Ulrich Melle's helpful introduction to these lectures.

[19] Husserl, *Vorlesungen*, 90; Husserl, *Einleitung in die Ethik: Vorlesungen Sommersemester 1920/1924*, Husserliana XXXVII (Dordrecht: Kluwer, 2004), 15.

[20] Alasdair MacIntyre reminds us that Husserl's phenomenological empiricism responded in part to Humean skepticism, including perhaps the stricture against moving from is to ought. See MacIntyre, *Edith Stein: A Philosophical Prologue, 1913–1922* (Lanham, MD: Rowman and Littlefield, 2006), 19–49.

[21] Husserl, "Fünf Aufsätze über Erneuerung," in *Aufsätze und Vorträge: 1922–1937*, Husserliana XXVII (Dordrecht: Kluwer, 1989), 5.

[22] "We could not be persons for others," he wrote, "if a common surrounding world did not stand there for us in a community, in an intentional linkage of our lives." Husserl, *Ideas Pertaining to a Pure Phenomenology and to Phenomenological Philosophy*, Second Book (Dordrecht: Kluwer, 1989), 201. The manuscript was redacted first by Edith Stein and later again by Ludwig Landgrebe. Sections Two and Three in particular offered lengthy, if somewhat preliminary discurses on empathy and otherness, the intersubjective constitution of the world, and the distinction between the naturalistic and personalistic attitudes, physicalist and spiritual realities. Husserl even asserted that worldliness precedes and grounds intersubjective otherness.

In a 1919 letter to the young philosopher Arnold Metzger, introduced and translated by Erazim Kohák for *The Philosophical Forum* XXI (1963), 48–68, Husserl acknowledged that in his early career he had "no eyes for practical and cultural realities." (56)

[23] See Husserl's July 8, 1917 letter to Roman Ingarden, in *Briefe an Roman Ingarden* (The Hague, 1968), 6–7.

[24] Husserl, "Fichtes Menschheitsideal," 268. See Thomas Nenon and Hans Rainer Sepp's "Einleitung" to that volume.

[25] Husserl, "Fünf Aufsätze," 35.

[26] Husserl, "Fünf Aufsätze," 33. Again this is a Fichtean evocation. See also Andrzej Gniazdowski, "Phänomenologie und Politik: Husserl's These von der Erneuerung der Menschheit," in Paweł Dybel and Hans Jörg Sandkühler, eds., *Der Begriff des Subjekts in der modernen und postmodernen Philosophie* (Frankfurt, 2004), 67–79.

[27] Donn Welton in *The Other Husserl* and Anthony Steinbock in *Home and Beyond* provide two recent interpretations of these articles.

[28] Husserl, "Fünf Aufsätze," 3.

[29] Ibid., 5.

[30] Husserl, "Philosophy as Rigorous Science," in *Phenomenology and the Crisis of Philosophy*, trans. Quentin Lauer (New York: Harper, 1965), 71–147. See also "The Dilthey-Husserl Correspondence" in *Husserl: Shorter Works*, eds. Peter McCormick and Frederick Elliston (Notre Dame: Notre Dame, 1981), 198–209.

[31] Husserl, "Fünf Aufsätze," 5.

[32] Ibid., 7.

[33] Ibid., 14, 16.

[34] Ibid., 20.

[35] Ibid., 36. Again, the Kantian vocabulary should not mislead us into seeing a Kantian project. Just as Husserl's phenomenological gaze took in more of human experience than Kant's transcendental, his imperative was not rooted in a formal concept of universal duty; instead, it found the particular universal within individual circumstances and experiences annealed in the cauldron of intuited possibility.

[36] Patočka, too, would highlight the liberating value of a dissatisfaction that rejected factual earthly bounds. See the 1950s essay "Negative Platonism," translated in Kohák, *Jan Patočka*, 193.

[37] Husserl, "Fünf Aufsätze," 32.

[38] Most remarkably, Husserl occasionally wrote of "die Gottesidee," the infinite potential contained within men though never attained in life. This divinity, as it were, formed a double, a "better I," toward which a person could orient herself. Only God, noted the Lutheran convert, could achieve true rational perfection. Ibid., 33–35. Again, Patočka, too, often celebrated liberating value of religion in orienting humans toward their higher freedom.

[39] Ibid., 38.

[40] Ibid., 39.

[41] Ibid., 45.

[42] Ibid., 22. This formulation first appeared in the *Ideas Pertaining to a Pure Phenomenology and to a Phenomenological Philosophy*, Second Book, trans by Richard Rojcewicz and André Schuwer (Dordrecht: Kluwer, 1989), 205. For a discussion of this formulation, see Hart, *The Person and the Common Life*.

[43] Husserl himself did not use this Greek-*cum*-Heideggerian term, but the *ethōs* is similar to an individual's "seelischer Habitus." Ibid., 23. For Heidegger's famous evocation of the term, see Heidegger, "Letter on Humanism," in *Basic Writings* (New York: HarperCollins, 1993), 256–57.

[44] Ibid., 48.

[45] Ibid., 52.

[46] Ibid., 57.

[47] "The humanity of higher human nature or reason," he later wrote "requires ... a genuine philosophy." Husserl, *The Crisis*, 291; (*Die Krisis der europäischen Wissenschaften und die transzendentale Phänomenologie*, Husserliana VI (Dordrecht, 1976), 338).

[48] Husserl, "Fünf Aufsätze," 58.

[49] Ibid., 55.

[50] Hart, "I, We, and God," 126.

[51] Husserl, "Fünf Aufsätze," 86. Cf. Patočka, *Plato and Europe* (Palo Alto, CA: Stanford, 2002). The concept of movement formed the core of Patočka's later phenomenology.

[52] In numerous essays, Hart and Buckley try to extrapolate political positions from Husserl's various pronouncements. While Robert Sokolowski's assertion, in *Introduction to Phenomenology* (Cambridge: Cambridge, 1999), 203–04, that phenomenology has not developed a political philosophy may be strictly accurate for Husserl, it ignores the political interests and implications operative in his work. It certainly does not hold for later phenomenological thinkers. For explorations in the political potencies of phenomenology, see Kevin Thompson and Lester Embree, eds. *Phenomenology of the Political* (Dordrecht: Kluwer, 2000).

[53] Husserl, "The Vienna Lecture," in *The Crisis of European Sciences and Transcendental Phenomenology* (Evanston, IL: Northwestern, 1970), 287–88.

[54] Aviezer Tucker argues that Patočka's late-1970s dissident writings owed more to his Husserlian inheritance than to the Heidegger he celebrated in his despair a half-decade earlier. Tucker, *The Philosophy and Politics of Czech Dissidence: From Patočka to Husserl* (Pittsburgh, PA: Pittsburgh, 2000), 58.

[55] By this time, of course, Husserl had already broached the topic of intersubjectivity in the 1910–11 lectures, *The Basic Problems of Phenomenology*.

[56] Welton, *The Other Husserl*, 319.

[57] Husserl, "Fünf Aufsätze," 48. See also pp. 4, 20.

[58] Husserl, "Gemeingeist I," 171, 174, 175.

[59] Husserl, "Gemeingeist I," 175–84.

[60] See not only the famous Fifth Cartesian Meditation, but also the extensive lectures and notes, dating from 1905, in the three *Husserliana* series volumes published as *Zur Phänomenologie der Intersubjektivität* (XIII, XIV, XV) The latter two volumes fall at or after the date of the *Kaizo* essays.

[61] Prague in particular afforded the chance to renew ties with Landgrebe and Patočka in the land of his "first teacher," Tomáš Masaryk, who fulfilled the "spirit of international humanity" that Husserl's philosophy endorsed. Letter from Husserl to the Austrian legal philosopher Felix Kaufmann, May 5, 1936 in Husserl, *Briefwechsel, Band IV: Die Freiburger Schüler* (Dordrecht: Kluwer, 1994), 224–25. His family suffered as well from the Nazi race laws. His son, a World War I veteran, lost his post as a jurisprudence professor, and his daughter was unable to secure domestic help. Husserl, Letters to Landgrebe, December 19, 1935; and the Dutch philosopher Hendrik J. Pos, January 17, 1935; in Husserl, *Briefwechsel, Band IV: Die Freiburger Schüler*, 343, 448.

[62] Husserl, Letter to Felix Kaufmann, January 5, 1934 in Husserl, *Briefwechsel, Band IV*, 201.

[63] Husserl, *Die Krisis der europäischen Wissenschaft und die transzendentale Phänomenologie* (Hamburg: Felix Meiner, 1996), 10; *Crisis*, 11.

[64] In his critique of these attitudes, Husserl remained a solid Brentanian. For Brentano's condemnation of modern philosophical skepticism and irrationalism, see his essays in *Die Vier Phasen der Philosophie und ihr augenblicklicher Stand* (Leipzig: Meiner, 1926) and *Über die Zukunft der Philosophie* (Leipzig: Meiner, 1929).

[65] As Guy von Kerckhoven has noted, the term lifeworld first surfaced in *Ideen II* in the early 1920s, when Husserl was rethinking his philosophical project along cultural and ethical lines. It was not carefully elaborated, however, until the end of his life. See von Kerckhoven, "Zur Genese des Begriffs 'Lebenswelt' bei Edmund Husserl," *Archiv für Begriffsgeschichte* 29 (1985), 182–203.

[66] In fact, Husserl ultimately abandoned his goal of publishing a German edition of the meditations due to dissatisfaction with their analyses. Several recent works have challenged this focus on empathy, showing that Husserl's late and incomplete "generative" phenomenology took a novel tack on questions of world, community, and intersubjectivity. See Steinbock, *Home and Beyond*; Welton, *The Other Husserl*. For an earlier work that makes similar claims regarding Husserl's analysis of history, see David Carr, *Phenomenology and the Problem of History: A Study of Husserl's Transcendental Philosophy* (Evanston, IL: Northwestern, 1974).

[67] He may, of course, have presumed that the problem was already dealt with by this stage in his career, yet even so there are tensions between the worldliness of the *Crisis* and the egology of the *Meditations*. As Steinbock argues, the Cartesian approach to subjectivity retained an egological core, for the Other always appeared as a second and subordinated self. Steinbock, *Home and Beyond*, 49–78.

[68] On the relation between Husserl and Heidegger, see Steven Galt Crowell, *Husserl, Heidegger, and the Space of Meaning*. I do want to stress, however, that Husserl's worldly inclinations can already be found in writings prior to the appearance of Heidegger's famous analysis in *Being and Time*, that they cannot simply be written off as a reaction to the ideas of his student-turned-apostate.

[69] A joint student of Husserl and Heidegger, Eugen Fink spent a career arguing that the transcendental origin of the world was the *Grundproblem* of Husserl's phenomenology as early as the *Logical Investigations*. See his essays "Die Spätphilosophie Husserls in der Freiburger Zeit," *Phaenomenologica* IV (1960; from a 1959 lecture), 99–115; and *Die Phänomenologische Philosophie Edmund Husserls in der gegenwärtigen Kritik* (Berlin: Pan-Verlagsgesellschaft, 1934). On Fink, Husserl, and Heidegger, see Ronald Bruzina, *Edmund Husserl and Eugen Fink: Beginnings and Ends in Phenomenology* (New Haven, CT: Yale, 2004).

[70] I do not mean by this statement to sweep aside Husserl's frequent resort to primal subjectivity even in his late works, most famously in the posthumously published Crisis Part III §54 on the *Ur-Ich*. But, because it is downplayed by critics of Husserlian egology, I do want to argue with James Dodd that there is a fruitful and influential "tension between, on the one hand, Husserl's development of the theme of history [as a fundament of extra-subjective worldliness] and, on the other, his unrelenting focus on the personal dimension of philosophical life." See Dodd's thought-provoking *Crisis and Reflection: An Essay on Husserl's Crisis of the European Sciences* (Dordrecht: Kluwer, 2004), 67. This tension lends the *Crisis* a greater openness and suppleness than many critics allow. For example, Husserl may have already foreseen the danger, which Adorno and Habermas later censured, of the excessive subjectivity inherent in transcendental egology. See Adorno's famous complaint: "The 'absolutely other,' which should arise within the phenomenological ἐποχή is ... nothing other than the reified performance of the subject radically alienated from its own origin." *Against Epistemology: A Metacritique*. Trans. Willis Domingo (Cambridge: MIT, 1983), 163; Jürgen Habermas, *The Theory of Communicative Action, Vol. 2: Lifeworld and System: A Critique of Functionalist Reason*. Trans. Thomas McCarthy (Boston, MA: Beacon, 1985). Dodd, in fact, interprets the *Ur-Ich* not as an individual subjectivity *in* the world but as the subjectivity *of* the world, a world subjectivity.

[71] Husserl, "Universale Teleologie. Der Intersubjektive, Alle und jede subjekte umspannende Trieb transzendental Gesehen. Sein der monadischen Totalität," in *Zur Phänomenologie der Intersubjektivität*, Part III (The Hague: Nijhoff, 1973), 594.

[72] Husserl, "Foundational Investigations of the Phenomenological Origin of the Spatiality of Nature: The Originary Ark, the Earth, Does Not Move," in Maurice Merleau-Ponty, *Husserl at the Limits of Phenomenology* (Evanston: Northwestern, 2002).

[73] Husserl, "The World of the Living Present and the Constitution of the Surrounding World that is Outside the Flesh," in Maurice Merleau-Ponty, *Husserl at the Limits of Phenomenology*, 118, 132–33.

[74] Husserl, "Foundational Investigations," 121.

[75] Husserl, "The World of the Living Present," 153.

[76] Husserl, *Crisis*, 14–16.

77 Lifeworlds (in some manuscripts separated into home- and alien-worlds) were constitutive experiential frameworks characterized by a unifying *ēthos*.

78 Husserl, *Crisis*, 1, 12, 7–103. Of course, Husserl's analysis is Eurocentric in both neutral and negative senses; it seeks the roots of phenomenological reason in the European tradition and then proposes Europe as the vanguard of humanity universally. The wide non-European interest in phenomenology from its earliest days suggests that its methods can be severed from the Eurocentrism of its founder. Patočka would later criticize his mentor for this lingering rationalist Eurocentrism, though he did not fully escape it himself. See his manuscript "Réflexion sur l'Europe" [untitled in the German original], published posthumously in French translation in the essay collection *Liberté et sacrifice: Ecrits politiques* (Grenoble: Jerome Millon, 1990), 181–213.

79 See his discussion of the "perspectival style" of world and experience in two manuscripts from the early 1930s, "Foundational Investigations of the Phenomenological Origin of the Spatiality of Nature: The Originary Ark, the Earth, Does Not Move," and "The World of the Living Present and the Constitution of the Surrounding World that is Outside the Flesh," published in Merleau-Ponty, *Husserl at the Limits of Phenomenology*.

80 Husserl, *Crisis*, 31.

81 Husserl, *Crisis*, 32.

82 Husserl, *Crisis*, 31–2.

83 Husserl, *Crisis*, 21–57. Cf. Husserl, "The Origin of Geometry," appended to the English *Crisis* volume, 353–78.

84 Husserl, *Crisis*, 59.

85 In this, his critique moved him beyond the Brentanian legacy of intentionality, which linked subject and object but nonetheless preserved a divide between mental and physical phenomena, between the mind and its environs.

86 The term comes from Bernhard Waldenfels, "Experience of the Alien in Husserl's Phenomenology," *Research in Phenomenology* 20 (1990), 19. For an example of his persistent late invocation of transcendental phenomenological abstention – his goal of becoming a "non-participating onlooker", a "mere spectator, or observing ego," an "impartial observer" of the "life-process in reduced form," see the "The Amsterdam Lectures <on> Phenomenological Psychology" in Husserl, *Psychological and Transcendental Phenomenology and the Confrontation with Heidegger (1927–1931)* (Dordrecht: Kluwer, 1997), 222–24.

87 "[W]orld is a validity which has sprung up within subjectivity, indeed ... within my subjectivity," he wrote in the *Crisis*, 96. His disciples would reject this configuration. For Heidegger, the self became an empty clearing, and Fink espoused an asubjective phenomenology.

88 Husserl, *Crisis*, 8.

89 Klaus Held, in fact, maintained that Husserl's commitment to philosophical responsibility was rooted in his vision of the Greek tradition. At its origins, Attican philosophy grew from two intellectual commitments: *theoría* ,an organized sense of human wonder and curiosity); and *lógos/lógon didónai*, a responsibility to explain or account for things (in words). See Held, "Husserls These von der Europäisierung der Menschheit," in *Phänomenologie im Widerstreit: Zum 50. Todestag Edmund Husserls*, ed. Christoph Jamme and Otto Pöggeler (Frankfurt am Main: Suhrkamp, 1989), 13–39.

90 Husserl, *Crisis*, 8. Again, see Patočka, *Plato and Europe*, for the next generation of this argument.

91 Husserl, *Crisis*, 9.

92 Husserl, *Crisis*, 17.

SAULIUS GENIUSAS

THE QUESTION OF THE SUBJECT: JAN PATOČKA'S PHENOMENOLOGICAL CONTRIBUTION

Oh, those Greeks! They knew how to live. What is required for that is to stop courageously at the surface, ... to adore appearance, to believe in ... the whole Olympus of appearance.
Nietzsche, Gay Science (Preface to the Second Edition)

ABSTRACT

Can phenomenology offer a meaningful alternative to the structuralist and the poststructuralist pronouncement of the death of the subject? I suggest that a meaningful alternative could be established on the basis of Jan Patočka's phenomenological revival of Antiquity. According to my central thesis, Patočka's notion of the "Care for the Soul" provides the phenomenological resources for a novel sense of subjectivity. To substantiate this claim, my chapter is divided into six parts. After sketching the central problematic in the first part, I turn in the second part to a description of the central reasons that underlie the death of the subject thesis. The third part shows how from Patočka's works one can unearth the *phenomenological* basis that underlies this proclamation. The fourth part inquires into the close ties between the "death of the subject" thesis and Patočka's asubjective phenomenology. The fifth part spells out how Patočka's revival of Antiquity, under the heading of the "Care for the Soul," generates a novel sense of subjectivity. On this basis, my concluding section suggests that Jan Patočka's revival of Antiquity provides the resources needed to raise the question of subjectivity in the aftermath of the "death of the subject" thesis.

1. Nothing has unified European philosophy over the last century more than the question of subjectivity. On the one hand, the phenomenological analyses of subjectivity arguably are the most profound and deep-reaching that we can find in the whole history of philosophy. On the other hand, the structuralist and the poststructuralist critiques of the subject are the most piercing critiques the subject has ever seen.

In what follows, I would like to turn to Jan Patočka because his works provide the needed resources to open up a dialogue between these traditions, which sometimes seem to be almost diametrically opposed to each other. I would like to suggest that Patočka provides the most forceful, robust and intriguing expression of the phenomenological standpoint in the context of the debates that surround the "death of the subject" thesis; and he does so by incorporating the philosophical insights that have found expression in the radical critiques of the subject. Thus in what follows, I will argue that the structuralist and the poststructuralist critiques of the subject notwithstanding, phenomenology has the resources needed to raise anew the question of subjectivity. Such a possibility of reconstructing subjectivity will lead us further (or back) to Antiquity; it will lead us to the *epimeleia tes psyches*, the care for the soul.

Yet before turning to this theme directly, I would first like to say a few words about the "death of the subject" thesis.

2. This thesis springs from the realization that *subjectivity is not autonomous* and that therefore, *it cannot be conceived as the ultimate source of meaning and intelligibility*. Subjectivity is not autonomous because it is always determined by the unconscious mind, by history, by the opacity of language, and by social power. According to the proponents of the "death of the subject" thesis, the autonomous subject that was, for instance, so forcefully defended by the Enlightenment thinkers, is nothing more than a utopian dream.

Yet here we find ourselves on slippery ground and we need to be cautious. From the very start, the "death of the subject" thesis faces two serious objections. First, the proponents of this thesis can all-too-easily be accused that, at best, they only build straw men. The history of philosophy is filled with examples of how a certain thinker rejects the conception of the subject defended by earlier thinkers and replaces the discarded conception with a new notion of subjectivity. For instance, Kant's transcendental subject emerges out of a critique of Descartes' ego, just as Hegel's notion of spirit is built upon a rejection of the Kantian subject. Similarly, Husserl's transcendental subjectivity is an alternative to Descartes' ego and Kant's "I think," just as Heidegger's notion of Dasein emerges out of a rejection of Husserlian subjectivity. One could therefore argue that the proclamation of the "death of the subject" can only address a particular notion of the subject and for this reason, it cannot help but must leave other notions of the subject intact. This means that, paradoxically, the subject can only "die" a number of deaths—all metaphorical, and thus incapable of bringing about the subject's demise.

Such is the first objection. Secondly, the very fact that the subject has been attacked from so many perspectives, and for so many reasons, makes it difficult to conceive what the proclamation of the "death of the subject" could possibly mean. For anyone who seriously aims to proclaim the death of the subject, the subject turns out to be, like Typhon, a monster with a hundred heads: it is a highly hybrid figure which embraces Descartes' *res cogitans*, Leibniz's monadology, Kant's transcendental subject, Hegel's Absolute and finally Husserl's intentional consciousness. The subject announced dead turns out to be so *obese* that in fact, it can no longer be considered a subject at all; in the aftermath of its demise, it is in no way clear that it is truly a *face* of the subject that was drawn in sand before, as Foucault has put it, it has been "erased at the edge of the sea."

Here I am reminded of a beautiful story told by the great Argentinean writer, Jorge Luis Borges. The story is called "A New Refutation of Time." In this story, Borges aims to extend the critical function of British empiricism to the problematic of time. Berkeley denied that there exist any objects independently of our perceptions; Hume took this argument further and claimed that any kind of subject is nothing more than a recollection of sensations; so Borges wants to take the matter even further than

Berkeley or Hume and argue that *there is no time*. Why? Here is Borges' argument: if man is nothing more than a collection of sensations, and if these sensations can be remembered, then the recollection of these sensations means that the same sensation can repeat itself at least twice. But this repetition breaks apart the linear flow of time. And thus, if there is no subject, there is no time.

Yet interestingly enough—and this is the reason why I turned to this story— Borges finishes his analysis with *a refutation of this refutation*. As he puts it, "the world, unfortunately, is real; I, unfortunately, am Borges" (Borges, 234). One might very well wonder whether we, in the aftermath of all the critiques of the subject, will not be drawn to a similar conclusion.

At least one thing is uncontroversial: *to give up the subject as the autonomous source of meaning and intelligibility does not yet mean to give up the subject in any sense you please*. Yet before turning to the notion of subjectivity that the "death of the subject" thesis leaves intact, it is proper to raise a different concern: What is it that motivates the proponents of this thesis to speak of subjectivity's *demise*? Echoing Nietzsche, one could say: these proponents of the "death of the subject" thesis—*they are interesting*! What do they really want? What is it really that always drives them in just *this* direction?

Arguably, what underlies the death of the subject thesis is the very fact that most of the critiques of the subject have not culminated in the subject's downfall. *The proclamation of the death of the subject is precisely triggered by the limits from which less radical critiques of the subject suffer.*

Consider in this regard Michel Foucault's *The Order of Things*. Like many other proponents of the "death of the subject" thesis, Foucault is well aware that not all critiques of the subject lead to the proclamation of the death of the subject. In fact, Foucault's proclamation of the "death of man" is not so much directed against the Enlightenment notion of autonomous subjectivity but rather against those discourses that defend the subject on the grounds of its *heteronomy*. For Foucault, anthropology, as an analytic of man, emerges out of a critique of the sovereignty of the "I think." Anthropology emerges precisely when the sovereignty of the Classical discourse on the subject reaches its limit, i.e., when subjectivity comes to be conceived as a living, speaking, and laboring individual. For this reason, for Foucault, biology, philology, and economics are anthropological disciplines *par excellence*. Foucault goes as far as to suggest that the emergence of these disciplines marks the *birth* of man.

This clearly means that it would be a mistake to reduce Foucault's proclamation of the death of man to a merely forceful turn of phrase, which does nothing more than call to abandon a particular conception of subjectivity. Less clearly, it would also be a mistake to confuse this proclamation with a critique directed only against transcendental notions of the subject. Foucault's proclamation of the death of man is first and foremost directed against the *anthropological* critiques of the subject, viz., those critiques, which merely aim to correct an illegitimate notion of subjectivity. Foucault's analysis is geared toward the realization that just as the anthropological narrative (conceived as the analytic of finitude) surpasses the Classical discourse (based on the primacy of representation), so the anthropological narrative must also be surpassed by the death of man. According to Foucault, anthropology inevitably leads to the anthropological sleep, because *the primacy of discourse is irreconcilable with the being of man*. As Foucault puts it,

> But the right to conceive both of the being of language and of the being of man may be forever excluded; there may be ... an inerasable hiatus at that point (precisely that hiatus in which we exist and talk), so that it would be necessary to dismiss as fantasy any anthropology in which there was any question of the being of language, or any conception of language or signification which attempted to connect with, manifest, and free the being proper to man. (Foucault, 339)

How does Foucault support this claim? At first glance it seems that his argument relies on a mere conjecture that the being of man and the being of language are incompatible: "The only thing we know at the moment, in all certainty, is that in Western culture the being of man and the being of language have never, at any time, been able to coexist and to articulate themselves one upon the other" (Foucault, 339). Yet a closer look reveals that this conjecture is further grounded in what Foucault sees as an *irreducible confusion of the empirical and the transcendental*. It is interesting to note that this confusion is nothing other than a particular formulation of what Husserl has called "the paradox of subjectivity."[1] Even more interestingly, while Husserl was full of optimism that phenomenology has the resources to resolve this paradox, for Foucault, its resolution is not feasible. Lacking a successful resolution, this paradox leads to the ultimate conclusion of *The Order of Things*: "Man is an invention of recent date. And one perhaps nearing its end" (Foucault, 387).

How exactly is one to understand the above-mentioned confusion of the empirical and the transcendental? On the one hand, anthropology reverses the Classical priority of the transcendental over the empirical; it discovers the irreducibility of life, discourse, and labor to any kind of transcendental narratives. Yet on the other hand, anthropology covers up its fundamental discovery, it masks the "grey space of empiricity" by doubling the transcendental function: "the man of nature, of exchange, or of discourse, [are made to] serve as the foundation of his own finitude" (Foucault, 341). Put otherwise, anthropology's great discovery of the primacy of discourse is thereby covered up by the reinstated primacy of subjectivity. Foucault sees only one possible solution to this irreducible confusion of the empirical and the transcendental: only the destruction of anthropology can awaken thought from the anthropological sleep (Foucault, 341–342).

As I have indicated above, this confusion of the empirical and the transcendental can be conceived as a version of the paradox of subjectivity. As Husserl had formulated this paradox in the *Crisis*, "how can a component part of the world, its human subjectivity, constitute the whole world...?" And as he went on to say, "the subjective part of the world swallows up, so to speak, the whole world and thus itself too. What an absurdity!" (Husserl, 179–180). How can I conceive of myself as a subject *in* the world and a subject *for* the world? Needless to say, Husserl's resolution of the paradox is quite different from the one that Foucault offers. For Husserl, the paradox leads to the realization that there is a good sense in which one could claim that the subject's worldly existence is an accomplishment of his own transcendental subjectivity. A mistake to avoid here is to resist the temptation to conceive the primal ego in terms of what Heidegger has called *Vorhandenheit*, i.e., to conceive this ego as an *independently* existing substance that brings about subjectivity's own worldly existence. Rather, Husserl's insight is that my self-understanding, as the understanding of my "empirical" existence, remains inadequate for as long as I conceive of it *independently* from transcendental accomplishments. More precisely, for Husserl, my understanding of subjectivity remains incomplete and distorted for as long as I do not take into account that subjectivity is irreducibly both transcendental and "empirical," both for the world and in the world.

A detailed comparison of the two resolutions of this paradox would take me too far afield. In the present context, it suffices to see that the emergence of what Foucault calls "the confusion of the empirical and the transcendental" does not necessitate the conclusion he is drawn to. It is therefore meaningful to once again return to the question I had posed earlier—what is it that motivates the proponents of the death of the subject thesis to speak of the *death* of subjectivity?

One could think of the matter in terms of the Oedipal complex: a new generation of thinkers needs to establish itself independently from earlier generations. Now the structuralists and the poststructuralists were brought up in the eras that were heavily dominated by phenomenology and existentialism. What better option do you have to "kill your father" within such a context than to proclaim the death of subjectivity? Yet if one were to stick just to this explanation, a feeling would continue to linger that one has done no more than swept the problem under the carpet: what motivates the death of the subject thesis still remains unexplained.

Here, with this problem in mind, I would like to turn to Jan Patočka. I, personally, feel indebted to this thinker for having shown to me the phenomenological reasons that underlie the death of the subject thesis.

3. In *Plato and Europe*, Patočka provides an intriguing phenomenological description of the situation in which mankind finds itself today. With Eugene Ionesco in mind (and Heidegger in the back of his mind), Patočka asks: would it be possible to find an expression of the entire mood that could capture present day humanity? From the response offered, one can also extract an answer to the question I posed in the previous section—the question regarding the motivating force that underlies the death of the subject thesis.

> And this mood is: a deep helplessness and inability to stand upon anything in any way solid. In the nineteenth century people still had the sense that they could somehow direct their fate, that humanity could control its affairs. This sentiment has completely abandoned us. Now we live with the opposite sentiment: something is carrying us away; and what is carrying us away is contradictory, it prevents us from taking a univocal position. We do not know what we want; nobody does.

As Patočka goes on to say,

> we are the victims of contradictory prophets; some proclaim the unleashing of instincts, others absolute discipline and obedience. Thus a deep helplessness and distress. Every human initiative or deed is socialized, controlled, and integrated into current affairs and carried off alone into the unknown. This is the sentiment of estrangement. What grows from it is surprisingly a will to power, but power that has no subject. Power is just accumulated and accumulating, and it does what it wants. Here is an awareness of a horrible trend toward the abyss.

The world in which we find ourselves no longer appears as the self-realization of reason. To speak of such a realization in the aftermath of two world wars and in communist Prague would be simply absurd. And arguably, some thirty-five years later, our situation is, at least in principle, not that different. We enter into wars and we do not know the reasons that lead us to them or the ways that can get us out of them. We find ourselves in an economic crisis, and all that remains is to guess what it is that has led us to it, or what it is that will lead us out of it. When we reflect on our past experience, we are bound to discover that at least some of the most profound decisions that have shaped our existence have been reached without any awareness regarding their value and significance. In a way, all our lives are decided for us, and not decided by us. Everything around us is arbitrary and contradictory. To posit a subject in such a context would mean to close one's eyes to this arbitrariness and these contradictions. Nothing in our experience warrants the trust in the subject. At best, what the subject can do is *try* to catch up with itself, nothing more than *try* to appropriate its own being. But what is this if not a fruitless attempt to catch up with one's own shadow? Along with Heidegger, one could say that the subject is always ahead of itself; yet in contrast to Heidegger, one should also add: the subject is not ahead of itself because *it* projects its own possibilities; *for this it cannot do*. The subject's possibilities do not belong to the subject itself; the subject's possibilities are always already projected for it.

If such indeed is the overwhelming mood in which humanity finds itself today, then the proclamation of the death of the subject is only understandable. One could thus say that what Patočka provides is a phenomenological description of the experiential resources that underlie the death of the subject thesis. Not only does this thesis reflect our present condition; there are good reasons to conceive of it as a thesis that *emerges* out of a reflection on this condition.

Now if phenomenology can be characterized as a philosophical reflection on lived-experience, then here we come across a new realization and a new demand. We come to the realization that phenomenological reflections remain sterile for as long as phenomenology opposes the death of the subject thesis. And thus we face a new demand, viz., the need to incorporate the death of the subject thesis into phenomenology.

Yet is this demand reasonable? How can the recognition of the death of the subject give rise to phenomenology? After all, as the textbook definitions suggest, phenomenology is the analysis of the structures of subjectivity. Yet Patočka's phenomenology is not its orthodox interpretation. While it is well known that what Patočka represents is *asubjective* phenomenology, it is often overlooked that asubjective phenomenology emerges out of the same grounds as the death of the subject thesis. Taking this into account, one obtains the means to interpret asubjective phenomenology as a phenomenological alternative to the death of the subject thesis.

4. As Erazim Kohák has pertinently remarked, Patočka "considers Husserl's philosophy a towering achievement, for having freed modern thought from psychologism, for having contributed to its concepts of eidetic intuition and intentionality, of *epoche* and reduction, of temporality and of the body-subject, of the intersubjectivity

of the realm of reason and the radical interdependence of world and man—and, fundamentally, for having provided a diagnosis of the crisis of modernity" (*PSW*, 97). This deep appreciation notwithstanding, Patočka is also convinced that Husserl's phenomenology suffers from a significant limitation, viz. from the illegitimate privilege it bestows upon transcendental subjectivity. Patočka's characterization of his own project as asubjective phenomenology suggests not only that phenomenology has the reasons to correct Husserl's misinterpretation (or at least imprecise formulations) of his own fundamental discoveries, but also that asubjective phenomenology has in fact found the means to overcome them.

So as to make sense of Patočka's critique of Husserl, let me turn to the fundamental distinction that underlies classical phenomenology: the distinction between the thing itself and its manners of appearance. This distinction is the dominating theme in Husserl's phenomenology. It is already operative in Husserl's early *Logical Investigations*; and in the last and unfinished *Crisis*, Husserl famously remarks that "the first breakthrough of this correlation between the experienced object and its manners of givenness ... affected me so deeply that my whole subsequent lifework has been dominated by the task of systematically elaborating on this a priori of correlation" (Husserl, 166). The distinction between the object and its modes of givenness also plays a central role in Patočka's thought. However, it is crucial not to overlook that for Husserl and Patočka, this distinction means something significantly different. It is precisely in these differences that one can discern the central reasons that underlie Patočka's call for asubjective phenomenology.

On a general level, one could say that the correlation of objects and their modes of givenness highlights the manner in which human existence is bound to the rest of what is. To clarify this point, let me draw your attention to something very commonplace: If you find this text interesting, you will be absorbed in it, and therefore you will remain *indifferent* to things around you. If you find this text tedious, you might very well direct your attention to the pen in your hand, the cup of coffee on your table, or the voices coming in through the open window. As a subject of experience, you can be either concerned with, or indifferent to things in the surrounding world. Inanimate things, on the other hand, can be neither concerned with us, nor indifferent to us. *They are fully cut off from us, while our existence is such that we always stand in relation to them, even when we are indifferent to them.* This difference is trivial, one is willing to say; yet at a closer glance, it becomes highly intriguing.

Its significance becomes clear as soon as one asks: What is it that allows for the subjects to be concerned with objects? That is, what is it that subjects have that inanimate objects lack? And the answer is the following: Things themselves *appear* to the subject. They do not appear to the pen in my hands or the cup of coffee or the table; yet they do appear to me. But clearly, this must mean that insofar as I am a subject, I am aware not only of things, but also of appearances. *I am a being that is not restricted to the domain of beings.* I am free from beings and it is this freedom that links me to them: things show themselves to me; things *appear*. It is this distinction between things and appearances that is so central in phenomenology.

But what is it that we really know of appearances? With this question, we find ourselves in an awkward situation: on the one hand, all that we can know about things depends on how they appear to us. If things did not appear to us, our being would be not that different from the being of inanimate objects that surround us. Our being would be even more restricted than the being of a person in a vegetative state, assuming, of course, that such a person is still capable of dreams. On the other hand, as soon as we raise questions about the appearance of things, we immediately end up reducing appearance to things. Appearances are, to borrow a metaphor from William James, like snowflakes caught in the palm of our hands: as soon as we catch them, they become something they are not—drops of water. So similarly, as soon as we start to reflect on appearances, we end up transforming them into something they are not: we transform them into *objects* of appearances.

As I have mentioned above, Husserl was the first philosopher to reflect on the irreducible difference between objects and their manners of givenness. Overstating the matter only slightly, one could say that the central ambition of Husserl's phenomenology has been that defending the autonomy of appearances from the natural tendency to reabsorb them into what they are not, i.e., from the danger of transforming appearances into *objects* of appearances. Yet according to Patočka, Husserl's battle has been successful only in part. Husserl has failed to notice that the autonomy of appearances needs to be defended on *two fronts*, for just as there is a *natural* tendency to reduce appearances to objects, so there is also a *philosophical* tendency to reduce them to subjectivity. Thus Patočka insists that Husserl's fixated defense of appearances against their naturalistic misinterpretations led him to reaffirm and even strengthen the philosophical illusion that appearances could be derived from subjectivity. On Patočka's view, Husserl did not liberate himself from the danger of reducing appearances to something they are not: he reduced them to the structures of subjectivity.[2] But this reduction is unjustified: appearances retain their autonomy not only from objects, but also from the subject. This is the reason that underlies Patočka's qualification of his project as *asubjective* phenomenology.[3]

One could characterize asubjective phenomenology as a type of phenomenology that restricts itself to the analysis of the structures of manifestation and does not reduce manifestation either to objectivity, or to subjectivity. Arguably, there are certain laws of appearances that are not reducible either to objects or to subjectivity. Asubjective phenomenology is meant to be nothing other than the analysis of these non-objective and non-subjective laws of appearances.

The phenomenon of the *world* is an apt illustration of such laws. On the one hand, the world is clearly not reducible to subjectivity. "We come into this world"; "we leave this world": the finitude of human existence calls for the recognition of the world's transcendence in regard to the subject. But if the world does not belong to the subject, does this mean that the world is an object of sorts? Here again, we have to answer with a No. I can see a number of objects around me, but clearly, I cannot expect to see the world as one object among others. This points to something curious: on the one hand, I cannot experience a single thing that does not belong to the world; on the other hand, the world itself is neither an object, not the subject. So how can the world still "be," despite these negative qualifications? Patočka has the resources to answer this question: the world belongs neither to the subject, nor to the object; it belongs to the domain of appearances.[4] And it prescribes a law to objects. This law states: Each and every object will be an object only insofar as it belongs the world.[5]

Such a refusal to restrict the source of meaning to either subjectivity or objectivity brings Patočka into proximity with the proponents of the "death of the subject" thesis. One could even contend that under the heading of asubjective phenomenology, Patočka takes the critique of the subject further than is done by the proponents of this thesis. The reason for such a contention would lie in the realization that here we are not facing a limit that has to do with discourse, interpretation or understanding. Patočka brings to light the need to give up the primacy of subjectivity at much more basic levels of experience—at any level, in fact, at which appearance is operative. Put otherwise, here the limit on subjectivity is not imposed "from outside"; here subjectivity is itself driven to the realization that it is always something derivative and consequential.

Yet Patočka's central contribution to the problematic of the subject lies elsewhere. It is indeed remarkable that the qualification of phenomenology as asubjective leads him to inquire into the notion of subjectivity that this phenomenology would not only legitimize, but also calls for.

To make sense of this, one can begin by raising some classical phenomenological questions: could there be appearances without something that appears, i.e., without objects? To this we are to answer with a No. Correlatively, could there be appearances without a subject to which appearances are given? That is, could there be appearances without anyone who "has," or experiences them? Again, we have to answer with a No. It thereby becomes understandable why Patočka would claim that "the structure of the phenomenon as the phenomenon renders possible the existence of—what?—the kind of beings such as man" (*PE*, 31). Thus having made a circle, we come back to the original and, arguably, central phenomenological question: what is subjectivity?

Initially, one can answer this question by saying that *subjectivity is the dative of manifestation*. Subjectivity is that to which appearances are given. However, for Patočka, such an answer would be insufficient. Patočka has pursued a number of different ways in which the question of subjectivity could be thematized.[6] I will address one of them—the one I consider most promising and most intriguing. Let me turn to Patočka's notion of subjectivity conceived in terms of the Ancient Greek notion of the *care for the soul*.

5. What is the soul, and what does it care for? As Petr Lom has pertinently remarked, just as Patočka himself is neither a mystic, nor a theologian, so for him philosophy is neither myth, nor religious consolation.[7] For Patočka, philosophy has exclusively to do with the ability and determination to seek the truth. And what is it that enables human beings to pursue truth? It is nothing other than what Patočka calls the soul. The soul is "just that which is capable of truth within man" (*PE* 36); the essential care for the soul is nothing other than "living in truth."[8]

It is interesting to note that the notion of "living in truth," which was to become so central in the Charta 77 movement, has its origins in Patočka's conception of the Idea of philosophy. It is no less interesting that for Patočka, the concept of "living in truth" has its origins in Husserl's notion of philosophical responsibility. Both of these themes point back to the birth of philosophy in Ancient Greece. In the Warsaw Lecture from 1971, Patočka interpreted Husserl's reflections on the birth of philosophy as an approach that singles out "living in truth" as the very "spirit of Europe":

> As Husserl sees it, what makes Europe special is precisely the fact that reason constitutes the central axis of its history. There are numerous cultural traditions, but only the European places the universality of evidence—and so of proof and of reason—at the very center of its aspiration. The vision of living in truth, of living, as Husserl has it, responsibly, emerges only in Europe, and only here did it develop in the form of a continuous thought, capable of being universally duplicated and of being deepened and corrected through a shared effort. (*PSW*, 223)

It is no exaggeration to suggest that for Patočka, the care of the soul, on which the possibility of living in truth rests, is nothing less that the secret axis of European thought. This concept has its historical origins in Ancient philosophy. Within this context, we find the care of the soul in Plato's reflections as well as in those of Democritus and Aristotle.

Patočka singles out three crucial ways in which care of the soul has been articulated in Antiquity. First, this care has been thematized as ontocosmology—as the search for understanding the world as the horizon of one's existence. Secondly, it has been also understood in the context of political life—as the search for a communal life that would open the space for human being's freedom. Thirdly, it has been also thematized in terms of the relation of the human being to her own mortality.[9]

When Patočka characterizes the present condition as that of a crisis, he means by this that the three elements of the care for the soul no longer find their resonance in our existence. For Patočka, this signifies a double crisis: the crisis of philosophy and of Europe. For Patočka, the future of philosophy and of Europe is dependent upon finding the means to reawaken the care for the soul. Hence the significance of the questions that we find in the introductory lecture: "Can the care of the soul, which is the fundamental heritage of Europe, still speak to us today? Speak to us, who need to find something to lean on in this common agreement about decline, in this weakness, in this consent to the fall?" (*PE*, 14)

If we ask about the reasons that underlie this crisis, we are in an interesting way led back to the death of the subject thesis. "What led us into this state? What brought Europe here? The answer is simple: her disunited and enormous power" (*PE*, 9). More precisely, it is the emergence of autonomous and sovereign states, when coupled with the powers of science and technology that brings about the inner dissolution of Europe. One can conceive of this dissolution as Europe's internal fate, or at least a consequence that stems from its inner logic.

But what exactly is this inner logic? On the one hand, it has to do with technical power; on the other hand, it has to do with the instrumental reason, that reduces nature and subjectivity to a mere tool for exploitation and domination. Yet when so much is said, a new question emerges: what exactly underlies the domination of this technical instrumental rationality? On this more fundamental scale, Patočka's answer points in the direction of subjectivity itself. Following Nietzsche and Heidegger, Patočka argues that the present crisis is engendered by modern subjectivism, which aims to derive all understanding from subjectivity.[10] The subject, conceived as the origin of all meaning and intelligibility, gives rise to the will to power as the only measure of conduct. Yet just as Cronus was overpowered by his own children and vanquished into Tartarus' domain, so this subject is overcome by its offspring as well: we are left with "something like a will to power, but power that has no subject. It is not that someone should want this power; it is just accumulated and does what it wants with us" (*PE*, 6).

At first glance, it might seem that the care of the soul and asubjective phenomenology are completely unrelated themes. And yet, for Patočka, these themes are simply inseparable from each other.[11] As he puts it, "the entire essence of man, the whole question of his distinctiveness and of his possibilities is connected to the problem of manifestation" (*PE*, 26). More precisely, for Patočka, our capacity to live a life in truth derives from the givenness of appearances. For this givenness of appearances, this mere fact that we are the datives of manifestation, means that we can stay truthful to the manner in which things appear to us. On Patočka's view, *our capacity for truth is nothing other than our ability to remain truthful to appearances*.

To make sense of this, let us ask: what is truth? As Patočka has remarked, we all know that people have died in the name of truth. To die in the name of truth is to conceive of one's life in terms of *responsibility to how things show themselves*. In the highest peaks of humanity, we encounter those remarkable individuals who refuse to reject the manner in which things appear no matter what the consequences of such a refusal might entail. What these exceptional individuals represent so forcefully is something not exceptional at all: In regard to manifestation, subjectivity is not free. The manner in which the world reveals itself to us has always already engaged us and imposed a responsibility upon us. In contrast to the long-established and habitual attempts to reduce manifestation to semblance and thereby oppose it to truth, Patočka brings to light that truth is nothing other than the manner in which things manifest themselves; that truth is the manner in which things are; or better, that truth is the manner in which the very nature of things shows itself to us.[12] "Man is the caretaker of the phenomenon .... Man is a creature of truth—which means, of the phenomenon" (*PE*, 35). If this is accurate, then it is no exaggeration to suggest that the fundamental possibility of humanity coincides with the problem of manifestation.

Thus for Patočka, the death of the subject is a highly ambiguous theme. On the one hand, it is a symptom of "the crisis in which Europe finds itself today." On the other hand, when thought through under the heading of asubjective phenomenology, it offers a possible cure to this crisis, that is, it is something that provides new resources to reawaken the care for the soul. To express this differently: on the one hand, the death of the subject is a symptom of the fact that the care for the soul no longer speaks to us today. On the other hand, it is also a sign that appearance is more original than subjectivity and that the primacy of appearance can provide us with the resources needed to reawaken the care for the soul. To express this duplicity yet in another way: on the one hand, care for the soul stands for subjectivity's search for self-unity; care for the soul means "to want to be in unity with one's own self" (*PE*, 189). Yet on the other hand, such a notion of subjectivity is an accomplishment that rests upon the lack of unity in question; or in Patočka's own terms, the care for the soul is rooted in the fact that "man originally and always is not in this unity with himself" (*PE*, 188).

So what is subjectivity? Subjectivity is the dative of manifestation, but not only that. Subjectivity also holds a responsibility in regard to manifestation. For Patočka, this responsibility in regard to manifestation, conceived as the possibility of living in truth, is what defines the history of philosophy. Therefore, for Patočka, the thesis of the death of the subject, when not coupled with the question of subjectivity's rebirth, is in fact nothing other than the thesis of the death of philosophy. Herein lies Patočka's fascinating contribution to the problematic of subjectivity. For Patočka, in the aftermath of the radical critiques of the subject, the task of philosophy should be that of raising the question anew: what is subjectivity?

6. In my concluding remarks, let me once again return to Borges, to his short story "A New Refutation of Time," to which I had already referred earlier. At the end of this story, Borges writes:

> And yet, and yet ... Denying... the self is an apparent desperation and a secret consolation. Our destiny is not frightful by being unreal; it is frightful because it is irreversible and iron-clad. Time is the substance I am made of. Time is a river which sweeps me along, but I am the river; it is a tiger which destroys me, but I am the tiger; it is a fire which consumes me, but I am the fire. The world, unfortunately, is real...
> (Borges, 233–34)

...*and for better or worse, so is subjectivity*. Some thirty years ago, at the peak of the structuralist and poststructuralist critiques, the question of the subject seemed to be foreclosed. The general consensus was that these critiques have put the subject to death. Could it not be so that thirty years later, *presently*, one of the central philosophical tasks should be precisely that of reengaging the question of subjectivity?

I hope to have shown that Patočka's phenomenology provides plenty of resources for a phenomenological reconstruction of subjectivity.

James Madison University, Harrisonburg, VA, USA
e-mail: geniussh@jmu.edu

NOTES

[1] See in this regard Edmund Husserl, *The Crisis of European Sciences and Transcendental Phenomenology*, pp. 178–186.

[2] Erazim Kohák has succinctly expressed Patočka's position in regard to Husserlian phenomenology: "the great value of Husserl's work is its recognition of the dependence of particular objectivity on interaction with the subject. Its weakness is its failure to recognize the other 'objectivity'—the ontological irreducibility of the world-horizon, given equiprimordially with the being of subjectivity." (PSW, 92).

[3] According to Patočka, Husserl's phenomenology succumbs to an imprecise formulation of its own discoveries. So as to correct these imprecise formulations, Patočka argues for the need to draw a clear and sharp distinction between the *epoche* and the reduction. Patočka identifies the *epoche* with the authentically phenomenological attitude of suspending natural theses; *epoche* is exactly what is needed to enter into the phenomenological domain. When it comes to the reduction, Patočka suggests that this theme falls outside the scope of phenomenology proper and belongs to something that could be labeled as

phenomenological philosophy. More precisely, the reduction carries the threat of the return to subjectivity at the expense of phenomenality.

4 As Tamás Ullmann has put it, "for Patočka the whole is not an ontological but a profoundly phenomenological term: it is the essence of appearing, appearance as such" ("Negative Platonism and the Problem of Appearance," unpublished manuscript).

5 As I have mentioned repeatedly, this domain of appearances is not reducible either to objects, or to subjectivity. It is an absolutely independent domain. One could say that what Patočka defends is a middle ground between Husserl and Plato. In regard to Husserl, he qualifies his position as asubjective phenomenology. I have explained this already: according to Patočka, Husserl's reduction of appearances to subjectivity lacks phenomenological legitimacy. On the other hand, in regard to Plato, Patočka qualifies his position as negative Platonism. This means that Patočka fully agrees with Plato that objects are determined by something that is independent of them. Yet this "something" in question, according to Patočka, does not point in the direction of ideas. Why? Because ideas themselves are objects of a particular kind. For Patočka, this transcendent domain, without which objects could not be objects, is not metaphysical; it rather is phenomenological, i.e., it is the domain of appearances themselves.

6 One such highly intriguing way has to do with the three movements of human life, the first of which Patočka calls *anchoring*, or *sinking roots*; the second one—self-sustenance, or reproduction; and the third one—self-achievement, or integration. See Jan Patočka, *Body, Community, Language, World*, trans. by Erazim Kohák, ed. by James Dodd (Open Court, 1998, 143–163).

7 See Petr Lom's *Foreword* to Jan Patočka, *Plato and Europe*, xiii–xxi.

8 "For Patočka, philosophy is not simply reflection about the meaning of life or the order of the world; it is a practice to shape the soul (the self) not simply in order to attain an abstract and eternal truth but to realize a true life: a life that is stable, is able to withstand the loss of meaning, of disorder, without closing the opening of freedom and receding into an ossification of social and human existence." (Arpad Szakolczai, "Thinking beyond the East-West divide: Foucault, Patočka, and the care of the self.")

9 For a concise treatment of this theme, see Petr Lom, op. cit., xvi.

10 See in this regard Petr Lom, op. cit., xviii.

11 Not surprisingly, therefore, in *Plato and Europe*, after providing a brief account of the current state of Europe, Patočka immediately turns to an account of asubjective phenomenology. See *PE*, Chapters 2 and 3.

12 Or as Patočka puts it, "how do we get to the nub of this most important thing—that thing on the basis of which only then can we have something like truth and error—because manifesting is the ground, without which truth and falsehood do not make sense" (PE, 25).

REFERENCES

Borges, Jorge Luis. 1962. *Labyrinths: Selected stories & other writings*. New York: New Directions.
Foucault, Michel. 1970. *The order of things: An archaeology of the human sciences*. London: Travistock Publications.
Husserl, Edmund. 1970. *The crisis of European sciences and transcendental phenomenology* (trans: Carr, David). Evanston: Northwestern University Press.
Patočka, Jan. 1989. *Philosophy and selected writings*, ed. Erazim Kohák. Chicago: The University of Chicago Press.
Patočka, Jan. 2002. *Plato and Europe* (trans: Lom, Petr). Stanford: Stanford University Press.
Szakolczai, Arpad. 1994. Thinking beyond the East-West divide: Foucault, Patočka, and the care of the self. *Social Research* 61:297–323.

SILVIA PIEROSARA

"HUMAN CREATIVITY ACCORDING TO THE BEING" AND NARRATIVE ETHICS: AN ACTUALIZATION OF ARISTOTLE'S ACCOUNT OF IMAGINATION

ABSTRACT

In the first part of the chapter, a couple of textual references from Aristotle's *De Anima* will be provided. According to the definition of imagination that can be found in Book Γ, imagination is not a sensation, but it is allowed due to sensation. In the second part of the chapter it will be shown that imagination has an intentional structure which can be assimilated to the teleological constitution of human condition. From this point of view, Aristotle's account of imagination has an intrinsically teleological structure: it can create either new events or new meanings only starting from the concrete limits of human condition. In the third part, it will be pointed out that, according to its hybrid nature, imagination, as a faculty, cannot be reduced neither to the plain reproduction of the existing order, nor to the radical invention of brand new features of human beings. In this being situated, the ontological quality of imagination can be discovered, or rediscovered. Human creativity (in the sense of *creation according to the being*) can be reached also through the innovative power of imagination. It is not a creation *ex nihilo*, but, rather, a way to project actions in order to testify a sense of the being itself. As a conclusion, an actualization of the theory of imagination as the condition of possibility of the contemporary revival of narrative ethics will be provided.

The following quotations, elicited from Aristotle's *On the Soul*, might be a good starting point to explain the proper nature of imagination:

> "Imagination is different from both perception and thought; imagination always implies perception, and is itself implied by judgement."[1]
> "Sensation is always present but imagination is not. If sensation and imagination were identical in actuality, then imagination would be possible for all creatures; but this appears not to be the case; for instance it is not true for the ant, the bee, or the grub. Again, all sensations are true, but most imaginations are false. Nor we say "I imagine that it is a man" when our sense is functioning accurately with regard to its object, but only when we do not perceive distinctly. And, as we have said before, visions are seen by men even with their eyes shut. Nor is imagination any one of the faculties which are always right, such as knowledge or intelligence; for imagination may be false."[2]

> "If, then, imagination involves nothing else than we have stated, and is as we have described it, then imagination must be a movement produced by sensation actively operating. Since sight is the chief sense, the name φαντασία (imagination) is derived from φάος (light), because without light it is impossible to see."³

According to these extracts, the Aristotelian definition of imagination can be divided into three main parts, each of them corresponding to a quotation. First of all, his definition is reached *via negativa*: he asks both himself and the reader what imagination is not. So, according to the first one, imagination is neither a perception nor a thought: it is situated "in the middle". It isn't thinkable without perception which is its condition of possibility. Moreover, imagination itself is the condition of possibility of judgement. In other words, imagination couldn't exist without perception; in the same way, judgment couldn't exist without imagination: it really is a middle term, just like in a mathematical proportion. The fact that imagination couldn't exist without perception means that the latter provides the former with "material", and it is, among other features, the most Kantian one or, finally, the very Kantian inheritance.

A very interesting definition of imagination is, in fact, provided by Kant in the *Critique of Pure Reason*: "*Imagination* is the faculty of representing an object even without its presence in intuition. Now, as all our intuition is sensuous, imagination, by reason of the subjective condition under which alone it can give a corresponding intuition to the conceptions of the understanding, belongs to sensibility. But in so far as the synthesis of the imagination is an act of spontaneity, which is determinative, and not, like sense, merely determinable, and which is consequently able to determine sense *a priori*, according to its form, conformably to the unity of apperception, in so far is the imagination a faculty of determining sensibility *a priori*, and its synthesis of intuitions according to the categories must be the transcendental synthesis of the imagination. It is an operation of the understanding our sensibility, and the first application of the understanding to objects of possible intuition and the some time the basis for the exercise of the other functions of that faculty. As figurative, it is distinguished from the merely intellectual synthesis, which is produced by the understanding alone, without the aid of imagination. Now, in so far as imagination is spontaneity, I some time call it also the *productive* imagination, and distinguish it from the *reproductive*, the synthesis of which is subject entirely to empirical laws, these of association, namely, and which, therefore, contributes nothing to the explanation of the possibility of a *a priori* congnition, and for this reason belongs not to transcendental philosophy, but to psychology."⁴

Imagination is the power to manipulate different products of perception without necessarily corresponding to real course of things. It can, for instance, put together perceptions which are not together in the reality or, on the contrary, it can divide what is not divided into reality.

As previously stated, judgement couldn't exist without imagination, which is a sort of "laboratory" for it. Like in a laboratory, in fact, judgement takes form from a frequent exercise of imagination, which puts together and project the concordance between a subject and its predication. According to this view, the word "judgement" is taken in the very Aristotelian meaning. Like perception for imagination, imagination itself provides "material" for judgment, and it is able to give a configuration to possible ways of judging, by representing them in the mind.[5] Therefore, according to the first Aristotelian quotation and, moreover, according to the *history of its effects*,[6] the first feature of imagination is to be located in a middle position between perception and judgement and, subsequently, to be able to allow communication between them.

The second quotation from Aristotle's *On the Soul* corresponds to the second part of the definition: imagination is not a sensation. If they were the same in actuality, then imagination, like sensation, would be possible for all creatures, and this is not the case. The most relevant point is that imagination can be experienced in a phenomenological way *only when there is no clarity of sensation*. When you clearly see, there's no need to imagine; when you clearly hear, there's no need to imagine a conversation nor a symphony; when you're tasting a flavour or touching, or smelling an odour, you don't need to ask your faculty of imagination for help. This is because sensation and imagination are not the same, from the point of view of actuality: they don't correspond to the same process; in their transition from potentiality to actuality their radical difference can be better pointed out.

They both are seen as a process, as a transition between a potential state and an actual one. It is exactly within this transition that one can take a very well shaped picture of their differences. On one side, sensation is always true; on the other side, imagination can be false; moreover, sensation is always present,[7] but, in opposition, imagination is not always present. It doesn't mean that sensation is always actual, but that imagination is actual, only when sensation is potential. Therefore, sensation can be re-activated through imagination in any moment of our life, even while sleeping. In this sense, if it is true that imagination cannot exist without having had a perceptual experience before, it is as much true that it is more pervading than perception. This statement can be demonstrated also if we think that we can even imagine of having a perception.

According to this second quotation, it is clear that the radical power of imagination consists of being able to move itself away from reality, only after (and thanks to) having experienced some elements of reality itself. In this way, the creative power of imagination confirms its being situated in a perceptual field, or, to express it in a hermeneutical way, in a perspective over the truth.

Another relevant feature of imagination can be provided by the third quotation from Aristotle's work. The quotation describes a meaningful analogy: as it is impossible to see without light, it is impossible to imagine without sight. In fact, the Greek word $\varphi\alpha\nu\tau\alpha\sigma\acute{\iota}\alpha$ (imagination) is derived from $\varphi\acute{\alpha}o\varsigma$ (light). First of all, it is just the case to quickly mention the primacy of sight over the other senses, and it is a typical Western philosophical paradigm, from Plato on.[8] A synecdoche can be easily recognized here, that is, a part is taken for the whole; secondly, sight is the condition of possibility of imagination. There can be no imagination without the actuality of seeing (that is, the actual side of sight).

Thirdly, the semantics of perception and the semantics of sensation in the English version both correspond to the Greek semantic field of αἴσθησις; from this last notation, a question (if not a contradiction) can be raised. If we accept that perception and sensation are the same, and we must accept it according to the original Greek text, it can be easily realized that Aristotles says that imagination presents itself only when perceptions are not so clear; but, later, he adds that imagination is impossible without sensations, like sight is impossible without light. Moreover, he tries to clarify this point saying that sensation can be either potential or actual, while imagination is given only when neither sight nor seeing are present. Presence here means either potentiality, or actuality.

One way to answer these questions and to overcome the contradiction is to consider once again the hybrid nature of the faculty of imagination and, what is more, to remember that imagination has a process form, which represents the tension between potentiality and actuality. Imagination, according to this point, is not always present, but it can always be made present. That is, unlike sensation, it is not easy to distinguish between its potentiality and its actuality, between form and content, because its borders are not so neat. It is a kind of perpetually potential faculty, whose function is to create dynamic images not *ex nihilo*, but rather according to the being.

In this sense, the products of imagination are the products of a peculiar kind of *human creativity according to the being*, whose semantic field evokes a conscious and aware teleology, which isn't able to create *ex nihilo*, but is able to create from a given world of beings, and to give them a sense, a direction. At this stage, one could ask which being should imagination be accorded to, and the answer is quite simple: imagination is a creation according to the sensations, either present or not, either past or present. In other words, sensation is the *inescapable framework*[9] of imagination: if ontology can be assimilated to an inquiry concerning the conditions of possibility, and if a condition of possibility can be represented as an inescapable framework, then it is demonstrated that sensations are the "being" of imagination, which can create only according to them.

Imagination is perhaps an emblematic case of *human creativity according to the being*, because its structure itself, as already stated, calls for the ontological *substratum* of sensitive data or phenomena collected by the sensitive apparatus. Sensation isn't itself properly a *being* (even if it could be considered a *being* whereas *being* signifies the relationship between potentiality and actuality: in this sense, one could say that sensation is *in being*, that is, *in fieri*); rather, it provides imagination with *beings*. Thus, imagination neither creates *ex nihilo*, nor accomplishes a necessary teleological "destiny"; rather, it uses images, sounds, smells, and so on, coming from sensation and it works on them, modeling or combining their parts together.

Therefore, imagination has a tight relation with human creativity, which indicates a *creation according to being*: teleology represents the conceptual connection between them, that is, imagination develops a form of life, a way to be moral, a new and original project over the future. But it performs all these mental operations only because of its being situated in an ontological historical perspective; moreover, this kind of mental operations is not necessary, but, on the contrary, free: the freedom of imagination is, thus, a situated freedom, rather than an absolute one. Imagination does not accomplish a written destiny nor carries it out an ontological necessity: it has the potentiality to "give ear" to the being and to carry out its potentialities, or, the other way round, it can betray the being by not carrying it out. It has the capability of reading the potentialities of the present, of a person, of a child, of a situation, and to make them either flourish or not.

As previously stated, teleology is an essential concept in order to explain the role and the richness of imagination. First of all, teleology represents a human hypothesis concerning a very relevant point: it might be the very human way to make sense of one's life. In order to make one's life meaningful, it is necessary to believe in the possibility, even in the presence, of a sense in history, and in every life story; and the most reasonable attempt to give a sense is to give an ending to the story: there is a close connection between teleology and *the sense of an ending*.[10] In a second sense, teleology describes human intentionality in a very proper way; intentionality can, in fact, be found either in the individual dimension of the personhood, or in the collective dimension of history. In each case, however, intentionality means the human tension toward a sense, and it signifies that humanity structures the life experience *as if* there be a sense. Thus, intentionality means openness to a sense and attempt to organize oneself's life in a teleological perspective.

The fact that humanity tries to organize personal life and history *as if* there were a sense, doesn't represent itself an empirical certitude, nor a radical denial of the presence of a sense in history. Rather, it indicates the ontological and moral engagement of each personal life. In this context, imagination can be assumed as a paradigm of the ontological and moral engagement of personhood. In fact, imagination can be considered as the faculty which best explains the teleology of life. It is situated in the space between ontology and ethics: as a standpoint for its creations, it can use the ontological background of sensation; as a teleological aim, it has morality. Its work can either grant or deny the possible harmony between ontology and ethics.

The agreement between ontology and ethics is not external in respect to imagination, but it concerns its own structure: it is the capability of personal imagination to represent to itself a different point of view, assuming the ontological (and "ontic") perspective of its own life as a standpoint, and transcending from the bare ontology in order to gain an ethical overview over the teleology of life. Imagination is, thus, structurally directed to an aim, and the quality of the aim depends on the ethical orientation of teleology. Imagination is teleological because it is intentional, and it can find a direction towards an ethical teleology. In a nutshell, imagination can be morally oriented: it can potentially become a moral faculty: we could also speak of *moral imagination*, as S. Lovibond, among the others, does.

As previously stated, imagination is able to show that intentionality can be both personal and collective; moreover, it can also be the connection itself. On moral imagination as a connection between individual and collective dimension, the American philosopher S. Lovibond thinks that moral imagination can be a critical scrutiny of existing institutions by "seeing new aspects, and – arising logically out of such scrutiny."[11] She goes on pointing out the relevance of language as argumentative ethical strategy: "The speculative construction of alternatives. The fact of syntactic structure in language ensures that as competent speakers about ethics, we can represent to ourselves moral justifications for replacing existing institutions by others – even though this competence is itself grounded in our personal history of incorporation in our personal history of incorporation into the existing institutions – imagination as a linguistic capacity."[12]

Moreover, the American philosopher postulates a close connection between imagination and expressive skills: "Even if no one within a particular community actually possesses this philosophical conception of moral and political conflict, outside spectators (e. g.) historians can still, where appropriate, describe the experience of that community in the terms which it suggests – in terms of a struggle, that is between those forces tending towards a breakdown of ethical substance and those resisting such a tendence."[13]

According to the quotations , it is easy to realize that the morality of personal relations must not be taken for granted, but they must always be imagined in a different way. This feature of imagination can be also called *empathy*, and it denotes the competence of a person to put herself in someone else's shoes and to somehow revive her inner feelings. Thus, without imagination, not only judgment, but also empathy would be impossible. The ascription of morality to imagination is of a particular kind: it is neither totally necessary nor completely accidental; once again, also because of its morality, imagination seems to have a hybrid nature. Imagination can be moral, and its constitutional *being* allows it to be ontologically oriented to morality. In other words, moral orientation is not taken for granted; on the contrary, it requires a practical engagement and a continuous reflection over personal and social bonds.

A parallelism can be made to better explain the grade of externality (or, vice versa, of internality) concerning the ascription of morality to imagination. In the epistemic field, imagination is essential because it connects an extreme variety of sensations (or perceptions) and allows to make hypothesis to define an object or a phenomenon: this kind of process hides a deeply teleological structure, because from some perceptual premises – through imagination – a mental object is constructed and a hypothesis concerning its reality is made. The teleological structure is precisely the attitude of knowledge to give a sense (or a form, or an ending) to the external *stimuli*: this attitude is intrinsically morally oriented, even in the case of the epistemic process of knowledge.

In the moral field, imagination has the same role as in the epistemic field: it joins sensations and images and can formulate different kinds of moral actions, by projecting them in the mind and trying to empathize with one's future life. Such a faculty is thus intrinsically moral, because of its power to orient human agency towards good. Suppose you have to make a choice which involves others, and that the result of your choice is influencial to others (to *your* others). What happens in this case is that you cannot but imagine the ethical implications of your choice, even if you are either an utilitarian or a *homo oeconomicus*; that is because our agency is situated among others' agencies, and our definition of good is always conditioned by our deep relations.

Imagination is thus intrinsically oriented to morality: the reason is that it is a relational faculty, a dialogical faculty rather than a monological one; it represents, in fact, personal intentionality, and personal constitutive openness to the otherness. Imagination completes partial moral images and moulds them in order to create new and different courses of action. It is the very human (and the only) way to create, an ethical-ontological way, whose claim should be the faithfulness to the potentialities of human kind. Like in the case of epistemic imagination, in the ethical field imagination shows its attitude to unify fragments and to give them an order, constructing mental courses of human action, which represent, for the ethical life, possible models of action. It provides moral judgment with *schemas* for judging, exactly as it happens in the relation among sensation, imagination and judgment according to Aristotle's *De Anima*.

Such a characteristic makes possible to define imagination as the *laboratory* of moral judgment, as Paul Ricoeur describes it: "This mediating function performed by the narrative identity of the character between the poles of sameness and self-hood is attested to primarily by the *imaginative variations*, to which the narrative submits the identity. In truth, the narrative does not merely tolerate these variations, it engenders them, seeks them out. In this sense literature proves to consist in a vast laboratory for thought experiments in which the resources of variations encompassed by the narrative identity are put to the test of narration."[14]

According to the Aristotelian definition of imagination, it is quite easy to point out the analogy between the teleological constitution of human personhood, which structurally searches for a meaning, and the intentional nature of imagination. The close connection between the notions of teleology and intentionality is a standpoint of this analogy; the meaning is an aim: the human personhood must be predisposed to get it, without necessarily possessing it. The analogy between teleology and intentionality is thus a crucial condition to fully understand the sort of creative power of imagination. The force of imagination is exactly its being situated: its limit is not an accidental feature among the others, rather it is its structural way of being. Imagination cannot but start from the limits of knowledge, of sensation, and it cannot but try to overcome them, moulding new possibilities and imagining new scenarios of *fullness*, because its aim is fullness "according to the being".

The fact that fullness is the aim of imagination may help to clarify its intentional structure; if we mean the fullness in each grade it can be given, it is not difficult to understand the relationship that it has with imagination: starting from a merely epistemic standpoint, fullness can be in fact intended as the fullness of an object that we only partly perceive, and in this case intentionality of imagination makes it possible to predict and to metaphorically see the complete object; moreover, in a moral sense, fullness can signify the perfect action we can choose only imagining and evaluating its consequences, giving them the possibility to live in our minds, also in this case, intentionality of imagination is crucial to get the "fullness" of the action, that is, it is essential to express the tension towards the "good" in action; finally, fullness may signify the human need for a meaning, the projection of this need out of human history and, indeed, its capacity to orient history itself; the intentionality of imagination is crucial in this case too, because it provides both the lexicon and the iconography to think a conciliated world.

Intentionality and teleology share the same structure. Firstly, they are both situated in a living perspective; in other words, they both start from a standpoint, that is the living life itself. It is clear that this being situated represents a unavoidable limit, an *inescapable framework*, but it is the proper human way to tend to the fullness and to highlight the tension to it. Secondly, they both tend to something: whatever this "something" is, it is in both cases able to orient the fragments of knowledge, or of morality, we can experience in our lives in a coherent and well-ordered account. Therefore, both intentionality and teleology share a kind of propulsive linearity, which does not produce necessity, but freedom, and which starts from a point and is directed to another one. The peculiar kind of relationship between intentionality and teleology could also be described in the following terms: intentionality (and, obviously, the intentionality of imagination) is the very human way to decline teleology.

If it is true, the concept of personhood is nothing but this unique way to live the tension towards an end (in the sense of an aim); this aim has basically two possibilities to be realized: the first one is to fit itself to the teleology of being and, by doing so, to reach "fullness"; vice versa, because of the freedom of intentionality, it can betray the tension and tend to fragmentation, rather than the unification of the self in a historical context. The core of human creativity is situated exactly here: according to its main meaning, it is quite close to imagination, thanks to the intentional structure of the latter. Imagination is, in fact, situated in a space, in a time, in a story of life; and therefore it cannot but work according to its being situated, and according to the being which the sensations provide to it. Imagination can get to know the being; it can, subsequently, choose whether either to be faithful or to betray it. In both cases, its intentionality is proved.

The fact that the intentionality of imagination can be faithful to the being must be better explained, because it is exposed to many objections. One of the most relevant critiques could be that, since imagination has the only function of "respecting" the being, of "carrying it out", of fitting itself to it, there is no space to its creative power, and any possible representation of good action as a new element is impossible, because every image is already determinate and almost necessary, every possible future can be found all along as a possibility in the being: the creative force of imagination drowns into the *great see of being*.[15] The reply to this powerful objection is useful to clarify the relation between imagination and being and, as a consequence, between imagination and morality.

The solution of this dilemma could sound as follows: as a premise, one can consider the fact that imagination is situated not only as a limit, but also as a resource; the possibility that imagination can be faithful to the being means that it can find morality into the being and is able to make it evident, to work on it in order to let morality emerge from the being and to orientate it. Imagination is thus valued for its creative power, which is able to discover and, what is more, to shape, a sense of morality which consists on an ethical direction to be impressed to the reality. Imagination has the capability to recognize a regulative ideal into the being: once having found this regulative ideal, it is able to indicate a direction to action, by transforming ideality into a concrete image. In a nutshell, there is no way to avoid or misrecognize the creative power of imagination, even if we take into account its limiting and narrow bond with the being; the creative power of imagination is the very human way to decline intentionality.

Another suitable way to analyze the relation between the *intentionality of imagination* and the *teleology of the lifeworld* is the notion of participation: as far as imaginative intentionality participates in teleology, a teleological structure can be ascribed to the imagination as its main feature. As previously stated, intentionality of imagination is the authentic human way to assume the teleology of life and to freely accomplish it. The fulfillment of vocation to teleology is thus supposed not to be a necessary feature of imagination, but an intrinsic possibility, which can be confirmed or denied by the moral agency of each person.

According to its being closely related to imagination, *human creativity according to the being* can be defined as intrinsically moral. In fact, it is not a neutral process, which would regard only an increase of being, but as a morally oriented process which testifies the dialogical structure of each either imagined or real experience. This dialogical and "opened to the otherness" structure imposes to considerate morality as potentially intrinsic into the being, even though not definitively necessary. The discovering of the morality of the process of creation according to the being can be equated to a radical paradigm shift: from the paradigm of quantity, which is interested in the bare increase of being, there is a transition to the paradigm of quality, in which any moment of being can be transfigured into a moral one: it deals with a conversion, always possible, but never necessary.

To sum up, imagination can be considered the human way to decline the teleology: from one side, it is essential to rediscover the moral potential of the being and to avoid the automatisms of a totalitarian ontology of necessity; from the other side, imagination is able to represent the only human way to create, that is the assumption of the "being situated" as a resource, rather than as a limit. The limitedness of imagination, which can create only taking into account its sensations or perceptions, signifies its unavoidable historicity, which determines the conditions of the being imaginable of events. But the historicity of imagination doesn't draw the line at its transformative power; on the contrary, it makes this transformative power realizable and discloses the radical concreteness of imagination.

As a faculty, imagination cannot be reduced neither to the plain reproduction of the existing order, nor to the radical invention of totally new features: its hybrid nature reflects the middle position of human personhood in the world, so that imagination can't be assimilated to a bare description with the "mind's eye", neither can it be assimilated to a prescription; or, in other words, imagination is both descriptive and prescriptive, because it can't escape the historical and ontological framework, but it can operate in history in order to transform it; its providing models to action is similar to the force of the examples.

A. Ferrara, who has recently dedicated a book to this topic, describes the force of example as follows: "Alongside the force of what is and what ought to be, a third force gives shape to our world: the force of *what is as it should be* or the force of example. For a long time unrecognized and misleadingly assigned to the reductive realm of the aesthetic, the force of example is the force of what exerts appeal on us in all walks of life [...] by virtue of the singular and the exceptional *congruence* that what is exemplary realizes and exhibits between the order of its own reality and the order of normativity to which it responds".[16] Just like the force of examples, the kernel of the imaginative power is to create models for action, and, moreover, to highlight a *congruence* between what is and what ought to be. Imagination, like examples, exerts a force on human action, especially in the ethical field.

Another relevant analogy between examples and imagination is the concept of *congruence*: the force of the example is based on a congruence between description and prescription. An example represents a sensed history, in which concreteness goes together with the possibility to be universally valid. The congruence here is between what is and what ought to be. The same congruence can be found in the field of imagination: it looks at the way in which imagination finds and aims to creatively reproduce an order according to the being; it describes, in the sense of a reproduction, the world, and, what is more, it is able to model a morally oriented direction to the life, not only indicating it, but also prescribing what to do in order to reach this goal.

As the example, which highlights a congruence between the *is* and the *ought to be*, imagination can show a congruence between being and his moral declination, a congruence between images and their moral tension and, finally, a congruence between what is (that is, the realm of images) and what ought to be (that is, what images could become through their being oriented). The core of the analogy is thus the congruence between what is and what ought to be. Imagination has the force to show the possibility of the being to be morally oriented towards a sense. Imagination is able to reply to the questioning of a sense of history, describing the reality and making efforts to read differently the reality itself. The power to differently read reality is not morally neutral, because, by describing, it projects new possibilities and indicates the way to reach them.

The next step of this article is to highlight the close connection or, to properly speak of, the implication between imagination according to Aristotle's definition of it and contemporary narrative ethics. In the quotations below, Aristotle describes imagination as a middle faculty, which is possible thanks to the perceptual field and which makes judgment possible. This characteristic of being median is what makes possible the re-actualization of a theory of imagination from the viewpoint of contemporary narrative ethics, because narrativity too has a median role. In fact, it is either a description, or a prescription. In this sense, examples, especially those elicited from literature, represent a privileged perspective from which it could be easy to focus on the analogy between narrative ethics and imagination.

Once the definition of imagination as the proper faculty of *human creativity according to the being* has been accepted, a textual reference from P. Ricoeur can be useful to confirm and to prove this definition. According to the French philosopher, the value of imagination consists of "a free game with some possibilities, in a state of non-engagement towards the world. It is in this state of non-engagement that we experience new ideas, new values, new ways of being in the world. But this "common sense" connected to the notion of imagination can be fully recognized only in the fertility of imagination is connected with that of language, which is exemplified in the metaphorical process."[17] It is worth to note that the state of non-engagement does not mean a radical disengagement towards the world; on the contrary, the state of non-engagement is the condition of possibility of a really authentic ethical engagement, in order to reach which it is necessary to exert the faculty of judgment in a situated and both free way.

Only this continuous exercise of going out from the self to leave place to the otherness, that is an empathic exercise, allows the impartiality in moral judgment. Imagination is thus the capacity to freely combine different images or situations and to project a different course of action, starting from an unavoidable situated perspective. The non-engagement is thus the real possibility of a true engagement. From Ricoeur's perspective, language is able to establish a link between imagination and narrative ethics, because every representation is immediately translated into language and configured in a (hopefully) meaningful story. Man cannot but represent himself in a linguistic way: it is therefore confirmed the intentional structure of language, which represents a common feature of imagination, language and narratives. In *Oneself as Another*, Ricoeur starts from the literary narratives and proves that they structurally imply moral judgment: "The pleasure we take in following the fate of the characters implies, to be sure, that we suspend all moral judgment at the same time that we suspend action itself. But in the unreal sphere of fiction we never tire of explaining new ways of evaluating actions and characters. The thought experiments we conduct in the great laboratory of imaginary are also explorations in the realm of good and evil."[18] Even if one starts from narratives, the crucial point is that imagination seeks out the morality of new possible courses of action.

Moreover, in order to show the close connection between imagination and narrative ethics as one of its possible actualizations, it is necessary to try to exactly define what it is currently intended with *narrative ethics*. An exact definition of this recent declination of ethics could be problematic, because of the plurality of meanings it has in the contemporary studies. I choose here to quote three main definitions, as they appear in a recent German book.[19] In her introduction to the volume, K. Joisten lists three meanings of *narrative ethics*. According to the first, the language of narratives (and of narratology) renders ethics a scientific discipline, providing it with some useful categories; the category of "narrative" provides ethics with a rigorous language which let it "employ narrative elements in order to describe moral phenomena. Of course it is able to come through this trial, once it is considered as a science especially to critically analyze traditional customs, and to formulate moral judgments in a scientific way."[20]

The second meaning refers to the possibility to extract moral *examples* from literary or life narratives: "in this case, the word "narratives" has got, as its field of application, the wide range of stories and narrations."[21] The third and last meaning highlights a quasi-transcendental essentiality of narrativity in the constitution of human beings, who cannot but live and experience the relations and the world in a narrative form, from the beginning to the end of their lives. Narrativity isn't, thus, an accidental feature of human life, but it represents a proper ontological constitution: "to be a man means to be constitutively entangled in histories, that is to be narrative, to demonstrate to be disposed to be constantly structured and re-structured in a narrative form."[22]

Each of the listed meanings is useful to prove the hypothesis of an implication between Aristotelian imagination and contemporary narrative ethics. The first one points out the capacity of the narrative categories to give an order and a scientific form to ethics: it is as if narrative categories could create a model of the world, in which characters would be free to act and the spectator (who corresponds to the subject of the future action) could reflect carefully to take the best choice, once having imagined a parallel world, with its own characters and relational dynamics which are the copies of the real ones. It is quite easy to note the essential role of imagination in this first meaning of narrative ethics.

The second meaning is close to Ferrara's account of the force of the example, and highlights the close connection between examples and narrativity. Examples provide models for moral action; they are almost narrated and for this reason they exhibit a teleological structure, that is they seek out a sense: they signify something universal through concrete situations: "a judgment that unites a focus on particulars and yet an universal scope in its claim to validity is possible insofar as it appeals to *exemplary validity*. Exemplary validity represents an alternative way of understanding how it is that we are able to identify single objects as instances of a certain type of objects [...] A central role in validity so conceived is played by the *imagination*. For it is imagination – "the faculty to make present what is absent" – that evokes in our mind examples that might apply to our case. Imagination allows us to join together, under the different modalities of determinant and reflective judgment, particulars with general notions."[23]

What imagination does, is to create new models by using old images and combining them. Each model manifests a tendency towards a meaningfulness; moreover, each model can be equated to an example. Therefore, firstly, examples are narratives because they present themselves as stories, as real narratives; secondly, examples are narratives because of their intentional constitution and the possibility for them to be assimilated to imaginative models which tend towards a sense. Like examples, imagination is able to connect the particulars with general cases; narratives have got the same capacity: they are able to refer to universality, even if they give an account only of the particulars. Examples can be properly pieces of narratives, or they can be translated into narrative categories, but this translation seems to be necessary, or at least quasi-transcendental.

The third meaning that K. Joisten lists is crucial: narrativity is not an accidental feature of the human being, but it is their essential feature. Narratives are the tissue of every life, from their very beginning to their end. Narrativity means in fact dialogical openness and relational constitution; the person is constituted by others and it is represented in others' words and stories. Narrativity is the tissue of the human experience, which can be thus described as an experience of radical heterogeneity, and it should engender a grateful attitude towards who has made possible everyone's life story. The intentionality of stories and, in particular, of every life story, highlights the tensional structure of the human personhood.

To sum up, the definition of narrativity which is the standpoint of the third meaning of narrative ethics can be described as a combination of four features: dialogical openness; relational constitution; linguistic translatability; intentional orientation.

According to the dialogical openness, mankind is originally open to the otherness, which precedes and constitutes his lifeworld; it is because (and this is the second feature) to be relational is not accidental for the human beings, but it is constitutive, which means that relations are the main tissue of humanity, without them humanity is not thinkable; this relational constitution does not automatically correspond to a moral orientation towards the ethical life, but represents its condition of possibility, without which the orientation towards ethical life would not have the space to actualize itself.

The third feature is represented by the linguistic translatability: this feature cannot be assumed as a datum, but rather as a problem, because it is not taken for granted that any human experience can be said, or narrated, or communicated: the experience of evil, for example, or that of extreme suffering, are not so easy to tell or to be linguistically configured. If, in fact, the language, especially the language of narratives, express an attempt to give a sense, it is always possible not to find a sense. From this standpoint, it can be equallly said that language is "onni-pervading", because it shows a tendency which is constitutively human. The dramatic (even tragic) feature of humanity is being intentionally linguistic, together with the absolutely not granted possibility to express oneself.

The fourth feature is intentional orientation: according to the foregoing notations, this feature allows the possibility to focus on the close connection between imagination and narrative ethics. Imagination has a teleological orientation towards an end, intended as an accomplishment, either of an object, an action, or even of a history; moreover, its being situated represents the very human condition: the power to start a new course of actions, having thought of it before, is the only way to create. Imagination is thus the faculty of *creation according to the being*. Narrative ethics is at the same time teleologically oriented towards an end (both in the sense of an end and in the sense of an aim); it is able to configure a possible future and to create a model to the ethical action, starting from a situated perspective: in this sense, also narrative ethics can create *according to the being*, even though this kind of creation is not necessary but ontologically possible.

Therefore, narrative ethics can represent a proper way to actualize the hypothesis of the ontological force of the imagination. Firstly, what they have in common is the intentionality which manifests a teleological structure according to the being; secondly, their common structure is signified also by the relational constitution; both imagination and narrative ethics are fitting metaphors for this relational feature, because they represent the impossibility to find a sense of life in a complete isolation from the others. They confirm, rather, that the relationships with others are the condition of possibility of one's ontological being and of the reaching of a sense of fullness and commonality.

In a few words, like imagination, also narrative ethics can operate a *synthesis of the heterogeneous*. It means that they are both able to order epistemic processes (just think of the recent studies of narrative psychology[24] and to the relevance they give to the narratives in the learning process), human actions and entire life stories, in order to render ethical the ontological teleology. Without this kind of ethical reflection, the teleology would only remain ontologically possible, not being able to actualize itself. What is more, without this kind of reflection, teleology could not be defined *ethical*, because it would be automatic and it would not imply the individual freedom to choose whether to be moral or not: that is, in a state of necessary teleology, no one would be good or bad, because no one could choose and people would remain in a state of indeterminacy.

As a conclusion, two theoretical cores are worth of mentioning. The first one is that both imagination and narrative ethics point out that the only way for men to create is to create according to the being. Whether a person thinks of a new course of action, an accomplishment of history, a life story to be morally oriented, it is necessary to recognize the essential role that imagination plays in these cases and to identify its contiguity to the ethical-ontological processes of creation according to the being. Imagination is implied in narrative ethics, as already seen, because it represents a state of non-engagement through which different ways to project common life are highlighted. Narratives, for their part, represent the configuration of these projects of different ways to live the community relations, if we intend with "community" the relational tissue of the human experience. Before having a story to tell, we ourselves are stories, concrete living match points of stories and dialogical living tissues.

The second theoretical core can be described as follows: imagination and narrative ethics (that is, the narrative tissue and structure of the ethical life) are useful to rethink and to reflect on the configuration of the bonds (from the familiar ones, to the social ones) both in a retrospective and in a prospective direction. In a retrospective way, imagination is crucial in order to make present what is absent; moreover, it is crucial because of its heuristic power to empathize with past characters and to go back over the history imagining the real course of events; finally, it is very relevant because it allows to think the hypothetical consequences of a different course of history, in order to correct the perspective over the future.

In a very similar way, narrative ethics has a both a retrospective and a prospective validity. The case of the retrospective validity can be easily shown because of one's account with the past, it can contribute to better understanding of the present, either according to the personal life story, or according to the collective history. To be disposed to be hosted in the stories narrated by the others (stories which are made of words) means to recognize the structural heteronomy of the personhood. The possibility to be differently narrated, with different voices, corresponds to the possibility of an ethical reflection on one's relationships; this capacity is very ambiguous if we think of the past and of its possibility to be manipulated. Lastly, the prospective validity of narrative ethics concerns its capacity to project imaginatively different ways to decline relations and new ideas of common life ethically oriented. Imagination as "human creativity according to the being" revives thus in the contemporary perspective of narrative ethics, confirming once again its being a teleological capacity to implement and to actualize human condition.

University of Macerata, Macerata, Italy
e-mail: silviapierosara@libero.it

NOTES

1   Aristotle, *On the Soul*, trans. W. S. Hett (2nd ed., Cambridge, MA: Harvard University Press, 1957), p. 157.
2   Ibid., p. 159.
3   Ibid., p. 163.
4   Immanuel Kant, *Critique of Pure Reason*, trans. J. M. D. Meiklejohn, T. Kingsmill Abbot, W. Hastie, J. Creed Meredith (Chicago: Encyclopaedia Britannica, 1952), pp. 54–55.
5   Malcolm Schonfield writes: "It is natural to assign such interpretative activity to the imagination. This is particularly the case where the interpreting is conscious." [Malcolm Schonfield, "Aristotle on the Imagination", in Martha C. Nussbaum, Amélie. O. Rorty, *Essays on Aristotle's* De Anima (Oxford: Clarendon Press, 1992, p. 259)]. Following this interpretative line, Dorothea Frede writes: "For it will turn out that there is a wide *gap* between the two, and that at least one of the functions of imagination is to fill that gap. This is not to deny that some *phantasiai* are 'mere appearance'; it is just to show that not all are. I will confine myself to a depiction of two main functions of *phantasia* in Aristotle's psychology: its role in the *synthesis* and retention of sense-perceptions, and its role in applying *thought* to objects of sense-perception." (Dorothea Frede, "The Cognitive Role of *Phantasia*" in Nussbaum, Rorty, op. cit., p. 282).
6   See, for instance, the following works on Aristotelian imagination: Jan J. Chambliss, *Imagination and Reason in Plato, Aristotle, Vico, Rousseau, and Keats: an Essay on the Philosophy of Experience* (The Hague: Nijhoff, 1974); M. Schofield, *Aristotle on Imagination* (2nd ed., London: Duckworth, 1979); Thomas Claviez, *Aesthetics & Ethics: Otherness and Moral Imagination from Aristotle to Levinas and to Uncle Tom's Cabin to the House Made of Dawn* (Heidelberg: Winter 2008).
7   It could be interesting to show the surprising modernity of this Aristotelian statement: just think of Maurice Merleau Ponty's *Phenomenology of Perception*, trans. C. Smith (2nd ed., London: Routledge, 2002).
8   The primacy of sight has been clearly pointed out by Hans Blumenberg, *Shipwreck with Spectator: Paradigm of a Metaphor of Existence* (Cambridge, MA, London: MIT Press, 1997).
9   This expression is used by Charles Taylor, *Sources of the Self. The Making of Modern Identity* (Cambridge: Cambridge University Press, 1989). Such an expression can be defined in the following way: "My underlying thesis is that there is a close connection between the different conditions of identity, or of one's making sense, that I have been discussing. One could put it this way: because we cannot but orient ourselves to the good, and thus determine our place relative to it and hence determine the direction of our lives, we must inescapably understand our lives in narrative form, as a 'quest'. But one could perhaps start from another point: because we have to determine our place in relation to the good, therefore we cannot be without an orientation to it, and hence must see our life in story. From whichever direction, I see these conditions as connected facets of the same reality, inescapable structural requirements of human agency." (Ibid., pp. 51–52).
10 See Frank Kermode, *The Sense of an Ending. Studies in the Theory of Fiction* (London, New York: OUP, 1967). Kermode writes: "The physician Alkmeon observed, with Aristotle's approval, that men die because they cannot join the beginning and the end. What they, the dying men, can do is to imagine a significance for themselves in these unremembered but imaginable events. One of the ways in which they do so is to make objects in which everything is that exists un concord with everything else, and nothing else is, implying that this arrangement mirrors the dispositions of a creator, actual or possible." (*Ivi*, p. 4). A few pages after, he continues: "The matter is entirely in our own hands, of course; but our interest in it reflects our deep need for intelligible Ends. We project ourselves – a small, humble elect, perhaps – past the End, so as to see the structure whole, a thing we cannot do from our spot of time in the middle." (Ibid., pp. 7–8).
11 Sabina Lovibond, *Realism and Imagination in Ethics* (Oxford: Basil Blackwell, 1983), p. 117.
12 Ibid., p. 194.
13 Ibid.
14 Paul Ricoeur, *Oneself as Another*, trans. K. Blamey (Chicago: University of Chicago Press, 1994), p. 148.

[15] This expression is from Dante Alighieri, *The Divine Comedy, Paradise*, trans. C. E. Norton (Chicago: Encyclopaedia Britannica, 1952), Canto I, line 113, p. 107.

[16] Alessandro Ferrara, *The Force of the Example. Explorations on the Paradigm of Judgment* (New York: Columbia University Press, 2008), pp. 2–3.

[17] This quotation is elicited from the Italian edition of Paul Ricoeur, *Imagination in the Discourse and in the Action*, in Paul Ricoeur, *From text to action*, trans. K. Blamey and J. Thompson (London: Athlone, 1991). The Italian edition, from which the quotation is taken, is: Paul Ricoeur, *L'immaginazione nel discorso e nell'azione*, in *Dal testo all'azione. Saggi di ermeneutica*, trans. G. Grampa (Milano: Jaca Book, 1989), p. 212.

[18] Ricoeur, *Oneself as Another*, p. 164.

[19] Karen Joisten *Möglichkeiten und Grenzen einer narrativen Ethik*, in Karen Joisten (edits), *Narrative Ethik. Das Gut und das Böse Erzählen* (Berlin: Akademie, 2007), pp. 9–21. See also Claudia Öhlschläger (edits), *Narration und Ethik* (Paderborn: Fink, 2009); these two books are not translated in English.

[20] Joisten, op. cit., p. 10.

[21] Ibid, p. 11.

[22] Ibid., pp. 12–13.

[23] Ferrara, op. cit., pp. 47–48.

[24] See at least the very complete study of Janoś László, *The Science of Stories* (London and New York: Routledge, 2008).

REFERENCES

Alighieri, D. 1952. *The Divine comedy* (trans: Norton, C.E.). Chicago: Encyclopaedia Britannica.
Aristotle. 1957. *On the soul* (trans: Hett, W.S.), 2nd ed. Cambridge, MA: Harvard University Press.
Blumenberg, H. 1997. *Shipwreck with Spectator. Paradigm of a metaphor of existence* (trans: Rendall, S.). Cambridge, MA, London: The MIT Press.
Chambliss, J.J. 1974. *Imagination and reason in Plato, Aristotle, Vico, Rousseau, and Keats: An essay on the philosophy of experience*. The Hague: Nijhoff.
Claviez, T. 2008. *Aesthetics & Ethics: otherness and moral imagination from Aristotle to Levinas and to Uncle Tom's Cabin to the House made of Dawn*. Heidelberg: Winter.
Ferrara, A. 2008. *The force of the example. Explorations on the Paradigm of judgment*. New York: Columbia University Press.
Joisten, K. (eds.). 2007. *Narrative Ethik. Das Gut und das Böse Erzählen*. Berlin: Akademie.
Kant, I. 1952. *Critique of pure reason* (trans: Meiklejohn, J.M.D., T. Kingsmill Abbot, W. Hastie, and J. Creed Meredith). Chicago: Encyclopaedia Britannica.
Kermode, F. 1967. *The sense of an ending. Studies in the theory of fiction*. London, New York: OUP.
László, J. 2008. *The science of stories*. London and New York: Routledge.
Lovibond, S. 1983. *Realism and imagination in ethics*. Oxford: Basil Blackwell.
Modrak, D.K.W. 1986. Φαντασία Reconsidered. *Archiv für Geschichte der Philosophie* 68:47–69.
Nussbaum, M.C., and A.O. Rorty. (eds.). 1992. *Essays on Aristotle's* De Anima. Oxford: Clarendon Press.
Öhlschläger, C. (eds.). 2009. *Narration und Ethik*. Paderborn: Fink.
Ponty, M. Merleau. 2002. *Phenomenology of perception* (trans: Smith, C.), 2nd ed. London: Routledge.
Ricoeur, P. 1991. *From text to action* (trans: Blamey, K. and J. Thompson). London: Athlone.
Ricoeur, P. 1994. *Oneself as another* (trans: Blamey, K.). Chicago: University of Chicago Press.
Schofield, M. 1979. *Aristotle on imagination*, 2nd ed. London: Duckworth.
Taylor, Charles. 1989. *Sources of the self. The making of modern identity*. Cambridge: Cambridge University Press.

SECTION X
SEEKING THE LOGOS IN DIFFERENT CULTURES

KATARZYNA STARK

THEOSIS AND LIFE IN NICOLAI BERDYAEV'S PHILOSOPHY

ABSTRACT

The concept of *theosis* constitutes a central theme in the Byzantine theology and, generally, Eastern Christian spirituality. In his mystical realism, Nicolai Berdyaev refers to this tradition. For him, the mystical experience reveals the specific status of man as created in God's image. In his creative life, man can be divinized and, consequently, participate in the divine community. Berdyaev analyses the process of *theosis* referring to the most perfect example of Christ. *Theosis*, in the Russian philosopher's view, constitutes the aim of human existence. It discloses the reality which, from the perspective of contemporary philosophy Anna-Teresa Tymieniecka in her phenomenology defines as "the Fullness in the Ex-tasis of life, in the Glory of the Divine".

Nikolai Alexandrovich Berdyaev began his philosophical quest by looking for truth in Marxism, next to such philosophers as Sergei Bulgakov, Siemion Ludvigovich Frank, and Mikhail Ivanovich Tugan-Baranovsky. Accordingly, Berdyaev was initially seen as a member of the group of thinkers recognised as the so-called legal Marxists who spread the Marxist ideas in the 1890s in Russia. However, likewise most "legal Marxists", in his search for truth he did not remain a Marxist but instead turned to Christianity.

Berdyaev became a religious thinker not through a sudden conversion but rather as a result of the evolution of his ideas. His main interests lie in the question of the existential dialectics of the relation between man and God. Most interpreters of Berdyaev's philosophy regard him as an existentialist. This existential aspect is especially stressed by Fuad Nucho.[1] Berdyaev claims that the authentic philosophy is the representation of man's existential experience. As such, it bears a direct relation to life. According to F. Nucho, the anthropocentrism of Berdyaev's philosophy stems from the fact that the Russian existentialist acknowledges as the most important in the formation of philosophy concrete human experiences which form the image of man as a person.

Berdyaev maintains that the Divine element in man is revealed in the experience of the actualization of the indestructible image of God, which acquires its definitive form in the perfection of the existence of person. According to his idea of God-manhood, man is the co-creator and participator in God's existence to the same degree as God Himself assumes an active part in man's existence. This ontic interdependence between the divine and human existence, the conviction that neither God can be fully Himself without man nor is man himself without God, implies their mutual openness. It leads to the "mutual co-existence of the human reality in God and of the God's reality in man", as Wacław Hryniewicz notices.² Berdyaev illustrates the essence of these dialectic relations between man and God in the following terms: "By realising in himself the God's image, man actualizes the human image, and by actualizing in himself the human image, he actualizes the image of God. This is the mystery of God-manhood, the greatest mystery of human life. Manhood is God-manhood."³

The concepts of image, likeness, and person constitute the foundations of all anthropological reflections in Berdyaev's theory. They aim at confirming the thesis of man's birth in God. Within the orthodox thought, the realisation of the image of God in man found its theoretical justification in the complex idea of *theosis*, viz. the divinization of man's existence. This notion is founded on the conviction common in the orthodox thought that man has a theophoric nature, guaranteed by the indestructible image of God which is present within him. The idea of divinization reveals in man his Divine image. At the same time, divinization constitutes a full realization of the image of God. As W. Hryniewicz explains, the idea of *theosis* "means a spiritual development of man in the likeness of Christ, which takes place in the Holy Spirit. This development leads to the achievement of the true manhood."⁴ As a result of divinization, man achieves the highest degree of the ontic realization of his existence – the spiritual state. Only the divinized state of human nature, which reveals God's image within it, can lead to the affirmation of the fact that man has become a person. Within the orthodox thought, this state can be achieved here in our life, which is symbolized by the "light of Tabor". That is why the orthodox anthropology is so optimistic. It does not regard the divinization and participation in God's nature as an unachievable ideal, but rather confirms the belief in man's ability to accomplish this purpose. Seen in this light, man is not the prisoner of sin but rather, as a partner and collaborator of God, is called to the divinizing participation in the life of God already on earth.

The eternal process of the coming-to-life of the Holy Trinity, which takes place in the dynamics of God's Nothingness, implies the revelation of both God and man within it. In Berdyaev's theory, there is no original ontic separation between God and man. God does not exist "prior to" man. The births of God and man take place simultaneously in the theogenic-anthropogenic process. Thus, the existence of man constitutes an irreducible element in God's self-determination, as well as a necessary moment of co-creating the essence of the Holy Trinity life.

This dynamic vision of the Holy Trinity does not allow for the understanding of creation as a single act achieved in the eternity by God. The creation must encompass three dialectic moments of the manifestation of the creative power which correspond to the properties of each of the Three Divine Persons.

In Berdyaev's philosophy, the archetypal ideal image of creation within the Holy Trinity assumes the following form: the emanation of the Father's Divine might contained in the idea of the perfect world; the actualization of this idea in the world personified in the Son, who is the "absolute norm" for the world because it is Him who restores the ideal image of the creation; the divinization of the whole universe by the Holy Spirit, which constitutes the universal fulfilment of the idea of God-man individually actualized in the world.

The idea of God-manhood emphasizes the fact that the rule of synthetic balance between the Divine and the human element must be applied in the understanding of the relations between God and man. As Wacław Hryniewicz puts it: "(...) the idea of God-manhood inherent in man aimed at overcoming both transcendentism and immanentism. It aimed at exposing the partial truth present in both of them and at contributing to the revelation of the harmonic synthesis between the Divine and the human in the structure of the created reality."[5] The question of mutual transcendence and immanence of God and man finds its definite solution in the values pertinent to the God-man being, the essential meaning of which is co-created by the non-exclusive and non-contradictory principles of divinity and manhood.

In order to maintain his thesis of God-man and God-manhood, Berdyaev must assume a critical position in regard to the ontological monism (which treats man as one of the manifestations of the life of the Divinity, i.e. as one of the transitory moments in the Divinity development), pantheism (which allows no place for human freedom and creativity), and trascendent dualism (which excludes the reconciliation of divine and human nature, and the cooperation between God's and human will). Instead, he clears up his own understanding of dualism. He namely advocates a creative Christian God-man anthropology, a position which allows for independent existence of two united natures: divine and human, as well as for the cooperation between the divine and human freedoms.

The author of the *Truth and Revelation* emphasizes the fact that cognition is an act directed toward the foundations of reality. The philosopher must first grasp the source-presence of reality, and only on the basis of such experience can he formulate judgements about the world. This type of proceeding is both subject-rooted and highly creative which is why philosophy in Berdyaev's system becomes more akin to art than to science.[6] The existential philosopher thus criticises the perception of metaphysics as an objective science which analyses the Aristotelian notion of *Being qua Being*, and treats its substantiality in static categories.

The quest for authenticity in Berdyaev's philosophy is connected to his search for truth. The truth about life is not a concept reached by the intellect or by abstract logical reasoning. Rather, the truth reveals itself in the manifestation of the original sources of life, in which the human being takes an active part. Therefore, the recognition of truth goes hand in hand with the recognition of life, and it can be said that man cognizes truth to the extent he knows life. That is why the truth constitutes a dynamic process in which entire man must take an active part: both his spiritual side, his psyche, and his body.

The active engagement in the quest for truth corresponds to one's faith that its essence and value will eventually become revealed. This is where the religious aspect of the experience is founded, since, according to Berdyaev, God constitutes both the ultimate meaning of the existence of the world and the reason for it. On the level of the knowledge that is certain, there is no difference between the acquired philosophical wisdom and religious revelation, because God, understood not as a supranatural being but as the unquestionable value of the whole existence becomes the only standard of the acquired knowledge adequateness.

The author of *The Destiny of Man* proposes an answer to the question of the personal self-knowledge and the method guaranteeing the actualization of the authentic life by introducing the idea of God-man. David B. Richardson points out a specific aspect of Berdyaev's existentialism, viz. the fact that it becomes truly Christian owing to the assumption that man affirms his existence in the authentic way only when he becomes God-man.[7]

The authenticity of personal existence is actualized only within the image of God-man. What follows is that man's divine-human nature reveals itself fully in the area of the deepest existential experiences. These experiences are connected with the response man gives to the Divine. It would be pertinent to quote Anna-Teresa Tymieniecka in this context, who observes: "Thus to our viscera the Divine calls, and we 'respond' in our inward vision of the world, life, the existence of all thorough infinite understanding, mercy, forgiveness, generosity, love. Our vision of our universe is lifted above abysmal suffering to exalted enchantment with all the 'gifts' that nature, our Human Condition, and imagination shower upon us."[8] The existence of God-man reveals the manner of person realization in the finite, limited, temporary, and historic world. Christ's self-knowledge and the divinization of His human nature, break with, and, to certain extent, nullify the solidity of the phenomenal world. At the same time, the optimism which follows from the certainty of the person full realization in Christ, leads to the ultimate conclusion related to the value of man and affirmation of the entire world.

In order to explain the complex relations between God and man, Berdyaev specifically goes back to the religious experience, which – according to him – is the best possible way of grasping the truth about God and man.

In Berdyaev's philosophy, the religious experience is founded upon the meeting of God with man, in which God's reality proves to be *conditio sine qua non* of the authenticity and irrefutability of this experience. In such experience, God reveals Himself as a person, as a personal love, that both requires and awaits man's response. Thus, the religious experience, according to the Russian existentialist, is of the existential-dialogical nature, it is an experience of the conscious "I – thou" relation.

The Russian existentialist assumes that intuition plays the major role in religious experience. It is owing to intuition that the cognition of God acquires its spontaneous character. Intuitive cognition transcends the purely intellectual scope because it engages those aspects of human being which are related to emotions and volition. This allows for the thesis that Berdyaev regards intuition to be the constitutive characteristic of the subject,[9] the property which not only enables the subject to recognize and define any being, but also allows him to co-create it by means of his participation. In other words, in the intuitional act man attempts to grasp and understand the meaning of the universal, i.e. the archetypal structures of the world, by reaching the original source of all being.

The fullest kind of religious experience is the mystical experience, in which the personal "I", while maintaining its uniqueness, at the same time experiences the unmediated presence, acting, and union with God. In the philosophy of the Russian existentialist it becomes clear that the mystical experience reaches far deeper than any discursive knowledge of God. Compared to the dynamics of religious experience present in the former, the latter turns out to be both shallow, static, and schematised. Within the mystical experience, in the meeting and dialogue between two loves, any judgements or explications prove needless.

The first step in mystical cognition, seen as the way of entering the domain of God, is to clear the idea of God from any elements borrowed from the world of phenomena. Then we waive the conviction of our knowledge of God by means of putting in doubt and negating His anthropomorphic or sociomorphic image created by the empirical ego's stereotypical thinking. Since discursive cognition turns out to be insufficient to satisfy human yearning for the full knowledge about God, man has to engage his entire being, all its aspects, in order to fully open his existence so that God can present Himself in his immense greatness. On this level of cognition, man must first master his knowledge of himself, so that through self-knowledge he can cleanse his ego and within this new acquired purity "create space" for God's acting. At this point, man realizes his unity with God gained through *theosis*. In this context, we can employ Anna-Teresa Tymieniecka's words to clarify Berdyaev's idea: "Only now in the apparition of the Witness in the soul, in his entire presence which affirms himself in his transnatural meaning, do we discover that our personal destiny, so much sought after, was transnatural, that the modality elaborated in our soul, was the modality of the sacred."[10] The mystical cognition accepts every existential state of experiencing contradictions, inner fights, heroic spiritual battles. Each of them is in itself the emblem of the great effort of man on his path towards the full-knowledge of himself. The purified "I", that has overcome the states of existential conflicts and contradictions, gains the assurance of reaching the full union with God, Himself the ultimate aim of human pursuance and its definitive fulfilment.

Berdyaev does not regard mystical experience as a passive cognitive process. Mystical cognition, which activates the supra-consciousness, expands the limits of man's cognitive abilities. It reveals the limitless noumenal world, the original source of existence. Mystical realism, the key postulate in Berdyaev's philosophical approach to the world, foregrounds man's active existence within the world – his creative continuation of the act of co-creating the reality with God. While emphasizing the multifaceted character of human creativity and activity, the author of *Truth and Revelation* also stresses the fact that the creative power itself is gained through the mystical experience. It is in the mystical experience that infinity (which inspires man to actualize the truth about himself as God's image) opens to him the opportunity to gain the unquestionable knowledge about himself, and to divinize both himself and the world in accordance with this knowledge.

Berdyaev also applies negative theology in his description of the states of the deification of human nature, in which man becomes God. The negative theology is based upon the experience of the direct presence of God who reveals Himself within the subjective depth of person.

The object of the apophatic cognition and contemplation of the Three Persons of God is not only God Himself, whose nature cannot be expressed in any human thought, but also man. One can only experience the essence of manhood in the region of spiritual "pure existence". In apophatic theology, cognition consists in transgressing the limited cognitive method based upon rational concepts. On the way of gaining the apophatic knowledge of God, one begins to gradually question and reject everything that is learned in a positive way through subsequent stages of cognition. Such purification of one's consciousness can lead to the direct union with God. The apophatic cognition is ultimately crowned by man's experience of Divine luminousness, i.e. the inner light of God that deificates the human nature. Yet, in the experience of God's light, the divine and human nature remain separate.

Apophatic theology, seen as "the spiritual interpretation of God's mystery", is for Berdyaev the method of man's entering the "realm of pure existentiality", both through the consciousness transformation and the acquisition of supra-consciousness. It is not until reaching this cognitive level that the revelation of God's mystery becomes possible: God who has no name in any human language.

The person is considered a fundament of the *theosis* realization. Broadly speaking, in the view of Berdyaev's philosophy the person is created by God in eternity as an idea which is to realise itself in the self-development of man: through his conscious acting in the temporary existence. Thus, the Russian thinker assumes that the person comes directly from God since the achievement of the heights of man's development presupposes the existence of something greater than himself.[11] In his writings, Berdyaev often relates the concept of person to the concept of God: "The personality is the image and likeness of God, and that is its sole claim to existence; it appertains to the spiritual order and reveals itself in the destiny of existence."[12]

Berdyaev's concept of God-man, i.e. of man becoming God in his personal existence, stems from the personalistic premises of Christianity. The process of the formation of a personal consciousness in man began in the Old Testament and Greek culture, and reached its final stage in Christianity. According to Berdyaev, the essence of Christianity can be summarized in its recognition of the absolute value of every human person as the image and likeness of God.

Berdyaev, a fervent advocator of personalism defends the thesis that the person in his unique existence constitutes the highest possible value as he focuses within himself the perfection of being as such. Christianity, which discovers and respects the absolute value of every person, takes for its example the figure of Christ in whom personal life reaches its absolute fulfilment. "Human being, says Berdyaev, in the word's true sense, exists only in Christ and through Christ because Christ is God-man".[13] Taking into account the difference between image and likeness, one should notice that in Berdyaev's theory Christ has fully realized the idea of likeness understood as dynamic progress, as growing in Divine spiritual self-knowledge and in God's life, with the participation of man's freedom.[14] Thus, Christ restored the possibility of being entirely like God, that is of constituting God's image and likeness.

In Berdyaev's philosophy, the person has a paradoxical nature. He has a changeless fundament – God's image that actualizes itself as likeness within the variable. Although God's image of man belongs to the world of noumenal eternity, perfection and freedom, it realizes itself in the domain of temporality and time, within the world of phenomena, ruled by the principle of necessity and specific laws. These laws determine and limit the manner of actualizing God's image.

It is vital to clearly distinguish within the philosophy of the Russian existentialist between the corporal-psychological and the spiritual man. Man as a corporal-psychological being belongs to the objectified world and, accordingly, he can easily become objectified himself, that is lose the authentic character of his being. On the other hand, the spiritual man is a person, a free subject who is radically different from psycho-physical manifestations of his objectified nature. Hence, he cannot be defined or described in empirical categories. Personal existence is deeply rooted in the spiritual world and "the spirituality stemming from the depths is the force which both creates and sustains the person in man."[15]

In Berdyaev's theory, the fact that man belongs to the noumenal world means that only person – thing-in-itself – exists in an authentic way and is the creator of non-objectified meanings. The personal spirit regards creative acting within the world of phenomena as his greatest duty. Through creative acting, the spirit overcomes all types of transcendental appearance, restores and reveals its true image, the source of its personal existence as a free subject.

There is a point of contact between man and God in the original indestructible image. The image is as though a common form which allows for the actualization of the contents of the fundamental God-human relationship. The perfection assumed "at the beginning" in the image of the potential state, turns out to be the task set before man, which he is obliged to realize within his temporary life, engaging all his creative powers and abilities in the process. The fulfilment of this task means the actualization of God's likeness, a pleromatic fulfilment. Only then, once man has fully realized his likeness to God, does he become a person. Man's existence as the person means that God's image and likeness present within himself become fully revealed.

The exceptionality of Berdyaev's approach towards the realization of God's image lies in the fact that in this process he includes the actualization of the universum within the person's unique life. The person is the spiritual centre where microcosm meets macrocosm in this process, the individual fate unites with the fate of the world, and the individual history with the history of the universe. In this centre, a mutual infiltration occurs which leads to the union and harmony, instead of fight with the outer, objectified nature. Berdyaev describes this state of personal integrating spiritual centre in a metaphor of the sun radiating from man: "The sun should be inside man – the centre of the universe – the man himself should be the sun of the world, around whom everything circulates."[16]

Although the spiritual essence cannot be expressed directly in any conceptual philosophical system, its attributes can be grasped and described. Among them Berdyaev quotes freedom, creativity, love, integrity, as well as intuitional cognition, the pursuit of knowledge and reconciliation with God's reality.

In the meaning related to human structure, the spirit is treated as the decisive factor in the authenticity of personal existence. Seen as such, the spirit does not act as impersonal pre-rule of existence or the ideal fundament of the world, but shows its concrete nature, revealing itself in the personal existence. The spirit is "both the subject and subjectivity, both freedom, and a creative act."

The spirit has the power to deificate the human nature because "through spirit man becomes a Divine image and likeness. Spirit is the Divine element in man; and through it man can ascend to the highest spheres of the Godhead."[17] Following Berdyaev's thought we could thus say that in man who has become a spirit incarnate, "spirituality is the highest quality, a value, man's highest achievement. Spirit is, as it were, a Divine breath, penetrating human existence and endowing it with the highest dignity, with the highest quality of existence, with an inner independence and unity."[18] The unceasing influence of the Holy Spirit on person's life manifests itself both in the achievement of the source depth of spiritual life and in the attainment of the highest level in the development of the spiritual life. According to the author of *The Destiny of Man*, spirit and the Holy Spirit turn out to belong to the same reality, which is actualized to various degrees in various people.

The state of divinization, in which man acts by the power of the Holy Spirit, is only attainable if the spirit has succeeded in the integration of the psycho-physical dimension of existence and has achieved firm dominance over it. However, should man's spirituality remain still hidden and not actualized, his development will stop on the level of such experience and such self-knowledge which will neither allow him the insight into the essential and immeasurable profoundness of the spirit, nor prepare him for living in accordance with the spirit's principles. Berdyaev confirms this position when he says: "The failure to fully reveal the nature of the Holy Spirit is in Christianity a failure to fully reveal the ultimate, overwhelming profoundness of the spiritual life; it is a constraint of the spirit by the soul, it is the lack of awareness that all spiritual life, all true spiritual culture, originate in God and in the Holy Spirit."[19]

Through the Holy Spirit man enters the sphere of pure spirituality. The Holy Trinity then turns out to be a symbol encompassing both God's and man's essence. God becomes incarnate in Christ, while through the acting of the Holy Spirit man surpasses his carnality and is incorporated in the mystery of God's inner life.[20] God's incarnation creates the opportunity for man to attain divinity. It is primarily Christ as God-man who neither diminishes nor annihilates His divinity, but rather maintains it entirely, completing and enriching it with the human element. Thus, Christ who is born in the spirit and delivers His spirit to people, elevates human nature to His own Divine-human dignity. People "are gods" by dint of Christ's actualized archetype of the self and they can actualise this archetype themselves. Accordingly, the "Christ-transformed" people who bear the image of Christ-Anthropos within themselves, through the divinizing influence of the Holy Spirit, can reveal the fullness of the divine life in the Holy Trinity that is inexhaustible in its manifestations.

As Marek Styczyński notices, in Berdyaev's thought Christ actualizes the spirituality which corresponds to the "everlasting man", and by doing so, he becomes the precursor of a new spiritual humanity. The author of *The Meaning of History* explains this idea as follows: "The appearance of Christ marks the beginning of a new human species, i.e. a Christ-governed humanity, spiritual humanity, born and reborn in Christ."[21]

As a new Adam, Christ stands in opposition to the old Adam – the man of nature. The beingness of Christ is formulated not as an abstract idea of perfection but as an attainable form of existence, which excels the way man of nature exists. Christ sets the example on how to overcome the old Adam's nature and how to reveal in his nature the spirit which has been dormant or restrained by the order of the phenomenal world. He also shows how to include the spirit in the process of reforming and divinizing the human nature. Hereby Berdyaev confirms that this type of idea of the spiritual man – unknown to the ancient world – is a specifically Christian concept.

Christ is the prototype of perfection. It is in Him that the universum finds its realization. It should be once again emphasized that this realization of the fullness of life in Berdyaev's idea of God-man is related to the idea of *theosis* – divinization – which stems from the stoic idea of all creation's participation in God's nature. Owing to Saint Athanasius the Great, this idea has become the central one within the Eastern soteriology. As the philosopher states: "Theosis makes man Divine, while at the same time preserving his human nature. Thus, instead of the human personality being annihilated, it is made in the image of God and the Divine Trinity. The personality can be thus preserved only in and through Christ."[22] Through the divinization of His human nature, Christ achieves the highest level of the realization of His being – the spiritual state. The divinization of Christ's human nature does not lead to its identification with God's nature, but rather it furnishes Christ's human nature with a new quality – not present in the nature of the old Adam. John Meyendorff very aptly describes the state of divinization when he asserts that Christ's divinized human nature never forfeited its human features. Rather, these features become even more real and authentic through the contact with the divine ideal, by power of which they were created. "In Jesus Christ, Meyendorff continues, God and man are one, in Him God becomes accessible not because He replaces or eliminates the *humanum*, but rather because He actualizes and reveals humanity in its purest and most authentic form."[23]

It is because Christ is also fully human, that the human nature becomes fully integrated within the life of the Holy Trinity. Man becomes God's Son when in the state of divinization of his nature he can ascertain without doubt that: "Christ does not exist outside of us, but in us, He is the Absolute Man in us, He is our participation in the Holy Trinity."[24]

The person achieves ultimate fulfilment when man has actualized his spiritual nature, and when he fully reveals the spiritually which at the beginning of the process of self-realization remains potential and unconscious. As a result of full realization of person's spiritual depth, man's existence is from then onwards judged and looked upon from the position of pure spirituality. At this level of existence, the person, as the highest realization of the spirit's essence, "can reveal the pure and original conscience, free from objectification, and sovereign in all matters."[25]

Berdyaev takes the heart to constitute the person's core. The notion of heart symbolizes the intangible centre of the essential spiritual unity of man who is substantially multifaceted. "The heart (...) is the seat of wisdom and the organ of the moral conscience, which is the supreme organ of all evaluation",[26] elucidates Berdyaev.

Theosis is actualized in life in which no radical separation between time and eternity exists. Time and eternity are not autonomous, absolute and contradictory values, but they co-exist and infiltrate each other. As Berdyaev puts it: "Eternal life is revealed in time, it may unfold itself in every instant as an eternal present. Eternal life is not a future life but life in the present, life in the depths of an instant of time. In those depths time is torn asunder. It is therefore a mistake to expect eternity in the future, in an existence beyond the grave and to look forward to death in time in order to enter into the Divine eternal life. Strictly speaking, eternity will never come in the future – in the future there can only be a bad infinity. (...) Eternity and eternal life come not in the future but in a moment, i.e. they are a deliverance from time, and mean ceasing to project life into time."[27] Berdyaev associates the concept of eternity in the first place with the perfect character of being, with the ultimate form of divinized life.

Within the context of the development of personal life, eternity is the synonym of the fullness of being together with the most intensive experiences of the authenticity of existence. It means an end to projecting life within the objectified forms of time, an end to the objectification of existence, as a result of which the spiritual sphere of existence is attained. Thus, according to Berdyaev, eternity encompasses spiritual life in its incessant and absolutely self-actualizing form. The hallmark of this fullness is the creative dynamism of spirit, which holds a decisive advantage over psychological and physical aspects of human existence and, through its power leading both these aspects to renewal and divinization, similarly to the "cosmic miracle of Christ's resurrection". This dynamism of the self-actualizing spirit leads Berdyaev to the claim that eternity is the eternal newness, eternal ecstasy of creation and dissolution of being within Divine freedom. From the perspective of personal self-realization, eternity is not just a state of happiness to come in distant future, but it rather reveals itself in every moment of the lasting itself which is experienced as the present.

Time has its source in eternity and, as such constitutes its manifestation, or, as Berdyaev puts it, its degraded form. D.B. Richardson, by discussing the meaning of time in Berdyaev's theory as epiphenomenon of eternity, paradoxically concludes that eternity is temporary. The positive value of time stems from the fact that in time there exists the potentiality to experience eternity. If this potentiality is actualized, then within the most secret depth of time the Divine image of man is revealed. That image constitutes the basis for the divinization of human nature and for his participation in God's eternity in its fullness. Man acquires this possibility thanks to the existential time. The existential time is subjective, it constitutes the immanent property of the subject, and is formed on the basis of changes, which through their dynamics decide on the form of personal existence. This category of time wholly depends on man who modifies it depending on the quality of his existence, the kind and intensiveness of experience.

As a "personal time", in which personal inner self-realization takes place, the existential time favours the realization of spirit in its source depth of eternal existence. It should be noted that only the person in their spiritual dimension can really affect the shape of that time, and only in this dimension time can be utilized and directed according to the person's own conscious intentions.

Close connection between the existential time, in which man experiences the fullness of being, and eternity itself is rendered in the Greek concept of *kairós*. *Kairós* literally means "the point which is selfsame to the attainment of the goal", as well as "the critical point", "appropriate moment", "convenient moment". This concept implies the cooperation between God and man, human positive answer to God's call, which man gives through the full use of the possibility of self-determination presented to him within the limits of his whole life. "Convenient time" spent in appropriate way gives man the chance to attain full existence, the crowning achievement of which is *theosis* taking place in spite of the enslaving limitations of the objectified time.

In a narrower sense, *kairós* means the precise moment, the unrepeatable "while", in which human perfection actualizes itself, thereby confirming that this "thought time" regarded in a broader sense – as the whole human life – has been spent fully. The perfection reached within this one unique moment of the present is the authentically experienced fullness of divinized life which participates in eternity. In this context, Czesław S. Bartnik's statement is very apt: "(...) eternity not so much 'is' at the end of time, but shows through the inside of each moment of time, each *kairós*. The point of transformation of all time is Jesus Christ, who is 'the fullness of time and eternity' through their personal connection."[28] Those "convenient moments", "points selfsame to the attainment of the goal", each time saturating and fully fulfilling the present "now" with their perfect content bring closer the ultimate realization of the idea of God-manhood.

According to Berdyaev, the process of self-realization, of the achievement of the fullness of existence in which the Divine-human unity of being is revealed in all its clarity, is tantamount to redemption.

Within the Orthodox tradition, redemption is understood as the ontological transformation and renewal of human nature, its deification and initiation into the participation in God's life. In accordance with the Orthodox thought, Berdyaev does not regard redemption as a gift, a reward or exculpation by God. Primarily, he sees redemption as the achievement of the perfection of life, and as a creative transformation of human nature, which leads to divinization and rebirth of all creation. The perfection of existence is not a gift from God, but rather a task which man must face in his great effort to create himself anew, in the process of his own creative self-perfection. Thus, redemption depends to a great degree on man himself, who grows to his greatness, i.e. the greatness of personal fulfilment.

In order to be redeemed man must activate the creative aspect of his nature. The question of creativity, extensively discussed by Berdyaev, stems primarily from the position that man is the image and likeness of the Creator. As W. Hryniewicz concisely puts it: "According to Berdyaev, the essence of God's image is to be found in man's creative nature. Through creativeness man both transcends and overcomes himself, his own duality, and he becomes most similar to his Creator."[29]

In Berdyaev's theory, becoming a person changes radically the relation of man to the act of creativity. The person in whom God has been born, is able to become a co-creator absolutely compatible with God's plan. Berdyaev calls this type of creation theurgy. "Theurgy, he writes, is man's cooperation with God. It is God-acting, God-man active creation. In theurgic creation, the tragic contrast between the object and the subject is ultimately abolished (...)"[30]

In his vocation to create, the person actualizes the plans of God who expects man to consciously undertake the trouble to continue Divine act of creation. In the theological language this continuation is referred to as "the eighth day of creation". Berdyaev identifies "the eighth day of creation" with time or the eon in history, in which man cooperates with God, thereby proving that the history of Divine-human existential relationships has not been determined once and for all, but rather it reveals itself in ever new manifestations of creativity. Divine co-creation with man will not end until the world reaches its perfection, or even until this perfection is constantly present.

The field of actualization of Divine-human plan is the world of persons, things, and phenomena. The diversity of the world's nature corresponds to various possibilities of the expression of man's creative abilities. Since Divine-human synergism encompasses all the manifestations of life, Berdyaev says that in theurgy – understood as universal acting – all forms of human creativity meet. The central role of the person, confirmed in the theory of the author of *Slavery and Freedom* by his call to create, reveals itself in his active life in the world, and not in a contemplative passivity. Man's creativity takes the character of work for the benefit of others, of the whole human and cosmic community, and it manifests itself in the spiritual openness to others, the world, and all supra-personal values.

From the perspective of person's individual life, eschaton requires that man through creativity thoroughly activate his Christological nature within the "power and fame" of God's might. It means that Christ will come only to the humanity that performs Christological act of self-revelation, i.e. reveals Divine might and fame in its nature. The condition to "see Christ's visage in might and fame" is to discover in the act of creation our own "might and fame", and only then will creativity, through which the true human nature becomes manifest, become the continuation and ultimate revelation of Christ-the Absolute Man.[31]

The Absolute Man in the age of God-Manhood does not emerge under the form of self-sacrifice, but under the form of persons who transcend their own pain in creativity. In this context, "the second coming" of Christ does not refer to His special "coming", His reappearance within history, but rather the "attainment" by humanity and the world – through the universally realized idea of theosis – of Divine-human reality, of the existence of resurrected Christ, who lives "in the body" of God Himself.

Department of Culture Studies and Philosophy, University of Science and Technology AGH, Kraków, Poland
e-mail: stark@agh.edu.pl

NOTES

[1] Cf. Fuad Nucho, *Berdyaev's Philosophy: The Existential Paradox of Freedom and Necessity* (New York: Anchor Books, 1966), p. 149.

[2] Wacław Hryniewicz, *Bóg naszej nadziei. Szkice teologiczno-ekumeniczne* 1966 (Opole: Wydawnictwo św. Krzyża, 1989), p. 168.

[3] Nicolai Berdyaev, *Existential Dialectics of the Divine and the Human* (Paris: YMCA-Press, 1952), p. 138.

[4] Hryniewicz, op. cit., p. 142, cf. also pp. 170, 279.

[5] Ibidem, p. 167.

[6] H. G. Gogochuri, one of the Russian critics of Berdyaev's thought, stresses this fact by showing that for Berdyav philosophy constitutes a specific unity of creativity and religious faith. He claims that Berdyaev transforms philosophy into a kind of religion. Compare also H. G. Gogochuri, *K marksistskoy kritikie religiozno-ekzistiencialisticheskogo ponimania specifiki filosovskogo znania (Na primierie filosofi N.A. Berdyaeva)* (Tbilisi, 1980).

[7] David B. Richardson, *Berdyaev's Philosophy of History: An Existential Theory of Social Creativity and Eschatology* (The Hague: Martinus Nijhoff, 1968), p. 135.

[8] Anna-Teresa Tymieniecka, *The Fullness of the Logos in the Key of Life, Book I: The Case of God in the New Enlightenment*, in *Analecta Husserliana, The Yearbook of Phenomenological Research*, Vol. C (New Hampshire: Springer, 2009), p. 253.

[9] In Berdyaev's theory intuition is no longer a cognitive method but turns out to be an ontic characteristic of the subject which encompasses "real religious experience", as is noted by W. A. Kuwakin in his *Kritica egzistencialisma Berdyaeva* (Moscow, 1976), p. 43.

[10] Anna-Teresa Tymieniecka, op. cit., p. 239.

[11] Cf. N. Berdyaev, *The Destiny of Man*, trans. N. Duddington (London: G. Bles, 1954), pp. 55, 248.

[12] Berdyaev, *Solitude and Society,* trans. George Reavey (London: G. Bles: The Centenary Press, 1938), p. 168. About man as God's image and likeness compare also: Berdyaev, *The Destiny of Man*, op. cit., pp. 53–55, and N. Berdyaev, *Spirit and Reality,* trans. by George Reavey (London: G. Bles, 1946), p. 33.

[13] Nicolai Berdyaev, *Filosofia swobodnogo ducha. Problematika i apologia christianstva* (Paris: YMCA Press, 1927–1928), Vol. II, p. 40.

[14] It is worth remembering, though, that in Berdyaev's theory man is not only God's image and likeness, but that in his nature there is also undetermined freedom. This freedom participates in a decisive way in either the realization of the image or in the efforts conducting to its destruction. Thus, the realization of the personal being places the Divine idea of man as well as the irrational principle of meonic freedom at the point of departure. Taking into account the existence of the meonic freedom we could say that in Berdyaev's philosophy the ideal fulfilment of the person includes also a permanent victory over the destructive energy of freedom.

[15] N. Berdyaev, *Existential Dialectics...,* op. cit., p. 165.

[16] N. Berdyaev, *Smysl tvorchestva.Opyt oprawdania chielovieka* (Moscow: G. A. Leman & S. I. Sacharow, 1916), p. 72.

[17] N. Berdyaev, *Spirit and Reality*, trans. George Reavey (London: Geoffrey Bles: The Centenary Press, 1946), p. 33.

[18] Ibid., p. 6.

[19] N. Berdyaev, *Filosofia swobodnego ducha. Probliematika i apologia christianstva* (Paris: YMCA Press, 1927–1928), Vol. I, pp. 83–84.

[20] A.-T. Tymieniecka, op. cit., p. 234.

[21] N. Berdyaev, N. Berdyaev, *Filosofia swobodnego ducha*, op. cit., Vol. II, p. 18.

[22] N. Berdyaev, *Spirit and Reality*, op. cit., p. 149.

[23] Cf. John Meyendorff, *Byzantine Theology. Historical Trends and Doctrinal Themes* (New York: Fordham University Press, 1979).

[24] N. Berdyaev, *Smysl tvorchestva*, op. cit., p. 254.

[25] N. Berdyaev, *Solitude and Society*, op. cit., p. 177.

[26] Ibidem, p. 169.

[27] N. Berdyaev, *The Destiny of Man*, op. cit., pp. 261–262.

[28] Czesław Stanisław Bartnik, *Chrystus jako sens historii* (Wrocław: Wydawnictwo Wrocławskiej Księgarni Archidiecezjalnej, 1987), p. 258.
[29] Hryniewicz, *Bóg naszej nadziei*, op. cit., p. 140.
[30] N. Berdyaev, *Smysl tvorchestva*, op. cit., p. 243.
[31] Cf. ibidem, pp. 312–314.

TSUNG-I DOW

HARMONIOUS BALANCE: THE ULTIMATE PHENOMENON OF LIFE EXPERIENCE, A CONFUCIAN ATTEMPT AND APPROACH

ABSTRACT

The concept of harmonious balance reflects the ultimate phenomenon of a dual structure operating in complementary contradiction through a cyclical progression to attain the ultimate balance of being and becoming amongst all things and events in the world. In the Chinese language, this phenomenon is written and pronounced as zhong huo, now abridged to one word, zhong. Upon the ascendancy of Confucianism as the state doctrine, as proclaimed by Emperor Wu in 136 B.C. during the Han dynasty (206 B.C.–220 A.D.), it became synonymous with the identity of China, zhong guo, following the presentation of the yin-yang correlation by Confucian scholar, Dong, Zhongshu, long after the death of Confucius (551–479 B.C.). The reciprocity of harmonious balance can be seen in the natural world as well as the foundation of moral order. It can be seen in human dynamics as well as the cosmos. Will the Confucian appeal and recognition of the import of harmonious balance become commonplace worldwide?

HARMONIOUS BALANCE: THE FUNDAMENTAL PRINCIPLE OF CONFUCIAN WORLD VIEW

The concept of harmonious balance reflects the ultimate phenomena of a binary or dual structure operating in complementary contradiction through cyclical progression to attain the ultimate balance of being and becoming amongst all things and events in the world. In the Chinese language, the phenomena is written and pronounced as zhong huo, now abridged to one word, zhong. It became synonymous with the identity of China, zhong guo, upon the ascendancy of Confucianism as the state doctrine proclaimed by Emperor Wu in 136 B.C., during the Han dynasty (206 B.C.–220 A.D.), upon the presentation of the yin-yang correlation by Confucian follower, Dong, Zhongshu, long after the death of Confucius (551–479 B.C.). In actuality, zhong is a metaphysical rendition of harmonious balance in the Chinese language, and should not be inaccurately interpreted as Middle Kingdom or center of the Earth, nor defined as its other interpretation of china, or porcelain.

Obviously, the basic ontological assumption of a binary or dual structure of the universe derives from commonsense observation of phenomena of the world and life experience. For example, human beings rely on light to carry on daily activities. The sun rises bringing daylight and then the sun sets bringing darkness. In a sense, the interaction of opposing phenomena of light and dark correlate with states of sleep and wakefulness operating in a continuum and reflect a harmonious balance. It serves to reveal the connectedness of past and future and passage of time. Within this premise, Indian Buddhists deny the existence of past and future, positing the only truly relevant time is the present, the here and now.

Operating in the same manner, a duality exists in human perception of our physical place and sense of location. We perceive left and right, front and back, up and down, forward and backward and in and out, among others. These characterize place and space.

In order for us to life, air is indispensable. The moment we stop having air to breathe or stop breathing, we die. Not only the act of breathing follows a binary structure of the opposing actions, inhaling and exhaling, to attain harmonious balance, but it also invokes the phenomena of the mind in its totality to harmonize the opposing functions of consciousness and unconsciousness for sustaining life. While consciousness may prompt us to seek fresh air as a manifestation of our will, the unconscious mind directs the breathing process even as we are sleeping; only through grave injury to the brain by intentional self destruction can this unconscious process be interrupted. The discovery of the phenomena of homeostasis in life deserves further enlightening.

Why are sunshine and air, for example, abundant on Earth and critical in supporting life? Science, the most reliable knowledge of human cognition, explains best the how versus the why. From the point of view of structuralism, Claude Levi Shaurr contends what makes us human is the minds' ability to reconcile those opposing phenomena in a binary structure to attain balance.

Long before Confucius, the legendary tribal chief Fu Xi is alleged to have created a symbolic representation based on this foundation. He used a solid line and an broken line (– and - -) to represent the primordial pair to imply a binary structure of the world, which subsequently evolved into the sequence of eight diagram used to calculate and predict the consequences of human action. This rendition of a broken line, representing yin, and the unbroken line, representing yang, later on became the foundation of the yin-yang duality and the Chinese written language. Even today's 0.1 digital unit can be allied with the yin-yang duality of the solid and broken line.

Following Fu Xi, King Da Yu of the Neolithic Xia Dynasty is reported in the *Book of History* (Shu Jim, Da Yu Mou: *Analects*, BK XX, Yao Yueh, Ch. 1) exhorted people to maintain a balanced approach to resolve world affairs (Yun zhi jue zhong). Later on, Confucius not only reinforced this concept of Zhong by insisting on upholding a neutral posture to keep up with the momentum of creativity (Zhi lian yong zhong) but also reinterpreted it as a timely change in order to assure survivability and durability (Qiong tung da bian) in the *Book of Changes*. In his view, change is inevitable, yet it is also necessary for existence. On an ontological basis, harmonious balance arises prior to imbalance and existence prior to change. Mencius acclaimed Confucius as a sage of change (Shen zhi shi). Too much or too little change violate the princple of Zhong, which means "just right."

In the spirit of harmony and unity, the *Doctrine of the Mean* (Zhong Yong,) purported to have been composed by Zi Si, the grandson of Confucius, to preserve the original ideas of Confucius on the concept of Zhong, declared that "balance is the root of the world, while harmony is the universal path for all. Let the states of balance and power prevail, heaven and Earth will be placed in their order and all things will be nourished and flourish" (James Legge, translator *The Four Books*, Paragon, N.Y. 1966, p. 351). Neo-Confucian Zhu Xi rendered it as a prime text for all Chinese since the Song Dynasty (960–1271 A.D.).

Naturally, the assumption of a binary structure symbolized in the primordial pair (yin-yang) interacting in complementary contradiction to attain balance in cyclical progression envelops the nature of polarity, the phenomenon of interdependence and interpenetration within interaction. Within this process of mutual generation, the aspect of complementarity evolves in a dominant state of balance or symmetry that constitutes the being of an object or event which is identifiable, concrete or constant. Conversely, a contradictory state evolves into a dominant state of disharmony, imbalance or asymmetry which brings about a change or transformation of the object or event, which constitutes the phenomenon of becoming. In such a relationship, the maximum development of the state of complementarity is also self-limiting. The polarity of a thing or event contains within it the seed of its opposite. Neither the state of harmony nor the state of disharmony is exclusive or absolute but an approximation of infinite magnitudes and levels. The dominance of one aspect over the other does not exclude the other nor annihilate the polar interaction process altogether. Every thing or event in the Universe contains a yin-yang. It is a continuum of cyclical progression. The *Book of Changes* concluded: "complete not yet." This process of being and becoming amongst world phenomena is self-evident. Commonsense tells us that it takes two either to balance or to disrupt the other. Neither of the primordial pair can exist alone. The phenomena of the world are ever-changing and fundamentally ever-lasting. For instance, the zenith of the sun is the beginning of its decline. In humankind, the unity of male and female leading to reproduction is the foundation upon which existence is based. In the study of science, experts are not certain why negative and positive forces exist in the world but know without this complementary yet contradictory interaction the world as we know it would cease to exist. It is difficult to contemplate the possibilities if this duality did not exist – would the world such as it is today continue to operate?

However, the implications of the non-causality and non-locality of nothingness as it evolves into the concrete world, as represented by yin-yang interaction and the neo-Confucian Taiji Diagram, make it difficult to reconcile with Newtonian physics and its laws of causality and the exclusion of the middle. When these scientific precepts are weighed against Confucian metaphysics, some posit that the emergence of modern science was hampered in China and contributed to the decline of Chinese civilization. Realistically, this criticism may have its merits. For example, China's state examination system established by Emperor Tai Tsung Li Shimin, 627–649 A.D., during the Tang dynasty was originally hailed as a significant advancement in selecting qualified scholars to be government administrators. Some even saw it as an effective implementation of the democratization of government. Yet, in practice, this system turned into a pure memorization contest of Confucian classics, known as the eight-legs style. Questions and answers were restricted to whatever written material was contained in the *Analects, Mencius, Great Learning* and *Doctrine of the Mean*. Since advancement was only possible through this rigid system, all creative and innovative thinking was eliminated by the process.

Nevertheless, the world continued and continues to change. Confucius called for daily renovation as a way of life, intending it to facilitate one's adaptability to constantly changing situations. In his view, knowledge as such produced by the human mind was approximate and provisional. To know what is unknown also constituted knowledge (zhi bu zhi wei shi ye). He admonished people not to be self- righteous and self-assertive, stating "if there are three people on the street, one could be his teacher."

Naturally, change occurs. The emergence and advancement of quantum mechanics in physics not only led to the "information age" but strengthened world unity through what R.I.G. Hughes claims as "the unassailable truth that Taoist-Confucian speculation on the universe has in common with quantum mechanics" (R.I.G. Hughes, *Quantum Mechanics*, Preface, Harvard University Press, Cambridge, MA 1989). As well, the surprising parallels between the identities between the DNA reproduction sequence as addressed by Kary B. Mullis (Kary B. Mullis, "The Unusual Origin of the Polemeras Chain Reaction," Scientific American, April 1990, p. 58) and the hexagram progression sequence of neo-Confucian Shao Yan as practiced in the *Book of Changes*. Both processes follow seven stages ending in sixty-four units and demonstrate what Niels Bohr, the father of quantum mechanics, declares to be the principle of quantum mechanics interacting in complementary contradiction within a binary structure following a cyclical progression. Bohr was so taken with the confluence of these two fields that he adopted the Taiji Diagram as his coat of arms. Quantum mechanics revealed that our world consists of two fundamental particles, quarks and leptons. Operating in harmonious balance, their perpetual cycle sustains the being and becoming of all things and events. Each particle has a corresponding anti-particle; matter has anti-matter; and positrons have electrons, among others. Particles such as electrons never cease to spin. What gives an object its shape and definition is its spinning particles operating in a binary structure to reach a state of symmetry. Eventually, this state of symmetry is altered and objects change. This state of symmetry and asymmetry perpetually cycle in a complementary yet contradictory manner manifesting in a non-directed, experiential, moving, spinning, microscopic uncertain world (yin) operating alongside a directed, experiential, relatively stable, predictable world (yang). In daily living, the macroscopic immediately observable world takes precedence over the microscopic one. For example, we experience the land and earth around us to be flat, but quantum mechanics indicates that space is curved providing a source for gravity (Eduardo Crueron, "Adventure in Curved Space Time," Scientific American, Aug. 2009, pp. 38–45). At present, in dispute is the extent to which dark matter and dark energy exist in the known and unknown world. Recent concepts of super-position and super-symmetry proposed by scientists, Steven Weinberg, Dan Hooper and Gordon Kane, seem to describe the ultimate phenomena of harmonious balance and tend to convey the long-held Confucian world view using different linguistic terms. (Gordon Kane, "The Down of Physics

Beyond the Standard Model," Scientific American, Sept. 2006, pp. 96.) In the same manner as Bohr's quantum principle, Confucian Dong, Zhongshu's thesis (xiang fan er xiang cheng) stated that a quantum bond or entanglement unites us all.

Since Confucius was reluctant to engage in theoretical issues and stressed actions according to what he considered as the prevailing world view to sustaining survival along with the enjoyment and fulfillment of the meaning of life, his views were based on the human beings' existence within the nature as it was observable then. Thus, the implications taken from the principle of harmonious balance operating within human nature became a significant focus among his followers. To what extent their approaches succeeded require further inquiry.

IMPLEMENTATION OF CONFUCIAN WORLD VIEW IN HUMAN BEHAVIOR: THE CONCEPT OF REN AND ITS PRACTICE

In the Confucian view, the greatest attribute of heaven and earth (nature) is to provide for and nourish life (tien di zhi da de yueh sheng); yet, the greatest enemy and predator of man is man himself. Confucius chose the concept of Ren (jen) as his principal focus as a way to harmonize human relations and realize the goodness in man. The profound meaning of Ren can be seen in the etymology of its Chinese character, consisting of two people.

Although Confucius was reluctant to engage in the ontological exploration of the sense of Ren, it did not deter his followers from their attempts to systematize this concept on metaphysical grounds. For example, Dong, Zhongshu standardized the criteria of Ren in terms of a correlation theory. Later, neo-Confucians searched for the self-realization of Ren through the analysis of the essence of the mind. As a result, the concept of Ren has evolved from its original more commonsense approach to ontological, psychological and anthropological renditions. Throughout, the principle of centrality (zhong) or harmonious balance in twofold complementary contradiction has played a critical role. This evolution of thought can be characterized in three stages: (1) Ren as the manifestations of the life process of love and reciprocity; (2) Ren as the prerequisite for evolution, the moral order and cosmic order as one; (3) Ren as the internalization and individuation of the original mind, or the creation of the universal mind.

REN AS THE REFLECTION OF HUMANITY: FILIAL PIETY IN RECIPROCITY

In answer to the universal question "What is man?" Confucius proposed that the meaning of man was the man of Ren (Ren zhi ren yi). In his view, human beings differed little from animals except for their sense of Ren. Confucius defined Ren in many ways but the fundamental one was man's ability to love mankind (Ren zhi ai ren). Yet, Confucius also recognized the duality of human nature, the emotions of love and hate. He admonished his followers that "to love those whom men hate and to hate those whom men love is to outrage the natural feeling of man, disaster cannot fail to visit upon one who does so." He rejected the Buddha's approach in the *Dhammpada* to practice self-humiliation to dilute hatred nor responding with love to neutralize hate. He proposed that man love what is Ren and hate what is not Ren (Li Ren, Book IV, Ch. VI) thereby esteeming nothing above Ren and disallowing anything not Ren to effect one's person.

Perhaps the human mind's plasticity and capacity to reconcile and harmonize the opposite poles of emotion for the benefit of humankind are what sets us apart from other living creatures. The end result appears not just to be peaceful co-existence but the enrichment of others and ever-developing civilizations. The strongest urges humans feel appear to be centered on the two poles of life-love and death-hate but are both necessary and complementary. One state exclusive of the other is nearly impossible and would create a severe imbalance directly affecting the survival of the species. If we accept that every thing or event in this phenomenal world has a beginning and an end; all creations will perish, one's attitude is more flexible and more able to endure the torments of disappointment and the suffering of disease and death. It is irrefutable that a negative and a positive path exist for all things.

Human survival depends on the cycle of life. Evolutionary biologists point to the significance of reproductive fitness. Organs, homeostatic mechanisms and patterns of behavior that increase reproductive fitness are selectively favored and those considered harmful or less attractive are discarded (Francis Ayada, "Teleological Explanation in Evolutionary Biology" Philosophy of Science, Mar. 1970, p. 8). It might have been this view of priority in reproduction that prompted Confucius to place the enduring love between parents and children above that between husband and wife. In a recognizable, experiential world procreation takes precedence over sexual liaison. A self-centered love is perceived as a sickness. The satisfaction of one's needs, sexual and otherwise, is dependent on one's contribution to them in a dual exchange process. Confucius presented this concept of Ren through the principle of filial piety with the practice of reciprocity as its starting point.

Zi Kung, a student of Confucius, asked "is there one word that can serve as the guiding principle for the conduct of life?" and Confucius is purported to have replied reciprocity. "Do not do to others what you would not want to be done to you" (*Analects*, Book V, Gong Ye Chang, Ch. VI). Reciprocity serves as the kernel of human relations and the foundation of humanity. Confucius further elaborated, "Desiring to establish himself., he seeks to establish others; desiring to succeed himself, he helps others; he endeavors to enlarge the lives of others to enrich himself and serve others more adequately." Confucius counseled his followers to practice Ren one should subdue oneself to return to propriety and possibly even sacrifice oneself for the overall realization of Ren. One was expected to acquire knowledge to sustain life and realize Ren but not to use that knowledge to deprive others of their livelihood or life. And, on the most fundamental level of reciprocity and what is most commonly viewed as filial piety, one's son should serve his father as he would expect his son to serve him.

In *Classics of Filial Piety* (Xiao Jin) three principles were set forth. The first principle called for the priority of reproduction. One must have a wife and then a son to carry on the family line. The second principle exhorted one to never disgrace the family. The third principle called on one to contribute to society. Paternity was not seen as the sole factor in establishing a bond between father and son. An exchange of ideas and support was critical to filial piety. The universality of love and affection between mother and child seems to be natural but this type of love does not subvert the guiding principles of filial piety and reciprocity.

REN AS THE PREREQUISITE FOR EVOLUTION: THE FIRST OR ULTIMATE CAUSE OF HUMAN EXISTENCE, THE TOTALITY OF HUMANITY

There is a Chinese saying that human beings are not wood and stones (ren fei mu shi) but a unique structure of blood and flesh (sue ru zhi qu) with the highest intelligence (wan wu zhi lin). From this universal Confucian vantage point, one can make three deductions. First, human beings are fundamentally part of nature, constituted from the most fundamental particles of the Universe, quarks and leptons. Yet, those particles are not natural elements since they have no "life" to them. Second, the vital energy which supports human life is supposedly derived from matter but it is still impossible to measure and quantify a human being's vital energy. Is an infant's energy solely dependent on its mother's milk? Third, becoming aware of one's existence in the world only is possible when one achieves a certain level of consciousness. While it may mean everything to the person, it is not measurable. Little is known about how consciousness comes to be or when it appeared in the evolutionary process.

In spite of growing exploration of cognition and scientific discoveries regarding the human brain, these advancements seem to raise more questions than provide answers. The more we know about an object, it seems, the less certain we are about it. Philosophers have been pondering these very questions over the ages as well. While scientists now feel the Big Bang theory answers questions about the origin of the universe and hence humankind, Taoists centuries ago envisioned "hung tung," or primordial chaos, as the beginning. The Confucian concept of yuan described by Dong, Zhongshu during the Han period or the Buddhist concept of emptiness or the zero state also speak to the primordial beginning. The question arises as to whether the evolution of the human species was a purposeful event or merely an accident. With that said, Taoists ponder why the development of civilizations and its many benefits has not also brought with it the ability for all humankind to live in peace and harmony. One wonders whether these developments have only served to further complicate the human condition.

As scientists delve deeper into all facets of human functioning and development, absolute belief in Darwin's theory of evolution through natural selection has been modified. Information regarding the role of RNA as well as DNA in molecular behavior has revealed its critical importance to metabolism and reproduction, hence evolution. These discoveries demonstrate that conscious-less molecules function with a specific orientation which is purposeful rather than purposeless (Thomas R. Cech, "RNA as an Enzyme," Scientific American, Nov. 1986, pp. 69–73).

If constituent molecules of an organism behave with a special purposeful orientation, then, it would logically follow that the most intelligent beings known on this universe, human beings, would have some intention in their evolution. The primary goal of human evolution, perhaps, lies in its survivability in a negative sense and reproduction in a positive sense. The universal worship of reproductive organs beginning with primitive cultures is an indication of this intentionality. Parental love and devotion among most creatures is universal. Sexuality seems to be a cosmic archetypal phenomenon constituting different expressions of one continuum. Maternal instincts and the determination to give birth, even in the most adverse conditions, serves as another indication of intentionality in human evolution.

While the evolution of each individual human being may seem to be a random and purposeless process, the evolution of the human species as a whole appears to have a unique group orientation bound together by a cooperative spirit intended to accomplish a common objective. According to Xiong Zi, human beings are not only social animals but are the most powerful ones. One-to-one a human cannot match the physical strength of a lion but its creative intelligence and cooperative efforts in organizational activities allow it to control a lion. Recent research indicates that human beings are even more group oriented than primates (Julian Jaynes, *The Origin of Consciousness in the Break Down of the Bicameral Mind*, Houghton Mifflin, Boston MA, 1976, p. 127). Interestingly, a thirsty baboon does not leave its group to seek water. It is only satisfied within the patterned activity of the group. Evolutionary history reminds us that what is best for the species may differ from what is best for its component individuals (Paul W. Ewald, "The Evolution of Virulence" Scientific American, April 1993, p. 86).

The reason why force, particularly military force, has been decisive in shaping the course of history and human life, may be due to its nature as the most effective organizational power in the human species. Positive human relations are critical for any organization to function effectively. The essence of Ren lies in its focus on the human relationships. Organizational activities involve the capacity for human beings to fuse reason with emotion for unity in action. This harmonious cooperation among individuals is essential for effectiveness. Intellectual commitment, belief in a common goal and enthusiastic support all come together among the constituency. Neither reason nor emotion alone will suffice. Their unity and will to action through the harmonization of individuals' altruistic and egoistic impulses lead to achieving a common goal. Confucians consider this realization of harmony in human existence to be Ren.

Military force, in its ultimate sense, fulfills this purpose of Ren. Xiong Zi, an avowed Confucian follower who was rejected by other Confucians, tells of a dialogue between Chen Xiao and the reknown military theoretician, Sun Qing Zi, in power during the Warring States period (403–221 B.C.). Chen questioned Sun about why his discourses on war were based on the principles of Ren and Yi, righteousness, when Ren meant to love man and Yi meant to uphold order, while war lead to disorder, struggle and killing. Sun was said to have retorted that Chun was ignorant because "to love man those who may injure men must be prevented from doing so, and in order to uphold order those who may lead to disorder must be restrained to prevent its occurrence. Thus, the essence of the armed forces is to suppress violence and establish order as a way of nourishing life." Sun assured him, "If the armed forces of Ren prevailed, it will bring divine transformation to the people, just as timely rain nourishes life in happiness."

No evidence has been found that directly shows that Confucius subscribed to Xiong's arguments but he did not dismiss the use of force in the *Analects*, unlike Buddha. He stated that military force was essential to government (Yan Yuan, Book XII, Ch. VII) and that a ruler of Ren must "employ the upright and put aside all of the crooked" (Yan Yuan, Book XII, Ch. XXII). It is the Confucian contention that harmonious cooperation among people through a self-regulated system designed to attain a goal is derived from the belief they are all part of a hierarchy of the unitary whole, Yuan (the source of Ren) as its totality of humanity.

This totality of humanity is reflected in the etymology of the character for Chinese, tien, which incorporates the symbol of person with two lines over the head indicating that an individual naturally exists within a two person relationship.

REN AS THE INTERNALIZATION OR INDIVIDUATION OF THE ORIGINAL MIND: THE CREATION OF THE UNIVERSAL MIND

Whether or not the phenomena of the mind can be analyzed without prejudice by our own mind does plague the rationality of such inquiry. Yet, in the cognitive process, there is no other way to elevate life except by elevating its ideas. It is the intuitive creativity of the mind which makes real those imagined forms or states that can or may exist for the benefit of the lives of others. In order to resolve this challenge surrounding the nature of the mind, we have discovered that this is not a hopeless situation; the world is knowable. Scientists are able to observe how the mind operates to reach approximate inferences even though they may not be exactly accurate. Our mind, as well, is most often able to accomplish our intentions.

In view of recent scientific discoveries on the phenomena of the mind, neither the Buddhist assumption of a universal mind existing unseen outside of the body nor the view of consciousness only comprised of ideals seem to hold ground. Even the prevailing contention by materialists of the existence of a mirror image of knowing is losing purchase because of the discovery that the mind does not only reflect a negative image, it does not operate in the same manner as a photocopier. As a consequence, the Confucian and neo-Confucian contention that with modification the mind, a unique attribute of the human species, has evolved from the unitary whole, or yuan, through the evolutionary process with its own unique structure and function, could be a viable alternative for the exploration of the phenomena of the mind. Cognition may be the unique manifestation of Ren, the totality of humanity (Tien da di da ren yi da). Zhu, Xi, the neo-Confucian from the Song dynasty, introduced the concept of Li (principle of reason) to explain the phenomena of the mind but insisted that it exist in Qi (the concrete energy or matter of the world). It is Zhu's contention that every thing or event in the universe must have a reason for its being and becoming. All Li originates from the same Li of the universe. Since human beings are part of the world, the human mind is derived from the Li of the unitary whole and exists in the human body. The assertion seems akin to that of science. In the first place, science has discovered that the basic material of the unity of human beings and nature is the same, quarks and leptons. Consciousness may depend on neuronal activities but awareness does not. The phenomenon of acupuncture analgesia still mystifies scientists but the existence of its effect has been confirmed. Secondly, despite the residual dispute regarding the classification of consciousness and unconsciousness of the mind, scientists generally have observed the fact that consciousness seems to arise from unconsciousness. Consciousness makes us aware of the self and ego, from which cognition springs. But is much of what occurs in the unconscious makes life possible because the autonomic nervous system manages all human systems fundamental to survival. Thirdly, although artificial intelligence has been developed to run computers, robotics and many types of instruments, it would not be possible without human development. The full spectrum of human cognition has not yet been replicated by computer scientists. The experience of pain and the broad array of emotions remain the privilege of the human mind.

The tremendous power of the human mind occurs on conscious and unconscious levels. It could be categorized as being managed by the autonomic nervous system and the self-conscious nervous system. Both exist within the same structure and demonstrate the mind's profound process of integration and harmonization. It seamlessly reconciles and synthesizes the complex phenomena occurring inside and outside the body both on an unconscious and conscious level. With each event a new cognitive event takes place. As we probe deeper into how things and events are perceived, or manifest themselves, we can see a common thread of contradictory phenomena, which can be categorized as primordial pairs. Quantum mechanics has revealed that the polar relationships of primordial pairs occurs even at the most fundamental level too. For instance, the mind perceives symmetric unity in asymmetric diversity, constancy in everlasting change and the finite in infinity, among others. It is the mind which resolves the contradictions of these two-fold world phenomena through coordinating unity; the outcome of which is better understanding and functionality. As perception is understood further, certain patterns of recognition and syntax appear to be genetically established in our physical makeup. It appears human beings have been endowed with certain cognitive, linguistic and inferential competencies. This constancy was highlighted by Mencius, when he described an innate ability to know or understand (Lian zhi) and the capacity to act (Liang neng) on what is right or wrong. How the link between RNA, DNA and the human mind's operations relate to the origins of the universe and evolution have yet to be fully understood.

What we have been able to ascertain is that the functions of the mind appear to have developed through a hierarchical structure from the most fundamental operations to higher level thinking and decision making, or intellect, and ultimately the integrative functions of understanding. If the mind achieves a sufficient level of organization along with the ability for focused concentration, it appears to lead to a creative intelligence through which the highest levels of consciousness can be attained. Whether or not a collective consciousness exists is still debatable. It is generally accepted that self-awareness goes through a process of refinement and is shaped by an individual's cultural values and traditions as well as the social environment which includes peer pressure.

Human nature reveals itself through the essence of cognition as the mind processes, integrates and harmonizes the complex feedback and information set before each individual. Only through reconciliatory openness or separation from the self by elevating one's mind above extremes and self-righteousness can we attain the freedom to resolve the double-bind of subjectivity versus objectivity in the creative process. Self-awareness may be the highest function of the mind, but it is not its essence.

In the Confucian view, educability is an innate human trait which can be fully developed to attain the utmost goodness in human existence. This is the goal enunciated in *Great Learning*. In *Analects*, Confucius implied that knowledge is Ren. "How can one attain Ren if one is not knowing?" (Wei zhi yan de ren) (Kung ye Chang, Book V, Ch. 18) He expounded, "Knowing can reach Ren and thus Ren can hold on to it," (Zhi ji zhi Ren neng sho zhi) (Wei Ling Kung, Book XV, Ch. 32) In Zhu Xi's rendition, the goal of cognition is to assure the unencumbered development of one's potential for the full self-awareness of Ren (Shi ren), the Universal mind.

CONFUCIAN ATTEMPT ON FAMILY-CENTERED AND COSMOPOLITAN PRACTICES

The spirit of Zhong that permeated Chinese culture may have contributed to the shift in identity among the Chinese people and, therefore the state, from a focus on race and nationality to that of ethics, or from ethnicity to the philosophical connotations of its actions. When one describes the Chinese people as Han Chinese, it is not an anthropological marker but political and culture identification with the characteristics of the Han dynasty. The true meaning of China in the Chinese language is Zhong Guo; most accurately translated as a state based on the Confucian principle of harmonious balance rather than the geographic description of the middle kingdom or center of the Earth. The traditional Chinese society was non-litigious with the absence of a civil code, legal profession and even police. When Han emperor Wu-ti elevated Dong Zhongshu's Confucian teachings to state doctrine, a correlation between man, nature and the universe was advanced. Dong firmly supported the continued existence of other schools of thought, thus Taoism, legalism among others coexist and continue to be influential.

For a period of time, the subtleties of metaphysical contemplations about life and the teachings of Mahayana Buddhism obscured Confucianism. Yet, the mentality of Confucian Zhong eventually diluted Buddhist metaphysics. Followers liberated themselves from their sole focus on the process of life and death, or the ultimate ending of the perpetual chain of sentient existence by nihilistic absorption into Nirvana. The doctrine of "all in one, and one in all" advocated by the Tien Tai and Hua Yan schools of Chinese Buddhism were ostensibly free from the severe polemic tensions that plagued other religions. The difficulty which proponents of Christianity encountered with the Chinese culture arose not so much from Chinese xenophobia but from the exclusivity of the practices of Christianity. For example, the celebrated case of the rites controversy among the Jesuits come to mind.

Since Confucius declined to speculate on issues such as the origin of the Universe or the origin of human beings and refrained from contemplating the supernatural phenomena of ghosts or the after life, he battled with the dictates of chance and fate. Confucians gradually shifted from their reliance on religious observation to the idea of sovereignty of the conscience through the power of the internalization of Ren as the means of salvation. For instance, a "clear conscience" is a reward in itself. This optimistic approach was derived from their faith in the rationality of human nature. In their view, rationality was a precondition of existence. Mencius described this rationality in human nature as reflective equilibrium or a normative principle prevailing both in human logic and in ethics. This faith in rationality naturally fostered a mentality of intellectual cosmopolitanism, which prevails upon the follower to recognize and acknowledge the diversity in the unity of world phenomena and to endeavor to search for this unity while preserving the diversity (Qiu tong cong yi) as a way to attain the benefits of mutual complementariness.

From the vantage of this intellectual atmosphere, they perceived that the ultimate goal of all religions was the same, to achieve the ultimate goodness of human nature (San jiao jiu liu, shu to tong gui). The practice by some to attempt to reach either an absolute idea or a classless society in which the two-fold contradictory process would cease, violated the Confucian principle of the unity in opposites. If contradiction is the root of movement, as Hegel claimed, then, the annihilation of contradiction by eliminating opposition would not succeed in ending contradiction because the world is not static. In human nature, the urge for love and the desire to prolong life as well enhance aesthetic appreciation appears universal. If human beings are part of nature, an individual should be able to act in harmony with the Universe; then, what one obeys is internally dictated by his moral conscience and not an external authority. In the past, there had been recalcitrant Confucian scholars but no Confucian missionaries. There was religious discrimination but no religious wars. Comparatively, Chinese Confucians appeared to have been less immediately dependent on religion than most other people.

Since human existence is a fundamental priority of Confucian concern, the process of reproduction and importance of its role in the structure of family life is of utmost importance to Confucian ideology. This is why the virtue of filial piety constitutes the basic Confucian tool for fulfilling Ren. In Chinese culture, the family is the main source of economic security, education, social contact, recreation and happiness as well as an organizational model. The ideal of such a social unit, in the past, was to have every generation reside in one great household with the father or a senior member of the family exercising the authority as its head. The importance placed on marriage was intended to solidify the family as an institution. Confucius declared that the Tao of a superior person begins with the relationship between husband and wife.

However, in the evolution of the human species, the reproductive process does not end with the immediate family unit but extends with greater complexity and diversity as it expands. In order to assure survival as a whole, there must be elemental rules of behavior for group activities which require obedience and the support of all the members of a given group. Under these circumstances, the fatherly role is extended to the entire group as a way to maintain the natural order and sustain life. Confucians considered this extension a necessity in order to confer the true meaning of the mandate of heaven. The ruler of a state is considered the father who discharges his heavenly responsibilities of providing for the livelihood of his people and leads them towards a moral life according to Confucian values. In the Confucian *Book of Li*, it sets forth the ideal of politics as a family-oriented welfare state:

> When the great Tao prevails, the world (government) belongs to all the people. Persons virtuous and capable were selected to serve the state with sincerity emphasized and peace cultivated. People did not love only their parents nor did they treat their own children as their only children. Provisions were secured for the aged, employment was given to the able-bodied, and the means raise the young were established. People disliked sing the natural resources underdeveloped and hated those who worked only for their own profit. . . .

Mencius reinforced the idea that the manifestation of the mandate of Heaven is through the will of the people, even though he failed to recommend free elections as a method for the peaceful transfer of political power based on general acceptance and the acquiescence of the people. Instead, the Confucians later on devised an examination system for selecting a centralized bureaucracy to perform government functions under the direction of an emperor whose power was established by military might. The emperor was viewed as the son of Heaven; while autocratic, emperors rarely claimed divine rights or advocated absolutism. In general, the examination system was administered fairly and without undue discrimination.

In such an atmosphere, cultural centrism or arrogance was unchecked. Yet, Confucius clearly defended his ideal in the sage rulers of Yao and Shun, who were not native Chinese but Yi, because his cosmopolitan doctrine allowed whomever was the most virtuous and qualified to govern China. He declared, "all are brothers within the four seas." The concepts of a nation state, citizenship, or even sovereignty were absent in past Chinese political vocabularies. Had it not been for the Confucian cosmopolitan attitude toward so-called non-Chinese, the Manchus might have become an independent political identity today. Peaceful absorption was the rule. Not one single word can be found in Confucian pronouncements which call for conquering for the sake of conquests. The fact that the Chinese did not colonize others at the height of their power was not so much because they were unable to do so but rather because they were not inclined to do so. Building the Great Wall was a defensive measure even though it was constructed during their mighty military power and the legalistic rule of the pre-Confucian Qin dynasty. The surprising Ming maritime expeditions were not motivated by colonization and even more perplexing was their sudden suspension. In the vicissitudes of Chinese history, whenever Confucianism asserted itself, the cosmopolitan tendency toward a highly civilized universal state prevailed under one ruler. Perhaps this is why the Chinese enjoyed a more prolonged peace than the rest of the world despite the fact that prolonged peace was typically associated with inertia and stagnation.

Confucius viewed harmonious balance as the most simple concept and, therefore, easily understood and able to be put into practice. Yet, he lamented that it was also hard for man over time to hold on to properly (zhong yun zhi wei de qui zhi mingy an ren yu xiu). The reciprocity of harmonious balance can be seen in the natural world as well as the foundation for moral order. Confucianism and harmonious balance are inextricably linked to China's civilization. The tenets of Confucianism may be seen as the reason why China has the most people on Earth and have contributed to its longevity and relative stability compared to other nations. One prescient Confucian edict set forth in the 8th century B.C. that no two people with same last name could marry enforced the healthy genetic separation of its people. Yet, it also had less desirable effects. Was it responsible for the subjugation of women, discrimination against merchants, enforcing degree worship, prevailing upon families to follow only one head, allowing its people to too readily accept foreign governments without stirring up nationalist feelings, discouraging free elections through teaching the "one hundred idiots remain idiots," promoting service as the sole function of government, reinforcing the idea that the "son of Heaven was the father of the people" and their rightful ruler and leader thereby inciting socialist leanings? Does the right to revolution exist?

Harmonious balance did not exclude any world phenomena. Its binary, dual structure of yin and yang continues to resonate with today's understanding of the cosmic and moral order. One wonders what the implications are for the latest trend regarding the growing number of Colleges of Confucius (at least 280) that China is establishing both within its borders and world wide. Will the Confucian appeal and the import of harmonious balance become commonplace worldwide?

Florida Atlantic University, Boca Raton, FL, USA (Retired)
e-mail: debdowh@aol.com

JONATHON APPELS

"DANCE: WALKING AND SELF-MOVING IN HUSSERL AND MERLEAU-PONTY"

Here I will treat a subject I first considered at the Radcliffe Institute for Advanced Study for the "Economies of Art" conference in June 2009. At that time I examined the relationship between the texture of a dance and its dissemination (whether through performance touring or through electronic image transmission). In this study I look at the lived experience of dance as it is sensed by choreographer, dancer, and audience. In order to do this I consider works by Edmund Husserl that examine both ordinary and unusual bodily activities (self-moving, walking to the edge of Germany, bodies falling off the earth, etc.). I combine Husserl's observations with ideas on the expressive body from Maurice Merleau-Ponty's *Phenomenology of Perception* in order to lay the groundwork for a theory of dance texture based on specific actions within lived expression. In this trajectory the forms of choreography are never "ideal", but always emanate from the aspirations of daily chores and daily gestures.

I will begin by examining Husserl's discussion of the body in the chapter "The Constitution of Psychic Reality through the Body" in his *Ideas Pertaining to a Pure Phenomenology and to a Phenomenological Philosophy. Second Book*. I then consider two short research manuscripts by Husserl. "The World of the Living Present and the Constitution of the Surrounding World External to the Organism" is particularly concerned with the constitution of a world as "external" and "present" only through the actions of self-movement, and indeed it can be argued that for Husserl the presence of the world (the surrounding world) can only be constituted through self-moving. I also discuss Husserl's "Foundational Investigations of the Phenomenological Origin of the Spatiality of Nature" which considers that the idea of movement of the body cannot be distinguished from an understanding of the idea of space, and the perception of space has a contingent relation to a subject's moving body. While these two research manuscripts of Husserl have been available in English with the publication of *Husserl, Shorter Works* there has been relatively little commentary on them.

What is of particular interest within these three texts to be considered is the tentative quality of Husserl's thinking, and how he is often able to stylistically mime the physicalized activity of cognition that he is investigating. For Husserl the body in motion does not appear as an essential component of his overall philosophical program. This is one reason, as we can see in his chapter "The Constitution of Psychic Reality through the Body," that the movement of the animate body is not discussed with specific focus, and is only part of Husserl's discussion on the functioning of the body's sensory processes. Thus, tactual, visual, and auditory descriptions both complement and intrude upon his discussion of the body's movement in this chapter. These sensations are linked with the movement of the body, suggesting an equivalence between them, rather than a dominance of the visual. It can be argued elsewhere that while Husserl does not discuss the movement of the body as an operative principle, the body's movement actually functions operatively within his phenomenological program.

In "The Constitution of Psychic Reality through the Body" Husserl writes, when lifting a thing, "I experience its weight, but at the same time I have weight-sensations localized in my Body" (153). The body is constituted doubly, having both a particular materiality, and also the sense and approach of things external to it: "warmth on the back of the hand, coldness in the feet, sensations of touch in the fingertips. I sense, extended over larger Bodily areas, the pressure and pull of my clothes" (153). The body belongs to the subject, and is a "field of localization of its sensations." The subject is an Ego that can "freely move this Body" (159).

In his manuscript, "The World of the Living Present and the Constitution of the Surrounding World External to the Organism," written in 1931, he discusses how the activity of walking is abstracted from the concept of "I am in motion in space" (248). Conversely, physical rest "is only experienced as rest through the power of those changes of appearance whereby physical movement is constituted" (249). Rest is therefore not necessarily the primary state.

Husserl defines walking:

Walking thereby receives the sense of a modification of all coexistent subjective appearances whereby now the intentionality of the appearance of things first remains preserved, as a self-constituting in the oriented things and in the change of orientation, as identical things. (250)

He then poses a particularly important question. Is the objective world constituted through "self-moving and having-moved"?

The third manuscript I will consider, "Foundational Investigations of the Phenomenological Origin of the Spatiality of Nature," was written between May 7 and May 9, 1934. The physical act of walking, combined with the "synthesis of actual experiential fields," and combined with the idea that "I have not paced off and become acquainted with what lies in the horizon, but I know that others have become acquainted with a piece further on, then again others yet another piece" creates from the act of walking itself, as well as the comprehension of the walks of others, the "idea of Germany" (222). This idea of the traversing of the physical earth and its 'boundaries' leads not only to the idea of a nation, but to the idea of the earth itself.

This process of combining the physical activity of walking with the mental and physical apprehension of what is "further on" is important for Husserl because the idea of the earth is thus not only physical, but the earth itself, as he states several paragraphs later, is a body. For "we Copernicans" realize that "The earth is a globe-shaped body," a synthesis of my perception and that of others, a "unity of mutually connected single experiences. Yet, it is a body!" The earth as a body is an "experiential basis for all bodies." Initially the basis is not "experienced at first as body but becomes a basis-body at higher levels of constitution of the world" through experience. Where is it that motion occurs for Husserl? Husserl writes, "Motion occurs on or in the earth, away from it or off it" (223).

He then considers aspects of motion as normative. Husserl begins with hesitation, "actual or possible mobility and changeability," then considers bodies "thrown into the air." Husserl continues, these bodies are thrown in the air "or somewhere or other in the process of moving, I know not to where–in relation to the earth as earth-basis." Even when motion ends there is the possibility of additional motion.

> Bodies moveable in earth-space have a horizon of possible motion and if motion ends, experience nevertheless indicates in advance the possibility of further motion, perhaps simultaneously with the possibility of new causes of motion by a possible push, etc. ... Bodies are in actual and possible motion and <there is> the possibility of always open possibility in actuality, in continuation, in change of direction, etc. (223)

We can now consider how Merleau-Ponty amplifies Husserl's sense of the circumlocution of the body. As Merleau-Ponty writes in *Phenomenology of Perception*, "I know that objects have several facets because I could make a tour of inspection of them, and in that sense I am conscious of the world through the medium of my body" (82). There is "a certain field of action" which is "spread around me":

> I do not need to visualize external space and my own body in order to move one within the other. It is enough that they exist for me, and that they form a certain field of action spread around me. (180)

The body acts as a fulcrum for its own form as it moves toward action and the gesture. Merleau-Ponty then conjoins the gesture and the body together within activity. Gestures have in themselves the action of the body form out of which they are composed. Furthermore, that action has a rhythm,

> One can see what there is in common between the gesture and its meaning, for example in the case of emotional expression and the emotions themselves: the smile, the relaxed face, gaiety of gesture really have in them the rhythm of action ... (186)

But for Husserl the body still has a functional purpose, for example, to determine whether something is an illusion. The body must move from here to there in order to know something about here and there, and that thing, that illusion there. Merleau-Ponty's sense of the body, on the other hand, knows itself already moving there, and thus there is no illusion over there in need of determination. What is over there is already known by my body moving there with it.

Merleau-Ponty outlines his project on motion as follows:

> The project towards motion is an act, which means that it traces out the spatio-temporal distance by actually covering it. (387)

Merleau-Ponty introduces the notion of transitions:

> If we want to take the phenomenon of movement seriously, we shall need to conceive a world which is not made up only of things, but which has in it also pure transitions. The something in transit which we have recognized as necessary to the constitution of a change is to be defined only in terms of the particular manner of its 'passing'. (275)

Indeed, "movement is a fact," Merleau-Ponty writes (277); this is a principle for Merleau-Ponty rather than a description.

In his lectures at the Sorbonne in 1957–58 Merleau-Ponty returns in more complete detail to actions of the expressive body in motion. He writes "Thus, in walking, the gaze spontaneously re-establishes the fixed line of the horizon and it is only when one pays attention to one's perception that one sees the landscape jump." The movement of walking is "the power of organizing at each step certain unfoldings of perceptual appearances" (164).

I'd like to now tie the discussion of bodily movement in Husserl and Merleau-Ponty to some of the activities found within Western concert dance. Curiously, by observing and sensing what the body is doing at the moment of its action, both Husserl and Merleau-Ponty assume characteristics of the dancer and the choreographer. How does the dancer sense self-moving? I have not treated here the subject of Husserl and memory, but for the dancer memory is key. The dancer would like to resist mnemonic notation, yet must apply observational skills in order to have a navigational system. These self-aware observations of the dancer are able to align remnants of the memory of danced movements. The dancer's body is both a mnemonic base and a self-moving system. Indeed, the dancer's observation of her or his movement propagates additional and continuous dance making.

The choreographer also searches for a notational system, concomitant with the observational system, but is more likely to find a memory system that is extended from the bodily fabric. For the choreographer, in a similar operation to that of the philosopher of bodily action, extends the body into a writing instrument of a differing grammatical index (whether a pen, typewriter, video camera, or a student or disciple, etc.).

We can consider at another time how the Greek tragediams were making choreography within their productions and how the Greek tragedies maintained a balance between writing and movement. The question remains, how does a writing system or a dancing system changeover to its extension? Likely this occurs when the world presents itself through action, such as how the decelerating train, in the moments before it stops, allows the body to glide through the air or hop off-kilter because of the sudden lack of forethought and lack of a projected mission. Also we see how the pedestrian signal lights at the intersection pull the body from this written page. Both of these examples do not consist of signs, but of extended writing and dancing functions.

What, then, is the role of the audience? The body of the audience member is not an indexically reduced subject situated at a mediating point between other artistic subjects. The audience member has the writing and dancing extension of the dancer, and those tools as well of the choreographer, and the philosopher of movement. The texture of the dance is the place where such extensions are viable, where such extensions comingle and propagate. Walking and self-moving are the basis of dance making, and therefore the basis of dance texture. The body of the audience member shares the knowledge of walking and self-moving, and this knowledge is shared within the dance texture.

The distinctions I have shown between Husserl and Merleau-Ponty are gradational, and are contained within the notions of sensation and expression. For Husserl, there may be the need to make a determination concerning an illusion "over there". This determination has a relationship to the vectors of movement. Thus the illusion could be understood by Husserl to be the audience itself. Yet Husserl's theories of walking and self-moving suggest that the illusion is already a subset of walking and self-moving. The illusion comes about, and is contained within the walking and self-moving. This shows that for Husserl the illusion is therefore not "over there" since the walking and self-moving bring us to the place of the illusion which is then erased through the activity of getting there.

For Merleau-Ponty the expressiveness of the body clarifies that the illusion is not "over there", but rather embellishes the quality of walking and self-moving. The illusion, allied with the body, works to destabilize the customary functionality of walking and self-moving. The illusion is a finesse of the imagination which enfolds the artfulness of the activity of moving. In this way the expressivity of walking is already a dance, and is already filled with dance texture.

Columbia University, New York, 10011 NY, USA
e-mail: jr2168@columbia.edu

REFERENCES

Husserl, Edmund. 1981a. Foundational investigations of the phenomenological origin of the spatiality of nature (trans: Kersten, Fred). In *Husserl, Shorter Works*, eds. Peter McCormick and Frederick Elliston, 222–233. Notre Dame, IN: University of Notre Dame Press.

Husserl, Edmund. 1981b. The world of the living present and the constitution of the surrounding world external to the organism (trans: Elliston, Frederick and Lenore Langsdorf). In *Husserl, Shorter Works*, eds. Peter McCormick and Frederick Elliston, 238–250. Notre Dame, IN: University of Notre Dame Press.

Husserl, Edmund. 1989. *Ideas pertaining to a pure phenomenology and to a phenomenological philosophy. Second Book* (trans: Rojcewicz, Richard and Andre Schuwer). Dordrecht: Kluwer.

Merleau-Ponty, Maurice. 1992. Husserl's concept of nature (Merleau-Ponty's 1957–58 Lectures). Compiled by Xavier Tilliette (trans: Leder, Drew). In *Texts and dialogues: Maurice Merleau-Ponty*, eds. Hugh J. Silverman and James Barry, Jr., 162–168. Atlantic Highlands, NJ: Humanities Press.

Merleau-Ponty, Maurice. 1962. *Phenomenology of perception* (trans: Smith, Colin). London: Routledge & Kegan Paul.

BRUCE ROSS

THE SONGLINES: DREAMING THE ANCESTORS AND SUSTAINING THE WORLD IN ABORIGINAL ART

Yet, in the East, they still preserve the once universal concept: that wandering re-establishes the original harmony which once existed between man and nature.
Bruce Chatwin, *The Songlines*[1]

By spending his whole life walking and singing his Ancestor's Songline, a man eventually became the track, the Ancestor and the song.
Bruce Chatwin, *The Songlines*[2]

At the point when sleep has not yet come and wakefulness vanishes, being *is revealed.*
SHIVA

ABSTRACT

This chapter is based on the November 2009 New York University exhibit "Icons of the Desert," a selection of modern and contemporary Australian Aboriginal art from Papunya, perhaps the most significant center for such art. Issues of the ancestor realm, Dreaming, singing, and the walkabout serve as a context for discussing Aboriginal art in ritual, narrative, and artistic manifestations. Among the oldest art traditions in the world, these paintings of semi-abstract and patterned spaces can be viewed as a living connection to the mythic realm or Dreaming that presupposes the very essence of basic survival and social continuity in the present as well as the moral order and fate that continues to sustain the present. The Aboriginal artist is immersed in Dreamtime, a mythic past eliding into the present, that challenges certain Western views of consciousness and basic reality.

The first attraction to modern Australian Aboriginal art may very well be its suggestion of modern Western abstract art. Aboriginal art is filled with abstract-seeming and semi-abstract patterns of lines, dots, and geometric forms, often with patches of often bright color filling in the forms and background, suggestive of Paul Klee, Wassily Kandinsky, and JuanMiró. Aboriginal art is, however, highly symbolic of a central theme of ritualized sacred space. If one recalls the Andrei Tarkovsky film *Nostalgia* in which a man believes he can redeem a fallen world by repeating a ritual gesture: walking slowly across a shallow indoor pool holding a lit candle, they would have a metaphor of what Aboriginal art is for the Aboriginal. It is moreover ontologically grounded. The idea of a spiritual pilgrimage to a sacred site, such as the Sufi poet Jelaluddin Rumi's tomb, in cultures worldwide or to an aesthetic site celebrated in poetry, as in the Japanese *utamakura*, combined give a good indication of what Aboriginal art is: a reference to a mythic time when the specific subject of the painting, a ritual encounter within an actual part or parts of a landscape, was thought to have been created. The artist in a ritualized act is engaging the *illud tempus* or time of origins. The deceptive abstraction and focus on sacred typography is carried over to the modern world in Charlie Tarawa Tjungurrayi's *An Audience with the Queen* (1989) in which a Kasimir Malevich-like white space encompassing twenty concentric alternating black and white circles is bordered by various colored squares but is actually a design of the palace area where the artist met Queen Elizabeth II.[3] Ronnie Tjampitjinpa's *Untitled* (2003) looks like a bull's eye target surrounded by maze-like formations but recounts an ancestor's visit to the actual site of Tjintjintjin.[4] Likewise, Shorty Lungkarta Tjungurrayi's *Mystery Homeland* (1972) with white concentric circles on a black or maroon background in compartmentalized rectangles,[5] Freddie Ngarrmaliny Timms' *Blackfella, whitefella* (1999) where four stacked small shapes hover in a black background,[6] or Minimini Mamarika's *Orion and the Pleiades* (1948) in which a horseshoe shape containing thirteen circles hangs above a capital "t" shape containing six radiating circles[7] contain symbolic meaning, connected to myth or contemporary issues.

Such symbolism appears in sacred body paint,[8] on sacred boards,[9] as sand mosaic,[10] or as totem display.[11] Wally Caruana, an expert on such art, offers keys to the symbolism: concentric circles "denote a site, a camp, a waterhole or a fire ... Meandering or straight lines may indicate lightning or water courses, or ... paths of ancestors and supernatural beings ... U-shapes ... represent settled people or breasts ... arcs ... boomerangs or wind-breaks ... short straight lines or bars ... spears and digging sticks ... fields of dots ... sparks, fire, burnt ground, clouds, rain ... ."[12] In Clarise Nampijinpa Poulson's *Flying Ant Dreaming* (1990)[13] the abstract-like complexity of pattern, with concentric circles, U-shapes, and squiggly lines represent a diagram of the Flying Ant ceremony taking place at a specific site. Some of the circles are termite mounds, housing a staple food for the Aboriginals, and the camps of those harvesting the food.[14] The connection of the contemporary Aboriginal world and the ritualized nature of that world is called Dreaming.

The Dreaming refers to issues related to the mythic past which define the present landscape and Aboriginal rituals. The Dreaming story, ritual, associated imagery, and land upon which the primal Dreaming took place are inherited and have the equivalent of a copyright.[15] Thus Peggy Napurrula Poulson, Maggie Napurrula Poulson, and Bessie Nakamarra Sims' *Possum Dreaming* (1988) incorporates the story of Possum Ancestor who created the local waterholes, the circles in a vertical line. The squiggly lines are the tracks of the nocturnal possums as well as the dance movements of the ritual associated with the Possum ancestor.[16] Likewise, Uni Nampijimpa Martin and Dolly Nampijimpa Daniels' *Fire Country Dreaming* (1988) relates the Blue-Tongued Lizard Man's punishment of his sons' ritual breach by burning the landscape, such as contemporary farmers do to revitalize the earth, and was followed in the myth by a revivifying rain storm. The tracks of the participants of the Fire Dreaming ceremony move around circular icons, which could also represent the collected rain.[17] The majority of the ritual Dreaming refers, according to Caruana, to the "activities and epic deeds of the supernatural beings and creator ancestors" such as the Rainbow Serpents, the Lightning Men, the Wagilag

Sisters, etc.[18] These figures occur in rock painting, such as the representational one of the Lightning Men at Katherine River in the Northern Territory.[19] More usually, in Dreaming ritual design and art derived from such design, the art is predominantly symbolic as to form. In Patricia Lee Napangarti's Miro-like *The Death of the Tjampitjin Fighting Man at Tjunta* (1989) an ancestral heroic battle is depicted in such form. The large inverted U-shape at the center is the hero facing off against the gathered U-shapes at the painting's bottom. The hero has brought his staff, a long brown line surrounded by white dots and sacred clan symbols, small bent brown shapes similarly surrounded by white dots, on his left and right. He had stopped for a drink at a pool, a round blue circle containing his footprints. The top represents his journey to the battle, a blue line following a drainage bed to an immense rock hole. The hole is bracketed by curved black, red, and white forms. This site is probably a ritual one as such structures seem important to ritual settings, as water is precious in the desert.[20] Perhaps the ancestor drew strength from a ritual setting before battle.

The epigraph attributed to the god Shiva describes the nature of altered states of consciousness. One prominent theory aligns such states in the spirituality of primal cultures with neurological patterns of mental imagery, such as dots and squiggles.[21] Two of the means of accessing such imagery are sensory deprivation and sensory overload. The rock hole in itself and when connected to a cave represents a passage to another dimension in many primary cultures. When Aboriginals enter a cave for a ritual or dance and sing in a ritually important setting, they are entering altered states of consciousness and a socially designated spiritual consciousness sanctioned by the Dreaming. The representation of such ritual is sacred to the Aboriginal and, except for an early period in modern Aboriginal painting, non-Aboriginals are not permitted to view such a representation directly. Tim Payungka Tjapangarti's *Cave Story* (1971) and Yumpuluru Tjungurrayi's *Cave Story* (1972) depict the consciousness of experiencing the Dreaming in a cave. This consciousness experiences both the external cave (the bottom of each painting) and the interior of the cave (the top of each painting). The first may depict ritual objects, elongated lozenge forms, within the cave. The second represents a rock hole dripping water, lines of white dots, into the cave. White dots in fact dominate both paintings as outlines of objects, as paths of water, or as demarcations of objects in the landscape, thus sacralizing the paintings indirectly.[22] Though three paintings are focused on a central concentric circle form, Tommy Lowry Tjapaltjarri's *Pintupi Medicine Dreaming* (1972) may be contrasted to Shorty Lungkarta Tjungurrayi's *Classic Pintupi Water Dreaming* (1972) and Old Walter Tjampitjinpa's *Rainbow and Water Story* (1972). The first painting, looking like a rudimentary Gingerbread Man surrounded by three ovals is a representation of an initiation ceremony with sexual overtones, the elongated arms and legs and the ovals representing male and female sexuality in a manner similar to the earliest cave and rock art of Paleolithic and primal societies. The other two allude to mythic water hole formation by snakes, the lines around and leading into the central concentric circle, in *Water Dreaming,* and by a lightning storm, the yellow thatch-patterned upper left corner, in *Rainbow and Water Story*. In the latter painting, the two arcs enclosing the water hole are simultaneously the rainbow and a design associated with the Water Men ritual.[23] Dreaming accounts for the mythic origins and ritual sustaining of flora, fauna, and meteorological conditions necessary for Aboriginal existence. Billy Stockman Tjapaltjarri's *Yala Dreaming* (1971) is an astounding Miró-like canvas of wild potato plants against a bright yellow background. It includes the curved forms of women harvesters and water holes, three of which are connected by footprints and identified as ritually significant by the white dots surrounding the lines between them.[24] Paddy Jupurrula Nelson, Paddy Japaljarri Sims, and Larry Jungarrayi Spencer's *Star Dreaming* (1985) relates to a fire ceremony celebrating the formation of the constellations. The central area of two concentric circles connected with red and black bars of various lengths seems a ritualized setting. Similar concentric circles border the left and bottom of this central space. Above are a profusion of star burst forms, the constellations that meld in the right border to sacred sites.[25]

Bruce Chatwin tried to account for the wondrous amalgam of geography and myth in *The Songlines*, summarized in the two epigraphs from that work. His 1987 work and the overly romantic view of the walkabout Aboriginal singing the universe into continuous existence is now considered fiction. Nicholas Roeg's 1987 film *Walkabout* likewise romanticizes the Aboriginal trek as a kind of Native North American Indian vision quest done while walking rather than sitting. The recent film *Australia* also romanticizes the walkabout, now considered an Aboriginal's general need to visit relatives and so forth. Yet the gist of Chatwin's placing song and dance in a sacred typography and the intricate knowledge of nature of the young Aboriginal in Roeg's film are close to an essential understanding of the Aboriginal consciousness and its spiritualizing the world, including patterns of Aboriginal behavior. In fact, even though Aboriginals share common concepts under different names, major distinct groups, in Arnheim Land in the north, Kimberley in the northwest, Victoria in the southwest, and the central desert area surrounding Papunya, have different respective typography and their respective art reflects this difference. Amazingly, the modern art of Papunya can be dated from the year 1971 when an art teacher encouraged the Aboriginal men to paint traditional imagery on the school walls. The subsequent individual and collaborative acrylic art work at first directly expressed the most sacred aspects of Aboriginal Dreaming. Later, it was felt that such expression was not appropriate for non-Aboriginal viewing and such art was modified and even obscured for non-Aboriginals, as in fact some of their publicly viewed tribal ceremonies and sacred earthworks were. Yet a sacred aspect in the modified paintings comes through as they are reflections of Dreaming.

Thus the paintings of the desert are dominated by water holes and the sacred history associated with the holes. Wimmitiji Tjapangarti's *The Artist's Country* (1989) represents the sacred typography where the artist lives: rectangular forms that are hills, meandering lines that are creeks, and round forms that are the water holes in a Jackson Pollack-like profusion. To the upper right are zinc white bird tracks associated with the ancestral Old Woman who turned into a bird.[26] Accordingly, Susie Bootja Napangarti's Kutal *Soakage* (1989) is centered on a water hole understood to be the dwelling of the Rainbow Snake that produces rainstorms and lightning alluded to in the multitude of the local dotted marbled stone forms which are all oriented to the central water hole, including serpent-like shapes.[27] Kaapa Mbitjana Tjampitjinpa's *Mikanji* (1971) depicts at its center the ritual Dreaming of a local water hole. The hole is bounded on the left and right by sacred poles and on the top and bottom by the primordial serpents that created the hole and bullroarers, ritual sound producing objects connected here directly to the hole. Across the bottom are ancestral kangaroo tracks, and at the extreme left and right are bands of water holes and ritual emblems.[28] The relationship of human sexuality and the water is evident: one produces life and one sustains life. This connection is dramatically illustrated in Uta Uta Tjangala's *Yumari* (1981) in which an ancestral male with prominent genitals runs through a landscape covered with water holes, images of which cover his body.[29] The ancestral male is probably the Old Man whose testicles have a life of their own and often go travelling by themselves. Long Jack Phillipus Tjakmarra's *Medicine Story* (1971) depicts this cartoonish improbability. The central cactus-like green penis is demarcated as sacred by white decorative dots. The ten red, brown, black, and white runaway testicles, perhaps on different journeys, are connected to the penis with straight red lines. The yellow, white, red, and black wavy lines are probably sperm.[30] Uta Uta Tjangala's *Medicine Story* (1971) repeats the central cactus-like form, here brown, connected to sixteen wandering brown and white testicles. At the bottom is a horizontal brown cucumber shape that is the Old Man lying down.[31] Other Dreamings relate to the origin of bush food and totems. Tim Leura Tjapaltjarri's *The Honey Ant Story* (1972) alludes to the ancestors who came to earth as honey ants and later turned into men. The central roundel is the honey ant nest imposed on a ritual shield-like board. Four rows of three vertical sacred stones are to the left and right. A ritual spear extends vertically between each of the two rows of stones. On the spears may be men dancing the Honey Ant dreaming.[32] Billy Stockman Tjapaltjarri's *Possum Dreaming* (1972) relates the wanderings of this mischievous totemic animal. The central wavy colored lines are the possum's main trail which is bordered by delicate going and coming possum tracks and concentric circles where the creature rested. In turn, all this is bordered on the left and right by pairs of sacred ceremonial stones.[33]

In addition to the thorough encompassing of the Aboriginal world in a sacred geography, it is ontologically bound more often than not in sacred time, particularly with regard to ritual acts and ritual objects. One of the most prominent objects in both enacted ritual and ritual allusion in painting is the *tjuringa*, defined by Caruana as "sacred and secret incised boards and stones ...."[34] These objects are stored in caves where ritual ceremonies are carried out. Thus Mick Namararri Tjapaltjarri's masterful *Big Cave Dreaming with Ceremonial Object* (1972) depicts the rock stratum of the cave with *tjuringa* in the largest stratum in the upper half of the painting. The lower half depicts ten men engaging in a ceremony to the right. An enormous *tjuringa* hangs from the cave in an expression of a trance state reception of its importance.[35] Similarly, Shorty Lungkarta Tjungurrayi's *Mystery Sand Mosaic* (1974) depicts a sacred sand painting, perhaps of a water hole entering a cave, the concentric circles in the center. Below the area are four larger-than-life *tsuringa*.[36] Clifford Possum Tjapaltjarri's (1972) *Emu Corroboree Man* (1972) represents the Emu Dreaming ceremony. Such ceremonies include singing, dancing, and musical accompaniment. Sacred *tjuringa* boards with clan emblems appear on the dancer's back and bracket him on all sides as do the emu and their tracks on his left and right. At the extreme bottom are two bullroarer instruments, a flattened piece of wood, often a *tjuringa*, swung by a piece of string in a horizontal circle to produce a whirring sound.[37] Sometimes this instrument is accompanied by clapping sticks and, rarely, the long *didgeridoo* wind instrument. Shorty Lungkarta Tjungurrayi's *Snake Dreaming at Lampintjanya* (1972) also focuses on bullroarer *tjuringa*, here accompanied by snakes and water holes, alluding to the mythic creation of the holes.[38]

Rituals and ritual objects occur in real time but are transformed into sacred time by the allusions evidenced in specific designs and movements. The actual paint color and design and the effect of light on these produces a trance effect that is regarded as an opening of sacred time, what Caruana refers to as "visual shimmer."[39] He further suggests that for the Aboriginal "designs embody the power of the supernatural beings, [and] they are intended to be sensed more than viewed."[40] The effect is not unlike that approached in the artistic intentions of op art. This effect is found in any number of Aboriginal paintings without support or easily discernable support of recognizable objects through an intensive treatment of design and compression of color opposition, such as Tim Leura Tjapaltjarri's *Bushfire Spirit Dreaming at Napperby* (1972),[41] where the stippling hides a central water hole and paths, perhaps streams or snakes leading to it, and Kaapa MbitjanTjampitjinpa's *A Small Snake* (1972), with similarly obscured snake tracks.[42] Two paintings completely obscure their subjects through inclusive miniaturizing of their compact patterning: Mick Namararri Tjapaltjarri's *Tjunginpa* (1991), associated with *Bettong* (kangaroo rat) Dreaming and the creature's tracks,[43] and Turkey Tolson Tjupurrula's *Straightening Spears at Ilyingoungou* (1990), a depiction of spear straightening in a fire.[44] Two other paintings reflect semi-abstract forms against op art-like backgrounds: Clifford Possum Tjapaltjarri's *Dreaming Story at Warlugulong* (1976), an explosion of brushfire related to an ancestral event with the black and white dots reflecting the burnt landscape,[45] and Anatjari Tjakamarra's *Yarranyanga* (1989), an ancestral allusion taking place among rock holes, a pattern of dizzying black and white concentric circles and their similarly colored connecting paths upon a variously colored stippled background, a claypan.[46] Incorporating many of the previous approaches, Anatjari Tjakamarra's *Pakarangura* (1972) surrounds a water hole and cave of concentric black, white, and red lines with four huge, intricately patterned water *tjuringa*. These forms are set against scalloped concentric semi-circles of alternating maroon and white lines suggestive of traditional Japanese depictions of waves but here probably desert effects.[47] The conjunction of water, a cave, and *tjuringa* with the optical effects reflects the ritually important impact of this Water Dreaming. It is not surprising that this painting is one of eight of the fifty featured paintings in Roger Benjamin's book *icons of the desert, Early Aboriginal Paintings from Papunya* that do not appear in the main text but rather in a supplement because of their especially sacred imagery.[48]

The concern, accordingly, by Aboriginal artists to hide sacred imagery in their work is one aspect of the seemingly obsessive use of dot fields while those very fields may reflect the tangible presence of the sacred. Caruana thus asserts: "Areas of dots may mask sacred designs, and they may be used to produce visually stimulating effects intended to evoke the presence of supernatural power in the earth."[49] The Shiva epigraph suggests that true ontology is elicited in a hypnagogic state. In the Aboriginal world as seen through a ritual trance state in a Dreaming precipitated by singing, dancing, music, body paint, sacred objects and their signs, sitting in a dark cave, and so forth, another true ontology is revealed to the Aboriginal. The visual stroboscopic effect of such elements carries over to Aboriginal painting. Looking at Emily Kame Kngwarreye's *Untitled* (1991), an infinite field of jumbled colored dots,[50] or Robert Ambrose Cole's *Untitled* (1994), an infinite field of orderly white dots almost covering its black background,[51] one gets a visual effect that may be suggestive of neurological imagery experienced in a trance state, a state that may certainly be part of the Dreaming. The fact that Aboriginal artists have been seen singing while they painted is not surprising. They were calling up the Dreamtime. The Dreaming is communion with the first beings and the ancestors through allusion in body paint, songs of the first time, dance, art, and sagas of the ancestors that define how the genders and their respective rites are arranged, how Aboriginals are connected to their landscape, to the heavens, and to their totem animals, and to the water and bush food they rely upon. Just as animism supports the worldview of Shamanism and Shinto, the Dreaming and singing orders the Aboriginal world in its tangibility and enlightens it in what is spirituality.

40 Manning Mill Road, Hampden, ME 04444, USA
e-mail: dr_bruce_ross@hotmail.com

NOTES

[1] Bruce Chatwin, *The Songlines* (London, England: Penguin, 1987), p. 178.
[2] Ibid., p. 179.
[3] *icons of the desert, Early Aboriginal Paintings from Papunya*, ed. Roger Benjamin (Ithaca, NY: Herbert F. Johnson Museum of Art, Cornell University, 2009), pp. 164–165.
[4] Ibid., pp. 172–173.
[5] Ibid., p. 59.
[6] Wally Caruana, *Aboriginal Art* (London, England: Thames & Hudson, 1993), pp. 181–182.
[7] Ibid., p. 79.
[8] Icons of the desert, op. cit., p. 35.
[9] Ibid., p. 41.
[10] Ibid., p. 33.
[11] Ibid., p. 35.
[12] *Aboriginal Art*, op. cit., p. 103.
[13] Ibid., p. 141.
[14] Ibid., p. 140.
[15] Ibid., p. 104.
[16] Ibid., pp. 137, 139.
[17] Ibid., pp. 136–137, 139.
[18] Ibid., p. 10.
[19] Ibid., p. 12.

20 Ibid., pp. 158, 160.
21 David Lewis-Williams and David Pearce, *Inside the Neolithic Mind, Consciousness, Cosmos and the Realm of the Gods* (London, England: Thames & Hudson, 2005), pp. 46–55.
22 *icons of the desert*, op. cit., pp. 90–91, 112–113.
23 Ibid., pp. 116–117, 122–123, 124–125.
24 *Aboriginal Art*, op. cit., p. 114.
25 Ibid., pp. 134–136.
26 Ibid., pp. 154–156.
27 Ibid., pp. 157–158.
28 *icons of the desert*, op. cit., p. 84 and supplement, p. 3.
29 Ibid., p. 34.
30 Ibid., pp. 86–87.
31 Ibid., pp. 88–89.
32 Ibid., pp. 104–105.
33 Ibid., pp. 102–103 and supplement, p. 7.
34 *Aboriginal Art*, op. cit., p. 102.
35 *icons of the desert*, op. cit., pp. 136–137.
36 Ibid., pp. 154–155.
37 Ibid., pp. 108–109, and supplement, pp. 10–11.
38 Ibid., p. 58.
39 Aboriginal Art, p. 60.
40 Ibid., p. 60.
41 *icons of the desert*, op. cit., pp. 138–139.
42 Ibid., pp. 140–141.
43 Ibid., pp. 168–169.
44 Ibid., pp. 43–44.
45 Ibid., pp. 156–157.
46 Ibid., pp. 166–167.
47 Ibid., pp. 134–135 and supplement, p. 13.
48 Ibid., p. 134. The statement on the black space for each of the eight is: "For reasons of its secret/sacred imagery, this image is reproduced in the supplement only."
49 *Aboriginal Art*, op. cit., p. 116.
50 Ibid., p. 152.
51 Ibid., p. 159.

SECTION XI
CONTEMPORARY RETRIEVING OF THE PRINCIPLES OF THE UNIVERSAL ORDER

MAIJA KŪLE

LOGOS AND LIFE: UNDERSTANDING OF RHYTHM

ABSTRACT

Human being exists in the flow of time and where there is interaction between time, space and energy there is some rhythm. Physical sciences tend to attribute to rhythm a mechanical overtone but phenomenology of life shows a rhythm in the context of logos and life. Phenomenology of life describes logos of life realizing in time, place and creative acts. Concept of creativity developed by Anna-Teresa Tymieniecka is similar to concept of energy recognized by Lefebvre. It means that we can investigate the problem of meaning of rhythm at the logos and life as creative experience. Rhythms appear as cosmic, natural, psychological, cultural, social and can be described as: (1) a repetition of movements, situations, acts; (2) cyclical processes of development or decay, (3) living beings birth, growth, decline, death, (4) philosophical ideas of cyclical time and eternal recurrence. Anna-Teresa Tymieniecka rhythm describes: (a) cycles, (b) pulsations, (c) circuit, (d) recurrence, (e) the swings of pendulum (analysis of literary works). Idea of eternal recurrence as cosmic/human being's rhythm mainly is developed at Nietzsche's philosophy.

The phenomenon of rhythm directly influences understanding of human life as a whole. Human being exists in the flow of time and where there is interaction between time, space and energy there is some rhythm. Physical sciences tend to attribute to rhythm a mechanical overtone but phenomenology of life shows a rhythm in the context of logos and life-world. Rhythm can be explained as a sequence of movement, changes, speed, pulsation; economists and social scientists speak about rhythm of economical periods, repetition of financial crisis, representatives of cultural and social studies – about eras, changes of civilizations, social cycles. Philosophers mainly did not include concept "rhythm" at the list of fundamental categories, only some of them have described cyclical development of cosmos, life and culture. Among them are ancient philosophers Heraclitus, Pythagoras, at 19th century – Hegel, Nietzsche, Schopenhauer, among contemporary philosophers – Gaston Bachelard and Henri Lefebvre with his idea of rhythmanalysis.

Stuart Elden writes about Henri Lefebvre concept of rhythmanalysis that French philosopher recognizes – everywhere where there is interaction between a place, a time and energy, there is rhythm. Phenomenology of life describes logos of life realizing in time, place and creative acts. Concept of creativity developed by Anna-Teresa Tymieniecka is similar to concept of energy recognized by Henri Lefebvre. It means that we can investigate the problem of meaning of rhythm at the logos and life as creative experience.

Lefebvre shows the interrelation of understandings of space and time in the comprehension of everyday life: music, the comodity, measurement, the media, political matters, city life etc.[1] Lefebvre's study includes a rhythmanalyst portrait – he [or she] listen to body, calls on all senses, experiences present moments, past and future images. The human being thinks with his body in lived temporality.

The rhythmanalyst has some points common with the phenomenologist – description of phenomena, body experience in the life-world and lived temporality, inner time consciousness. Analyst learns rhythms first from his personal body which serves as a metronome. The difficulty is to perceive distinct rhythms from the personal body, they can damage body existence if differ from universal natural or societal rhythms.

Rhythms appear as cosmic, natural, psychological, cultural, social. They can be described from the philosophical point of view as: (1) a repetition of movements, situations, acts; (2) cyclical processes of development or decay, (3) living beings birth, growth, decline, death, (4) ideas of cyclical time and eternal recurrence. Rhythm can be individual, particular and universal, it represents the lives of individuals or groups and appears as cosmic, natural order. Rhythms unite with one another or disunite creating a chaos. Unity of rhythms means polyrhythmia, disunity – eurhythmy, break of rhythm – arrhythmia.

The philosophical idea of the logos of life means recognition of the possible unity of rhythms – cosmic, natural, human and spiritual. Anna-Teresa Tymieniecka recognizes the creative act as the point of contemporary phenomenological access to the human condition. It means the radical change of the classical phenomenological perspective and gives a new interpretation of man as the creator and his specific telos. Logos of life has been interpreted within the creative inwardness. Description of pure conscious mechanisms of Husserlian phenomenology has been changed to the grasp of the rules of creative effort. Phenomenology of life opens a wide horizon of explanation of cosmic forces. Anna-Teresa Tymieniecka writes: "The mind is incarnate in living nature, finds in its processes and its generative forms a destiny parallel to its own. Thus arises a network of connections, which assigns its place to each phenomenon after having orchestrated all of them in the same symphony, to use the image dear to Leibniz. Living nature and fabricated nature bear the stamp of universal designs and have a role in the cosmic symphony."[2] The concept of "cosmic symphony" is very characteristic for phenomenology of life – it means orientation to the harmony of cosmic and human life rhythms, to the correspondence of cosmic, natural and existential life dimensions, *symphonic* polyrhythmia.

Henri Lefebvre classifies rhythms by crossing the notion of rhythm with those of the secret and public, the external and internal:

"(a) Secret rhythms: First, physiological rhythms, but also psychological ones (recollection and memory, said and the non-said, etc.).
(b) Public (therefore social) rhythms: calendars, fêtes, ceremonies and celebrations; or those that one declares and those that one exhibits as *virtuality,* as expression (digestion, tiredness, etc.).

(c) Fictional rhythms: Eloquence and verbal rhythms, but also elegance, gestures and learning processes. Those which are related to false secrets, or pseudo-dissimulations (short-, medium-, and long-term calculations and estimations). The imaginary!
(d) Dominating-dominated rhythms: completely made up: everyday or long-lasting, in music or in speech, aiming for an effect that is beyond themselves."[3]

Phenomenology of life does not emphasize classification of different rhythms but include rhythmanalysis in the creative orchestration of beingness within the Human Condition.

Phenomenology of life developed by Anna-Teresa Tymieniecka characterizes reality as the objectivity of the life-world. She describes: (a) cycles, (b) pulsations, (c) circuit, (d) recurrence, (e) the swings of pendulum (analysis of literary works). Reality is governed by logic of real facts. "Its stability in the *ever-recurring cycles of life* (it is characteristic feature of rhythm, M.K.), as well as in the seemingly foreseeable future progress within each cycle and above it, is naturally assumed by us to be *grounded in unchangeable rules and laws of our existence within the world of beings and things.*"[4]

Phenomenology of life does not classify rhythms as a movement of life physiological and social process and development of inner/outer experience. Rhythm exist in life cycles, it can be described as a vital pulsation and circuit. "In fact, the innumerable acts which we perform and which carry our vital progress (for example, acts of pulsation, instinctual acts, acts of sensation, feelings, desires, volitions) and which express our vital or as it is usually said 'animal' phase of existential progress, and express our specifically *human circuit* of experience as well, and which begins with the entrance into play of our cognitive, valuative, aesthetic, etc. faculties – that is, our fully developed human acts – are tempered in their respective intensities by the entire circuit within which they participate."[5]

In "Tractatus Brevis" Anna-Teresa Tymieniecka writes about the swing of the pendulum characterizing fictional rhythm.

Similar characteristic of rhythm gives Russian literary critic and philosopher Mikhail Bakhtin. He writes: "Rhythm is a value regulating internal (implicit) givenness, availability. Rhythm does not express experience, which is not well founded within it, it is not an emotive-volitional reaction to an object or meaning – it is a reaction to this reaction..."[6] Culture is an expression of rhythm. If rhythm may be likened to music, culture would be its lyrics. Understanding functions according to a certain rhythm.

Phenomenology of life recognizes that human beings are organized beings and we exist in a relatively stable world and are not fragments of dissolved chaos thanks to the system of recurrent order. Rhythm belongs to the phenomenon of recurrence.

The idea of rhythm is closely connected with the metaphor of eternal recurrence. Eternal recurrence demonstrates how an originally mythological sensually concrete image has been transformed into a philosophical idea. Friedrich Nietzsche is one of the originators of this idea[7] at a peculiar time when a trend of post-classical thinking emerged, concerning values and human life that radically altered the contemporary cultural orientation. Nietzsche himself broke the rhythm of Western culture by his "dynamite" style of philosophizing. It is noteworthy that at such crucial periods revived cultural phenomena as cyclical time, recurrence, previously discarded as unacceptable or incomprehensible, give an impetus to human life. There appears a diffusion of mythologico-poetico-religious terminology in Nietzsche's philosophical texts.

Alphonso Lingis interprets Nietzsche's idea of eternal recurrence in the context of the process of culture: "For Nietzsche the problem of the possibility of a culture today is not that of whether the lessons of the creative epochs of culture can be recalled today, but whether the forces of the creative ages of culture could recur today, whether the very feelings, the very dreams, from which the cultural forms which stand issued as monuments of the great festive moments of humanity, could recur in the late-born, civilized, rationalized life of today, that is, in the philistine life produced by our civilization that answers only to the need for comfort and for security".[8]

Idea of cycles and recurrence at the phenomenology of life is connected with the same motif – lessons of the creative experience.

The idea of eternal recurrence interprets time without beginning and end as a circle. In believing that man is experiencing each moment of his life for the first time, the human being is deceiving himself. But in reality, explains Nietzsche, man has already experienced it before and its passing is only apparent, since it will reappear in the future *ad infinitum*. Of course, it is hard to believe in the absolute recurrence of the human being's life and of the processes taking place in the world. But it is not difficult to believe in the existence of rhythm and eternal creativity of logos of life.

However, the notion of eternal recurrence is not a simple mythological image, a fancy idea that each moment of my life (as well as that of all the other people's) will recur again and again like a line of a song on a worn record. Eternal recurrence has nothing to do with the idea of a spoiled world. On the contrary, it is the idea of the fulfillment of the world, a peculiar symbol, a mystery incorporating deep archetypal statements concerning the circle (ring), rhythm, eternity realized in time and the value of everything that exists. Eternal recurrence symbolizes polyrhythmia as an unification of rhythms in cosmic harmony and normal human existential everydayness without catastrophe.

Nietzsche has said that if but one single moment in the world were to return then all the other moments would have to return too. This idea is not only a meditation on the flow of time and its direction. It is a strong belief in the orderliness of the world, in its changelessness and in the human potentialities for sufferings and attitudes towards them, belief in the appearance of value and its affirmation, the interconnection between a single moment and eternity, movement, changes and peace.

Recurrence as a symbol of cyclical time means ever returning of creative acts and existence of logos. In a word, it is a strong belief in the stability, value and firmness of all that exists, which manifests itself not in duration but in rhythm as reiteration and affirmation.

Friedrich Nietzsche has found a new way of affirmation that is no longer utilitarian, pragmatic or teological. The same but in alternative way has been done by the phenomenology of life with the concept of "logos of life". To affirm value of something does not mean to grasp its referentiality or utility. Value is not interpreted with the view of something else, or with the view of the human being. Affirmation is an end in itself, and not a performance for some reason within us. Forces are found in some sphere beneath human consciousness – in the creativity of life and especially in the human being's creativity. Stability in the world is retained in spite of the flow of time and not irrespective of it. The idea of eternal recurrence is originally a mythological image but in the context of the present-day cultural process it can be interpreted as a content-saturated metaphor and even a philosophical idea, which demonstrates the meaningfulness of rhythm and original affirmation of values.

Eternal recurrence contains the following significant dominant moments: rhythm and affirmation of meanings, which are accompanied by cosignificant elements – figure as circle, circuit, movement as time.

Moving along a circle and forever returning to the beginning (though – to be more exact – there is no beginning or end to a circle) means moving in a steady rhythm. Ancient cultures saw the mythological unity of the human being and the Universe and expressed this unity in a sense of rhythm.

The world's movement according to a rhythm is not alien to Latvian mythology, it finds expression in the language forms: *ritums, ritējums, aprite*. The Latvian female name **Rita** is rooted in Indo-European mythology. Latvian folksongs – *dainas* – tell how the contiguity of the Sun and the darkness begin revolving the eternal wheel of life.[9] They represent the ancient mythological sense of rhythm as a basis for creativity and stability against chaos.

The most distinctive dominant feature manifested in rhythm is the circle (ring, wheel). The return to the beginning of the curve of time locks up, as it were, into a circle. That is a universal symbol of eternity in the mythology of many peoples. To ancient people the circle signified the orderliness and fulfilment of the world, a uniform rhythm, which characterized the firmness and stability of everything that existed. The circle is known as the symbol of the Universe, all the movement in the Universe proceeding in a circle.

Circle, uniform rhythm as well as a specific understanding of eternity form the frame of eternal recurrence. It would not be correct, however, to reduce the idea of eternal recurrence to these forms. Not every concept, a feeling of the world, or an idea that admits of the circular movement is identical with the concept of eternal recurrence. There might occur similarity of form, yet, not of content, because the most essential notion of eternal recurrence is obtained by posing the question: what is it that returns?

In its primordial form the idea of eternal recurrence exists in its mythologically cosmological variant. Dominating in the above is the mythological image in combination with the cosmological interpretation of the Universe. The teaching of the eternal recurrence of the Universe dates back to ancient times as manifested in the world outlook of the ancient Greek philosopher Heraclitus, the Pythagoreans and the Stoics.

The purpose of eternal recurrence is the affirmation of the return itself. So that the questions: What is the sense of revolving? And what is the human's role in it?

do not apply. Rhythm exists in itself. These questions lie outside the logic of the idea on which the metaphor of recurrence is based. Antique philosophy represents the cosmocentric model of the world, in that it regards the human being as a natural part of the Universe, which picks up all the processes of the Universe and does not set himself apart as the subjective ruler of the world. The phenomenology of life developed by Anna-Teresa Tymieniecka recognizes similar structure of this model: the human person as the all-embracing functional complex and the transmutation center of the logos of life. The question is only about the meaningful role of human being.

Martin Heidegger shows how the human being in the course of civilization has lost understanding of the Being *(Sein)* and changed the rhythm of all living beings. This changes the rhythm of things because they become enslaved to the human being's rhythm of civilization.

As part of the Universe the human being has no purpose outside its rhythm. Though being transient and irrelevant, human actions will command the future, since every human deed bears eternity. Therefore the affirmation of oneself in the circulation of the Universe becomes so important. Heidegger interprets it as hearing of the voice of Being. The question about purpose does not apply either because its posing is based on the logic of a different idea, which acknowledges that the evolution of the world may have an aim outside its existence.

The dominant moment of rhythm is the circle. The typical characteristic of modern culture (time, history, the uniqueness of personality, way of value affirmation and etc.) is a "straight line" and feeling of historicity. The difference between the two significant moments – the circle and the straight line – is best laid bare in the understanding of time: there is cyclic time and linear time. The mythological sense of rhythm, which is based on the cyclic understanding of time, differs from the standpoint of Christianity, which creates a new (different) sense of the world, by postulating a historic dimension, irreversibility and the linear flow of time from the past to the future.

Linear time is a potentiality of historical thinking and a system of record. Mircea Eliade in his book *Le mythe de l'éternel retour* writes that the difference between the outlook of the human being of archaic society and that of modern society brought up in the Judaic-Christian tradition lies in the feeling of an intimate link with the rhythm of the universe and seeing one's own essence closely linked up with history.[10]

The idea of eternal recurrence portrays the clash of the cyclic and the linear time. It is a clash between the rhythm that affirms place and the rhythm that pushes forward. The first is represented by eternity going into depth; the second finds itself in prolongation. Friedrich Nietzsche interprets historical sense as a disease of his time. But the roots of historicity are much deeper. When The New Testament came in conflict with the views of antique philosophy advocating the cyclic rotation of the world, a clash of ideas and the opposition of the different sense of the world became unavoidable. Ridiculing the cyclic view St. Augustin wrote that it looks "...as if, for example, the philosopher Plato, having taught in the school at Athens which is called Academy so, numbers of ages before, at long but certain intervals, this same Plato, and the same school, and the same disciples existed, and so also are to be repeated during the countless cycles that are yet to be – far be it, I say, from us to believe this. For **once** (underscored by me – M. K.) Christ died for our sins; and rising from the dead, He dieth no more [...] And that too which follows, is, I think, appropriate enough: *'The wicked walk in a circle'*; not because their life is to recur by means of these circles, which these philosophers imagine, but because the path in which their false doctrine now runs is circuitous."[11]

The rhythm that pushes forward presupposes value affirmation between the positive and the negative values. But the cyclic rhythm appears without value distinctions, beyond good and evil.

The Russian specialist of Byzantine literature S. Averintsev writes: "If the world of Greek philosophy and Greek poetry is *cosmos*, i.e. a law-governed symmetrical spatial structure, then the world of the Bible is *olam*, i.e. a stream of time process carrying all things within itself. Inside the *cosmos* even time is given in a spatial modus: indeed, the teaching of eternal recurrence patently or latently present in all Greek conceptions of being, both mythological and philosophical, robs time of its inherent characteristic, namely its irreversibility, and lends it symmetry, which is only conceivable of space. Inside the *olam* even space is given in the mode of time-dynamics as a receptacle for irreversible events."[12]

Time connected with eternal return is spatial, i.e. spatially structured in the form of a circle obeying the principle of symmetry. The linear time, in its turn, does not allow events to recur. In this respect St. Augustine's statement is excellent: "Christ died but once". And that is all there is to it. The cultural paradigm has undergone a change from *always* to *but once*. Now the testimony of value is in singularity. A special place is assigned to individuality, to the unique, peculiar, matchless. The idea of personality as a unique, singular and deeply individual being, strictly speaking, is only possible in the paradigm of contemporary rhythm of life, for a personality is rooted in its history.

Man's uniquely life process as man's self-interpretative individualization describes phenomenology of life. Anna-Teresa Tymieniecka writes about self individualization: "It is the element of constructive differentiation from life-conditions while transforming them into *his conditions* [human being's, M.K.] of the life-world."[13]

The teaching of logos of life, creativity, individualization and rhythm acquires a new dimension in phenomenological philosophy. It is no longer the ancient cosmological idea, but a notion based on the phenomenology of creativity, on the hermeneutical understanding of cosmic life. Thus, a new interpretation of the idea is conceived – one that while retaining the features of an ancient understanding of logos – the moral sense of life –, emphasizes to a much greater extent the problems of subjectivity, morality, values and sense as against the problems of cosmologically by neutral rhythm. For phenomenology of life to believe in the logos of life, is to believe, that all creative and moral possibilities, what were once possible in humanity are still valid and in force in each individual, and at each moment of history.

In the context of modern civilization the idea of creative individualization resounds in the form of an appeal to become part and parcel of the life-functioning of the world, to appreciate the importance of the moment and not to exaggerate the role of history in the formation of the human being's life. It urges to remember the stability and order of the world which is not at all the making of human will. We see the horror of individual existence and yet we do not despair. The consolation cuts us off from the sphere of the changing phenomena. The struggle, sufferings, extinction, moral sense seem necessary now in the endless variety of forms resulting from life. Notwithstanding the fear and compassion we are all happy to be alive, yet not as subjective individuals but as everything alive with which we are inseparably linked.

The rhythm of creative process is a phenomenon uniting the world. It is a borderline state, which most often arises when culture is dissatisfied with pluralism, inner chaos, when the world is too divided and culture has become relative. The road from pluralism to monism is well known in the history of the world. One of the main questions of philosophy arising when the world stands at a crossroads of pluralism, disharmony, is how to substantiate the value and existence of separate individual things. One of the ways is to attribute value to things themselves; the second is to attribute value to them within the entirety of the world which affirms itself returning or locking itself out of the relative flow of time and including them into the development of logos of life. The individual being is not senseless. Namely, in the individual it is not the abstract form of humanity, but all the vital forces of all individuals that keep returning.

The anthropological line, which characterizes the idea of eternal recurrence, is described in Nietzsche's work *Unzeitgemässe Betrachtungen.* Nietzsche writes that the past and the future is the same, namely, it is something that in its obvious variety is typically uniform, and representing a constant return of unchangeable types it is essentially an image of eternally equal importance and changeless value. "If you are to venture to interpret the past you can do so only out of the fullest exertion of the vigour of the present: only when you put forth your noblest qualities in all their strength will you divine what is worth knowing and preserving in the past. [...] When the past speaks it always speaks as an oracle: only if you are an architect of the future and know the present will you understand it."[14]

That is the sphere where we can talk of rhythm – the world of values, sense and meanings. Meanings in the world of the human being exist only insofar as they are continuously affirmed anew. Without eternal recurrence human life is impossible. That is the eternal love, which returns from generation to generation, that is the eternal recurrence of likes and dislikes, of friendship and hate as long as there is human companionship. Thoughts, ideas, meanings return when they are thought out and comprehended anew. The values common to all mankind, a stable world order and a meaningful life cannot exist without it. Its precondition is rhythm characterized by stability. When the human being living in a world has detached him from the order of the cosmos, quite a specific problem arises as to how should the human world be put in order. The rhythm plays the role of a regulator then, for it provides an appraisal of every moment one has lived through, every action and every thought. The moral inherent in the phenomenon of rhythm appears to be even more ruthless than any other rigoristic moral. For phenomenology of life not the notion of being functions as a principle which sustains what there is. The main principle is "beingness", which means the principle of individualization and through which as a vehicle, life expands in its rhythm.

Institute of Philosophy and Sociology, University of Latvia, Riga, Latvia
e-mail: maija.kule@gmail.com

NOTES

[1] Stuart Elden, Rhythmoanalysis: an Introduction to the Understanding of Rhythms, in: Henri Lefebvre, *Rhythmoanalysis. Space, time and everyday life.* (London, New York: Continuum, 2004), p. vii.
[2] Anna-Teresa Tymieniecka, Logos and Life: Creative Experience and the Critique of Reason, in: *Analecta Husserliana* (Kluwer, 1988), Vol. XXIV, p. 81.
[3] Henri Lefebvre, *Rhythmoanalysis. Space, time and everyday life.* (London, New York: Continuum, 2004), p. 18.
[4] Anna-Teresa Tymieniecka, Logos and Life: Creative Experience and the Critique of Reason, in: *Analecta Husserliana* (Kluwer, 1988), Vol. XXIV, p. 307.
[5] Ibid., p. 416.
[6] Михаил Бахтин, *Эстетика словесного творчества* (Москва, 1979), с. 110.
[7] Joachim Köhler in his book *Who was Friedrich Nietzsche? Thoughts in a Centenary Year.* (Bonn: Inter Nationes, 2000) comments that Nietzsche's idea on eternal recurrence appeared as his hallucination figure, when he turned to opium. Such a bizarre notion is well known in the spheres of hallucination and epileptic attacks. But such an explanation of the genesis of the idea, to my mind, does not mean that the idea of eternal recurrence has no philosophical sense.
[8] Alphonso Lingis, Mastery in Eternal Recurrence, in: *Analecta Husserliana* (D. Reidel Publishing Company, 1986), Vol. XXI, p. 93.
[9] See: Vaira Vikis Freiberga (ed.), *Linguistics and Poetics of Latvian Folk Songs.* (Montreal: McGill-Queen's University Press, 1989).
[10] Mircea Eliade, *Le mythe de l'éternel retour.* (Éditions Gallimard, 1969).
[11] Saint Augustin, *The City of God.* (New York, 1950), Bk. XII, 13.
[12] Сергей Аверинцев, "Порядок космоса и порядок истории в мировоззрении раннего средневековья", *Античность и Византия* (Москва, 1975), с. 269–270.
[13] Anna-Teresa Tymieniecka, Logos and Life: Creative Experience and the Critique of Reason, in *Analecta Husserliana* (Kluwer, 1988), Vol. XXIV, p. 399.
[14] Friedrich Nietzsche, *Untimely Meditations* (Cambridge: Cambridge University Press, 1997), p. 94.

CLARA MANDOLINI

LIFE POWERFUL FORCE BETWEEN VIRTUALITY AND ENACTMENT

ABSTRACT

The chapter discusses the *peculiarity* and the *continuity* of Tymieniecka's phenomenology of life and philosophical enterprise, in relation to the renovation of the ancient concept of *logos spermatikos*, seminal reason. The theme is introduced by focusing on the central interrogative issue of philosophy as inquiry about life and its developmental features and conditions. Concepts, such as force and energy, receive a new relevancy in phenomenology of life, according to the demanding attempt to grasp the way how life develops itself and how it generates different living beings and their own unfolding.

WONDER AND CLASSIFICATION

A main problem lies under the whole human philosophical enterprise, as desiderative tension for knowledge: to uncover the veal of error, doubt, delusion that is always beyond human experience. The approach to reality, thus, whereas on the one hand is originally made possible by experience (which attests directly, even if not without already fallacious mediations, existence and the many characteristics of surrounding world), on the other hand it invokes the enterprise of the cognitive, analytical, transcendental peculiar ability of human thought. There are two in some way *a-symmetrical* poles, then, in the human cognitive attempt: *reality* itself, and *thought*, which tries to shape, reflect, narrate and penetrate reality, in turn grasped as a nucleus of beingness which is paradoxically "always-there", around us (but also inside us or physically far from us), still again out of reach for us and, nevertheless, "in sight". There is, so to say (continuing mentioning some ideas that have been at length assumed in the philosophical *koiné* of all times, despite the opinion of those who think that the philosophical thought could be *weakly* exercised), a continuous dialogue, as perennial challenge of human thought to reality, of which thought claims the authentic view, the unambiguous comprehension, the "embodiment" in his own "glance". But in this dialogue a problematic element appears soon. In fact, where is reality, or where experience shows that reality would be likely to be, man realizes that there was already something "behind", which does not appear immediately or plainly, which does not reveal itself in that "presumed" reality, and which is able to invalidate, falsify its pretension of completeness, immediacy, simplicity.

We come in this way to the birth-point of a most radical conceptual couple of the philosophical thought: reality/appearance, which is correspondent, even though not coincident, to the couple truth/illusion: we can see that, typically in the non-coincidence between these two conceptual couples, lies the *proprium* of phenomenology, as well as its novelty as a philosophical methodology and as a discipline of reasoning. But, at the same time, reality, as "thought" and "said" (*logos*), establishes itself as field that presents a gap in relation to its wrong, distorted mirroring: the "said", therefore, becomes the genetic field of appearance, of appearance intended as illusion and – let us recall Plato here – the misleading and deceptive (or, at least, incorrect) opinion (*doxa*). In this sense, at the birth moment of the inquiring thought (the philosophical thought in particular), reality is, no more, not only an experience which is in itself meaningful, but also an image to which the interrogating subject can ask for the credentials for exhibiting itself as truth or, on the opposite case, as illusion. Nietzsche, then, posed himself rightly in the core of this problematic connection when, in his criticism of "truth" and in his appreciation of *perspective*, he saw that reality, originally connected and incorporated to the living as such – that is, as *alive* and thus self-perceiving – becomes illusion as soon as it is established as *truth*, or, better, as soon as a representation of it is crystallised and identified with a supposed *true* reality. In simpler words, and even before that Nietzsche underlined it with strong emphasis, truth hides itself as such when an image of reality is declared as an unequivocal and faithful representation of reality.

Here we can understand why the theme of mutation and that of cause have become so crucial: every image of reality, while declaring itself authentic, falsifies itself, not being able to bear the evidence of its stability: the focus on the gap between reality and illusion, in this regard, recalls and requires, as a complementary and unavoidable theme, to focus on the great problem of mutation and becoming.

On this second level of reflection, however, the gap reality/illusion appears once more, even if provided with a dynamic connotation. We find that, maybe, falsification is not inherent really to the representative procedure itself, but rather to the constantly dynamical nature of what is, in any case, a cause of experience, of the experience that witnesses, to man, mutation, variation, mutability, ageing. Using a synthetic term to represent the passage between these two phases (not necessarily put in succession) of the philosophical attempt, we could say that, where the analytical level stops to the verification (or not) of representation, there is a "spirit of *classification*" acting, whilst, where one wants to discover the reality of mutation itself, there is a "spirit of *wonder*", as a key feeling and thinking attitude at the birthpoint of philosophy and even of science. Now it seems to us that phenomenology could be said a crowning synthesis of both those two attitudes of "classification" and "wonder", generating *method*. Phenomenology, in fact, does not conceal the representative aspect of experience and thought, but at the same time, through the Husserlian discovery of intentionality, it wants to be faithful to the experienced, lived reality, embodied in the sentient and thinking being itself, in the interrogating living being, namely the human being.

In phenomenology, classification and wonder, now fused, or, rather, put in an ever-fusing dipole, receive a fecund improving, because thought there becomes effectively an "interroga*ting* glance", which is not impersonal, simply ordering juxtaposition of the subjective "lens", but rather its reality in its "being-shown" and self-sharing, still remaining a partial and perfectible glance, however not separated from the universal whole. And the reason for this resides rightly in the fact that reality is not negated by phenomenology in its appearance, an appearance which is fatally consigned to a destiny of negation; but that it is approached as phenomenon which shows, even if partially (and it could not be otherwise), its own credentials of reality. In the words of Anna-Teresa Tymieniecka: "Phenomenology remains a path of inquiry focused on the very *sense* of phenomena, on what makes them "phenomena" for the acting and cognizing subject, what maintains articulation and order amid the fleeting, ungraspable appearances in which the real manifests itself and so grounds our vital, psychic, and mental existence".[1] In this way, phenomenology makes room, at the same time, both to the exigency of ordering, experiential and categorical[2] rigour, and to the vital necessity of fidelity to a reality that, if true, is "real", real in the infinitely fecund manner of life.

From this original attitude and novelty of the phenomenological approach – that here we cannot specify, limiting ourselves to refer to its father Husserl – a fundamental consequence, or better an interesting opportunity for thought comes out. This opportunity has been intellectually caught and developed by the whole philosophical work of Anna-Teresa Tymieniecka: we should underline the originality of her reflection together with its full adherence to the instituting question of perennial Philosophy, through a new elucidation of the nuanced implications of the couples reality/appearance and truth/illusion, balanced on the new barycentre of *life*. But she grasps a further element in those couples, by characterising logos both as a feature, organizational principle of vital becoming, and as an essentially interrogative logos.

APPEARANCE OF LIFE AND BEYOND

In Tymieniecka's fully phenomenological attempt, the dichotomy reality/appearance is stepped over (but not ignored or misunderstood) since it overpasses the level of an inquiry that is conceived as mere analysis and mirroring; rather, it seeks to "penetrate" reality in its making-itself as such, as, also, appearance and apparition, that is to say as "phenomenon": "The logos that humanity has been pondering for centuries and which we cannot fail to encounter all over again now through phenomenology we may seek to pursue either in full light or by unearthing it from thus far inaccessible locations as it radiates through the entire sequence of life and beingness-in-becoming pointing to further areas through the relevancies of each segment".[3]

However, Tymieniecka gives one more, new hint to phenomenology, a contribution to its further implementation rightly in the direction of the continuation and perfecting of its original characteristic of synthesis and conjunction between the inquiring rigorous approach and the intuitive, penetrating attitude. And this synthesis actually manifests itself as the true counterbalance to the rationalistic drift of some transcendental philosophy. Essentially, this element consists in the inquiring focus centred on life: this one is not assumed simply as object of a sectorial discipline (like in the case, for example, of biology), but almost as a counter-concept in relation to that of reality.

Life is reality not just in the sense of a "real" that man can find already-produced, in a sort of still-nature at hand, ready-to-be-seen; it is, rather, the *source* of an organised reality which is full of powers of change, a source of dynamisms, qualities, relations, energies, constructive and perennial virtualities and "voices" that ontologically "call" the response (a constructive, active, fecund or, differently, disruptive response) of other living beings. Life is phenomenon of genesis, production, growth, vitality, creativity of the real, since it is dynamism, process, becoming: thus, it is not just phenomenon or appearance, but source and generating process of appearance and phenomenon itself (not remaining simply noumenon, so to say). In this regard, life, as a concept, contains much *more* than the simpler – and, let us say, opaque, unqualified, "grey", *abstract* – concept of reality.

Even from this observation we could grasp the phenomenological continuity of such a philosophical approach with questions that (not without an extreme simplification, here needed for brevity) we were mentioning before: phenomenology of life is conceived as able to restore a reality that is *true* (real) inasmuch as – and because – it is process of its making itself a generative, creative and at the same time disruptive, transformation process: *"Life is the conveyor of beingness"*.[4]

The gravity centre, put on life, of this new phenomenology, overpasses then the focus of a philosophy centred on reality and on the idea of a rational mirroring of reality through the analytical lens of human intelligence. In this approach, life, in fact, is still reality, but it is not reduced to the traditional notion of reality: the idea of life makes up for classical aspects of philosophy, as inquiry oriented to causes, to the Principle (*arché*), to the innovative and welding element of multiplicity that is attested by experience; nonetheless, it gives to it new and determinant features. The fundamental idea that the focus on life contains is that the principle and *ratio*, the sense of reality as phenomenon is creation of itself as life, as generating dynamism of beings that, in turn, are creative participants in a becoming, in a sensed becoming constructively oriented. Life, in other words, is not static reality, so that an image, a representation of it, an enunciation declaring a state of affairs could adequate it in order to remain truthful intellectual experience; on the contrary, it is creative and fecund process and interaction of aspects, potentialities, realizations, but above all irreplaceable beings. To that corresponds the idea that logos is not merely a formal principle, but – we could say – virtual, ordering and self-conferring material, temporal, operative devices that are necessary for its own deployment in any form of life, what Tymieniecka includes in the concept of "individualising beingness-in-progress".[5]

From there the dynamic characterisation of logos follows, in the sense of an evaluation of its original connection with life: "Decisively, the nature of this course has been envisaged principally in terms of its formative, constructive progress, which implies forces and energies at work: it implies a self-prompting, that is, inner, dynamism".[6] But dynamism, in this conception, is not synonym of vitalism or chaotic aggregation of energetic drives: in fact, we are speaking about "logos", this is about an ordering principle: "The force of the Logos manifests itself in the logos' effusion of life. In acquires "shape" in its performance and is then intuited through that performance, from the inside, as it were. First of all, logos, the reason of reasons and the sense of everything, is not simply a set of principles articulating "matter". It is above all a force, a *driving force* that through its modalities is accountable not only for the incipient instance of originating life in its self-individualizing process but also for the pre-origination, pre-ontopoietic ground and for the subsequent striving toward the abyss of the spirit. Life, as the ontopoietic progress of the logos' drive in the self-individualization of beingness, emerges then as a manifestation of the ontopoietic process".[7]

Here is at stake a different vision of mutation and becoming, that tends to assume in deep the sense of reality as *evolution*, thus approaching in an interesting way some acquisitions of the biological and cosmological sciences. Moreover, it appears fecund just as a key to overpass the rationalist cage of thought, by thinking reality and beingness in the key of an entelechial order inside life's unfolding, which lets the "new philosopher" interpret the variety, fantasy, variability of the forms of life in the balanced view-point of both the particularity of concrete vital realisations and devices, and the general design underlying to the passage from virtuality to enactement. Tymieniecka highlights then the presence of a necessary spatio-temporal structure of life's becoming, and an order in which she rehabilitates the precious ancient concept of *entelechy*, rightly to explain the manifestation of order, enactement of virtualities, together with the selection of life's strategies and possibilities. But this is "order" not only in the sense of a classification, of an arrangement principle of already-made elements, beings or matters, but in the sense of a constructive order, capable of establishing hierarchy, order, arrangement in the (quite different) sense of becoming and growth.

The constructive design of the entelechy is not a mere formal blueprint. It is above all a set of selective virtualities – forces and energies endowed with propensities toward intergenerative fusing as well as toward entering into these fusings with appropriate elements such that a pattern of growth will be spontaneously outlined by their release. It is from this time-conditioned constructive project that spring forth constructive means, constructive postulates: inner/outer, and present/past/future. In other words, it is the inner postulates of growth that brings forth what we call the spatiotemporal schema of life.[8]

SEEDS OF LOGOS

The decisive point, then, in the phenomenology of life by Tymieniecka, is the new relation which is assumed and acknowledged between the creative principle of the fecund dynamism of life, its prompting and creative drive, and life itself as appearing reality, that is to say, at the same time, as object of the phenomenological "glance", and source of becoming manifestation itself: a manifestation which is both gnoseological and possibly representative on the one hand, and effective deployment and continuous genesis of reality on the other hand.

In fact, logos is conceived, in this perspective, not as abstract principle of knowledge or of subjective cognitive activity, but as concrete, effective, consistent "constructive track", "drive" that "carries the entirety of the givenness discovered"⁹ on the road of life manifestations and of a wonderful order: the two essential features of "logos of life" are, in the words of Tymieniecka, the twofold fact that "it [the logos] harnesses the universal becoming into the genesis of self-individualizing beingness as it both participates in the universal flux of life within the world, constituting it, and simultaneously makes it present to itself in innumerable perspectives".[10]

This means that logos is essentially "logos *of life*", first, in the sense that it provides life with its constructive recognisable sense and beings' and beingnesses' differentiation, interaction, variation, and so on, and, secondly, because it makes life capable of recognition and "vision" to itself, to at least some of its forms (and eminently, the human one). In this sense, Tymieniecka writes about an ontopoietical level of life – a level of life in which life assumes its creative and, at the same time, structured multiplicity –, on which also phenomenology, as interrogation of logos about logos (in life), becomes re-conjunction with life itself, by means of its inquiring and "re-cognitive" attitude. In this sense, Tymieniecka argues that life, both as ontopoietic and logoic reality (and now understood in its being manifestation of logos), can be said the really ultimate "being the yield of the very last reduction".[11]

In this perspective, that, as we can see, recuperates and renovates the philosophical and specifically phenomenological enterprise, one aspect in particular can be noted and analysed of life. It deals with the peculiar real (or, better, generative) and displaying (from the point of view of thought) dual characteristic of life. Tymieniecka, in other words, poses herself not *over* thought, nor *on this side*, but rightly on the bridge that originally connects the inquiring subject to the vital reality, from which it yet comes and in which still remains immersed, embodied. This aspect, really crucial from a gnoseological point of view, can be grasped in its peculiarity (and we should observe that it is at the basis of any reading of Tymieniecka), noting that it establishes the centrality of logos not beyond, over, or "at the margin" of life, but in its very heart, at the heart of its deployment *as* life: "Life as life, life in its emergence, let us emphasize again, is not merely an articulated line of construction, but on the contrary this rationale of the self-individualization is "animated".[12]

We thus jump to the question of the link between logos and life, or of the way in which logos should generate itself from life and then we proceed assuming that it can be found in life because life itself is logos. But we should be careful not to confuse this affirmation with a re-presentation of a quasi-Hegelian position. Of course, we must observe that logos does not coincide here with reason, as an abstractedly intended idea. On the contrary, it means that life is dynamism manifesting logos, principle, orientation, dynamic creativity. It means that, in the heart of reality, even of the inorganic one (which does not mean a field excluded from life), there is a kernel which produces itself as logos and "logoic force".

One more peculiarity of phenomenology of life, conceived as interrogative movement of thought and as follower of life's own genetic movement, consists in the *constructive* connotation of logos, what let us approach the peculiar content of the conceptual renovation obtained in the main concept of "seminal logos": "It is not though anticipating its furthest constructive results, such as human consciousness, and not by assuming an outside realm beyond it, but by laying out intuitively the logos' own life involvement and its realization in concrete life development that we may get to its ultimate constructive roots. They lie with the nature of the logos, which crystallizes its virtualities in projecting life".[13]

This aspect can be better clarified by going deeper into the original re-elaboration by Tymieniecka of the concept of seminal reason, or it would be better to say, with the Greek words, of *logos spermatikós*. In particular, we should grasp the double "direction" of this concept, which has been living throughout history of philosophy in a Kars line connecting the Stoics, Early Christian theologians, and later on Leibniz: life as carrier of sense and meaning that can be individuated in logos, but, also, logos as life provided with an own poietic, operative and coordinative directionality, as prompting force, as constructive *élan*.

The logos of life articulates itself into an organised network of forces, drives, elements, thus trespassing the status of a chaotic, simply magmatic aggregation of energies and opaque forces. It provides reality – and now we understand the essential difference between the concepts of reality as simple *be-there*, and reality as living beingness or *becoming-real* – with organisation, coordination, effectiveness regarding a sensed becoming: "Life is, then, a dynamic flux, but is far from a wild Heraclitean flux, for it articulates itself".[14]

Life, in other words, carries with itself constructive ontopoietic patterns, in virtue of which the flux of becoming starts being connoted with living forms, coordination and organisation, not only in the sense of intentionality, but, differently, in that of creativity, as constructive and creative drive in the constitution of reality:[15] "*The logos of life is not an uncommitted stream of neutral force; on the contrary, it exhibits a shaping force*".[16] This idea of a "shaping force" connotes Tymieniecka's idea of "logos of life" of its activating power, of the really life-essential feature of modelling energy, the same energy as that which lets the *seed* become a *three*. This aspect of double powerful and shaping force of logos is clearly at the stake in the conception of ontopoiesis as the real way-to-become of life.

And we know, as Tymieniecka observes, that this organisation and this structured order, even if totally variable and creative, manifests itself in the order of time (with its "timing"), of space, but also in the other dimensions that include the "remembering" aspect of the living forms. It comes to light, then, the question of force, in the sense of life as force-connoted: "Form and force appear prima facie to be factors of life most intimately enmeshed with each other. Can we disentangle their respective roles, or are they irremediably fluid? Where does the inquiry into the formal delineation of the deployment of life stop and the inquiry into the force carrying this deployment begin?".[17]

Tymieniecka underlines that this becoming is not unqualified, indifferent, brutal force of change or disruption; rather, it is force of production, energy which "calls" the beings in their ontological structure, which is tuned in the deep nature of beings and puts it into action, enacts the being' forces so to generate new ramifications of life.

The fundamental change in perspective that this new sense of logos brings, therefore, in the point of view of phenomenology of life, is linked with the dynamic connotation – in the sense of the *dynamis*, of the force – of life and of logos of life. The change in perspective is declared as follows: "To understand life it is indispensable to envisage from two perspectives: one may take in its surface phenomenal manifestation in a formal, structural, constitutive fashion, or one may peer into the depths of the energies, forces, dynamisms that carry it relentlessly onward".[18]

In this second sense of the inquiring attitude, posing ourselves, so to say, "on this side" of life as generative and patterning propulsion, we discover that life, as ontopoiesis, has the possibility to produce and change the world because its structure is based on internal *virtualities* to be put into activity and effectivity.[19] These virtualities recall, in our reading, the ancient notion of seminal reason, since they represent the connection point – in the sense of dynamism – between the evolution and enacting process of life's becoming, and its ordered structure.

We must underline the "germinal" and ontopoietic connotation of these virtualities which are at the same time vital and logoic, that approaches directly to a renovated sense of the idea of seminal reason. The germinal and seminal characteristic of the logos of life indicates that life is equipped with a inter-chained and phased structure even in its smallest corners and kernels, by which life shows continuity in creativity in any single being, and a peculiar "leaning forward" its development.[20] There are, hence, "germinal virtualities", virtualities that "shape"[21] the ontological material in the direction of life.

There is, here, a deep renovation of the idea of reason, that carry a genetic and preparatory order *in fieri* in the considered thought. And it is not haphazard that in this fecund field of explication, *enactement*, play of life-informed forces is called by Tymieniecka, with a highly evocative language, the "womb of life";[22] in it occurs a "virtual intergeneration",[23] which preludes to the incarnation of life, as able to modify itself in relation to the circumambient conditions.[24]

Here we arrive, finally, to one of the most interesting peculiarities of Tymieniecka's phenomenology of life in its innovative revisiting of ancient philosophy: "germinal" is virtuality and force that promotes and articulates life, but it is all other than a mere physical force, mechanically posed in relations of static balance and cause-effect. On the contrary the "germinal" attribution refers here to what is susceptible to inform in an original way the internal composition of beings; moreover, their *interior* composition, interior because proper of *a* living being.[25]

Università degli studi di Macerata, Macerata, Italy
e-mail: gilberto100@libero.it

NOTES

[1] Anna-Teresa Tymieniecka, *The Fullness of the Logos in the Key of Life*, Book I. *The Case of God in the New Enlightenment* (Dordrecht: Springer, 2009), "Analecta Husserliana", Vol. C, p. 14.
[2] She specifies this, still in one of her most recent texts, carrying on her long-term close confrontation with the Husserlian philosophy; Tymieniecka, *The Case of God in the New Enlightenment*, op. cit., p. 13.

[3] Tymieniecka, op. cit., p. 15.
[4] Tymieniecka, op. cit., p. 3, our italics.
[5] Tymieniecka, op. cit., p. 34.
[6] Ibid.
[7] Tymieniecka, op. cit., p. 33.
[8] Tymieniecka, op. cit., p. 6.
[9] Tymieniecka, op. cit., p. 27.
[10] Ibid.
[11] Ibid.
[12] Tymieniecka, op. cit., p. 6.
[13] Tymieniecka, op. cit., p. 30.
[14] Tymieniecka, op. cit., p. 11.
[15] Tymieniecka, op. cit., p. 25.
[16] Tymieniecka, op. cit., p. 30.
[17] Tymieniecka, op. cit., p. 34.
[18] Tymieniecka, op. cit., p. 35.
[19] Tymieniecka, op. cit., p. 37.
[20] Ibid.
[21] Ibid.
[22] Tymieniecka, op. cit., p. 39.
[23] Ibid.
[24] Tymieniecka, op. cit., pp. 41 ss.
[25] Tymieniecka, op. cit., p. 38.

PATRICIA TRUTTY-COOHILL

VISUALIZING TYMIENIECKA'S APPROACH WITH STRING THEORY

ABSTRACT

This chapter argues that to better understand Tymieniecka's thought, we might look to outside philosophy to the aesthetics of string theory. Brian Green's explanations and Tymieniecka's prose evoke a vibrating dynamism that call for a multidimensional approach to and understanding of the world whose "order may not be visible from any single dimension."

I am an art critic, so when I read philosophy, I test it in terms of the art I know. My philosophy books are annotated with lists of images, Anna-Teresa Tymieniecka's, the most heavily. For example: The fourth book of her *Logos and Life* quartet,[1] which discusses the internal dynamic of the Logos of life as a process of ontopoiesis, brought my mind to Leonardo da Vinci's brainstorm drawings (British Museum, Figure 1) for his revolutionary compositions of the Madonna and Child with St. Anne (the cartoon at the London National Gallery and the painting at the Louvre).[2]

The snare of lines in his preparatory drawings visualized Professor Tymieniecka's evocation of the emergence of the logos of life, which comes neither in a linear nor logical nor a commonsensical sequence "but in all its generic as well as functional operations, all the connective strings properly tied to a specific vibration, in a spontaneous effort that happens all at once."[3] The perspective of each medium allows, if not an experience *sub species aeterni*,[4] a more vibrant awareness of both and an example that understanding can be found through relationship, as in the fashioning rather than imagining a Mobius strip. It also calls for a leap into the *zeitgeist*, a territory of sense and non-sense very tempting to art historians.

The passage cited above is from Professor Tymieniecka's first book of her *Logos and Life* quartet, dedicated to the creative experience and the critique of reason. It was published in 1988, a critical point in the history of science for it is the time of the string theorists' attempt to unify two great incongruous fundamental theories: quantum mechanics and general relativity. By the early nineties the then many string theories were finally unified, to produce one of the most beautiful and elaborate physical theories ever invented. The diagram in Figure 2 from Wikipedia shows the underlying principle that underlying all "matter" are not points as we earlier imagined, but looping, vibrating strings of energy. Such a model unifies the static and dynamic theories of everything that have characterized Western philosophy from the time of the Presocratics.

The theory was the physicists' attempt to account for the flexibility and transformability evident in the cosmos as we know it. According to Brian Green, if we examine the particles that were thought to be elementary, we would find them not to

Figure 1. Leonardo da Vinci, *Sketches for Virgin and Child with St. John and St. Anne*, Chalk, pen and ink, stylus lines, on paper, London, British Museum. The full sheet contains small sketches that indicates some of Leonardo's varied thoughts of the composition possiblities. On the verso is his tracing of the composition from the stylus tracings

be points has been imagined but tiny loops, fields of energy, "vibrating and oscillating filaments, physicists call 'strings.'"[5] I can only claim to appreciate string theory as Brian Green has elegantly described it. My enthusiasm is for its aesthetic, just as I use Professor Tymieniecka's aesthetics to better appreciate works of art.[6] The ultimate beauties of both systems, the mathematics of string theory and the broad scope of Professor Tymieniecka's philosophy, are for keener minds than mine.

Nevertheless—it's always nevertheless that allows a speaker to discuss matters for which he is not well qualified—I find interesting analogies between the two who both allow wonder and imagination back into the world. Not only are the dangling, looping strings vibrating, but so are the multidimensional membranes (discovered in the nineties: called p-branes). And the dimensions number 10 or 11.[7] Professor Tymieniecka's explanations fall short here, for she only organizes her arguments into three "membranes" as it were. She explains her method in the opening of Book I of *Logos and Life*. She wrote, she says, a three-paneled work, a triptych like Hieronymous Bosch's famed *Garden of Earthly Delights* (Figure 3), not merely a continuous monolayered text. And she explains why; I will stitch a few statements together here for the sake of brevity:

Why, instead of following the usual way of composing a learned work, did I choose to make a presentation with no forcible direct continuity of rational argument [a single plane to make a point],

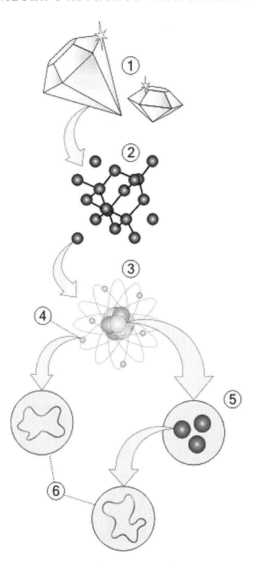

Figure 2. String theory diagram from Wikipedia: Levels of physical reality of matter: (1) Macroscipic level: matter (*diamond*); (2) Molecular level; (3) Atomic level of photons, neutrons, electrons; (4) Subatomic level: electrons; (5) Subatomic level: quarks; (6) String level

... but rather in the guise of a **triptych?** Their unity is not that of a continuing argument, but that of numerous significant threads, which ... maintain interconnections among various issues, various analytic complexes, and various dimensions which are projected by the great themes in question. In this interconnectedness resides that with which we aim to reveal: the workings of the creative condition of man (*LL1*, 8).

Anna-Teresa Tymieniecka sought a phenomenological recognition of all types of experience "without any forced connectedness or dubious speculative nets of unity."

Figure 3. Hieronymous Bosch, *Garden of Earthly Delights*, triptych opened. 15043-04. Madrid, Museo del Prado. My argument about the meaning of the form of the triptych does not depend of the interpretations of the iconography. All interpretations depend on the synergy of the parts

She emphasized that what may seem to be discrete, discontinuous and disordered from a single point, might not be so if viewed from multiple viewpoints: "Its order may not be visible from any single dimension" are her exact words. She therefore used the model of a man-devised and man-made object, as an analogy to her method.

The point of a triptych is that no part is complete individually. The whole can only be understood in terms of each part, the interaction of the parts with each other and with the whole, both inside and outside. The usual planar method of argument is not sufficient for the reality Tymieniecka describes. Her method is indeed her message.

Tymieniecka's aim, unlike that of the cognitive approaches, is to penetrate the "infinitely complex and differentiated web"—but think of this web as having *n* dimensions—a shadowy network of significant links that relate the poetic, the intellectual and the intuitive at the very least. Her terms are those of the experience of art: "the symphony of life" and the human creative endeavor. Only such a complexity can allude to life as lived. That these realities in their uniqueness cannot be verified by repetition, as string theory cannot be verified experimentally, is part of the charm.

In focusing her work on man's self-interpretation in existence she does not pin the metaphoric butterfly into a case in order to examine and describe it but rather she chooses to "film" it in the multiple dimensions of its existence. Her multidimensional camera is creativity. This way she hopes to find the true pattern of reality-in-becoming. Her language is all about encompassing, orchestration, dynamism, prompting, propulsion, passional strivings, abysmal depths, surgings. She warns that dealing with creative ciphering is a tortuous process. That our meandering course of understanding is chaotic merely demonstrates that the single lens of reason is not a sufficient tool for understanding the Logos of life. Leonardo da

Vinci had begun on such a path when he invented the exploded view of machines and anatomies, views in which the relationship between the various layers and the numerous directions of organization were revealed. Tymieniecka—as Laurence Kimmel characterized so well—does not "aspire to a god's eye view of reality [as the closed Bosch triptych, Figure 4, shows], but develops a view from within; it takes the course of an immersion into the creative mix that constitutes the total experience of existence—the cognitive, emotive, and volitional activities of human mind and culture ... she searches out the web of relations that together form the living tissue of a changing world."[8]

Tymieniecka's actions follow her thought and her writings, for she has established the World Phenomenology Institute which welcomes the insights of other scholars, who more and more, offer varied insights into the issues she has raised, and often provide case studies of the problems she has raised. As has been said of string

Figure 4. Bosch, *Garden of Earthly Delights*, triptych closed

theory, far from being a collection of chaotic experimental facts, these are the manifestations of one feature: the resonant patterns of vibration—the music rises from the notes the loops of string can play.

We strive who study the arts are in the challenging position of reckoning with the Pandora's box that Professor Tymieniecka has opened (*L&L1*,11). Or perhaps it is the other way around, we present her with escapees of the tradition of rationalist aesthetics, and see if she can find a place for them in her orchestration.

I plan to continue this sort of exploration into parallels between the New Physics and Tymieniecka's approach for it seems to me that her philosophy has more in common with that of the twentieth-century scientists—not its technicians—than is realized. These men were astounded by the wonders of the universe, wonders that Newtonian science could not even imagine. And so let me give Niels Bohr the last word: "Those who were not shocked when they first came across quantum theory cannot possibly have understood it."

I'm working on staying shocked.

Creative Arts Department, Siena College, Loudonville, NY, USA
e-mail: ptrutty@siena.edu

NOTES

[1] Anna-Teresa Tymieniecka, *Logos and Life: Impetus and Equipoise in the Life-Strategies of Reason, Book 4*, Dordrecht, Boston, London: Kluwer, 2000.

[2] Patricia Trutty-Coohill, "The Ontopoiesis of Leonardo da Vinci's Brainstorm Drawings" in *Analecta Husserliana* 92, 3–12.

[3] Anna-Teresa Tymieniecka, *Logos and Life, Book 1*, Dordrecht: Kluwer, 1988. Hereafter &*LL1*, in text.

[4] See Sibel Oktar's "The Place Where We See the World" in the current volume for a discussion of Wittgenstein on this subject. For a recent example of the method see my "Visualizing Tymieniekca's Approach to Originality," 275–294 in *Art Inspiring the Transmutations in Life*, *Analecta Husserliana* CVI, especially the comparison of a Tymieniecka text with an Georg Braque's analytical-cubist portrait *The Portuguese*, 277–279.

[5] Brian Green, *The Elegant Universe, Superstrings, Hidden Dimensions, and the Quest for the Ultimate Theory*, Norton, 1999. It was visualized beautifully on NOVA, available in streaming video on http://www.pbs.org/wgbh/nova/elegant/program.html.

[6] Telling of my own point of view is my initial "reading" of the motto of the "official string theory site" as "It's the 21st century: Time to feel your mind." I substituted "feel" for feed, probably because I had just absorbed Manjulika Ghoush's presentation! http://superstringtheory.com/index.html, consulted 11August 2010.

[7] For a demonstration see http://www.pbs.org/wgbh/nova/elegant/resonance.html.

[8] Laurence Kimmel, "Logos: Anna-Teresa Tymieniecka's Celebration of Life in Search of Wisdom" in *Thinking Through Anna-Teresa Tymieniecka's Logos and Life*, *Phenomenological* Inquiry XXVII (October 2003), 22–23.

NIKOLAY N. KOZHEVNIKOV

UNIVERSAL PRINCIPLES OF THE WORLD AND THE COORDINATE SYSTEM ON THE BASIS OF LIMIT DYNAMICAL EQUILIBRIUM

ABSTRACT

Our conception is fundamentally different from the former one because we start from the fundamental relative equilibrium, as though we shield all the previous levels of the world, about which we know nothing. All existing processes are balanced by the appropriate "limiting boundary surface." Beyond the unknown world remain the fundamental equilibrium quantities that are responses to that part of the world that is around us. The formula of the natural coordinate system is as follows. "One should identify two unequal parts in the real world. On the one hand, these dynamic equilibria are united in interrelated chains. The equilibria were created by the same laws at all levels of organization of the world. On the other hand, one could identify all the other non-equilibrium processes and phenomena". Any natural formation interactions with the dynamic equilibria are based on the fact that all natural formations tend to limit the fundamental equilibrium that is unreachable. In the article we consider the ontological levels of the world, which are comparable to "vacuum", "inertial systems", "thermodynamic equilibria", "spirituality", equilibrium parameters of which are well identified. "A man of networks" – is the nearest stage of human and mankind evolution, this is the only way for the mankind to maintain its real existence.

1. The deepest idea realized by mankind during its existence – is the idea of God. However, the very term "God" as a designation for "the great mystery of the world" takes all the complexity of the problem in the interpretation of choice and the specific details of interpretation. In the modern globalized world the philosophical understanding of this idea is becoming increasingly important. The religious interpretations of "God", because of their dogmatism, are one-sided and difficult to fit into the ongoing social and cultural processes, as well as their respective scientific and philosophical representation. Currently, there is a need for synthetic concepts representations of "God", able to absorb focused on the universal principles of the universe – the ideas of Plato, Kant, Fichte, Schelling, personalism, existentialism, world unity, cosmism and others nearest to them by the spirit of philosophical and religious systems. The bases for this synthesis are the planetary civilization shells, formed in the present and the associated with them web network: cultural, social, economic, etc.

The problems associated with analyzing the idea of "God" and correlates with the universal world harmony can be investigated only by relying on some ontological layers-bases. Some of the most important philosophical concepts are based on the fact that a person has to deal with, at least three worlds: objective, subjective, and transcendent. The simultaneous involvement of all these worlds in study of the problem can only confuse the initial concepts and approaches used. Attempts to highlight the above mentioned fundamental ontological layers of the world have taken over all periods in philosophy, science, religion, literature. The clear example is the four layers of N. Hartmann: inorganic, organic, psychological, spiritual. Various models of the hierarchy of nature and their corresponding levels of structural organization are emphasized in certain sciences, specifically in physics, chemistry and biology. Exceptionally clear layers being represented in literature ("The Divine Comedy" Dante Alighieri, "Rose of the world" Leonid Andreev).

However, our conception is fundamentally different from the former one because we start from the fundamental relative equilibrium, as though we shield all the previous levels of the world, about which we know nothing. All existing processes are balanced by the appropriate "limiting boundary surface." Beyond the unknown world remain the fundamental equilibrium quantities that are responses to that part of the world that is around us. As a frame of reference of self-organizing nature, which develops "turn by turn" – from one equilibrium state to another one, relative equilibrium are most natural. Exploring nature, society, consciousness, we often use the fundamental ideas and pushing away the fundamental equilibria, from "Silence of the universe". In science they are represented in the form of inertial systems of Galileo-Newton, thermodynamics' quasi-static processes. In religions they are the basic dogmas, such as the commandments of the New Testament. But the interest in science, philosophy, and art far more often had been focused on the active principle, more accessible and comprehensible in its manifestations. For instance, historical studies devoted to revolutions, wars, their leaders, whereas the underlying processes that shaped the course of history were determined by completely different reasons.

The fundamental dynamic equilibria associated with numerous private dynamic equilibria are forming their extensive network built on the same principles. The stability of these equilibria at all levels of nature, society, and mankind are defined by their connection with the three limits: (1) identificational, (2) communicational (network) and (3) full lifetime of the natural system – all three of which are unattainable for each individual system. As a result, a natural coordinate system of the world is being created based on these limiting relative equilibria. The system perfectly correlates with the basic, fundamental natural sciences and philosophical systems, dogmas and symbols of faith of the world religions.

2. The formula of the natural coordinate system is as follows. "One should identify two unequal parts in the real world. On the one hand, these dynamic equilibria are united in interrelated chains. The equilibria were created by the same laws at all levels of organization of the world. One could identify all the other non-equilibrium processes and phenomena on the other hand".[1] At every level of the world organization, these chains combine the fundamental equilibria with their other types: relative, limited, metastable, etc. These links are caused by the common origin of interactions between different natural processes, the likeness of their structures at all levels of the world. The very specific dynamic equilibrium is created by that part of

the energy that can be balanced in these specific natural systems, resulting in "cell dynamic equilibrium" common to all natural processes and coordinate system that are formed by a similar algorithm for all structural levels of the world. "The limiting boundary surface" is composed of many such equilibrium cells. The remaining unknown to us hidden processes form the surface of the identification limit of these cells, and their mutual balance within this surface – a communicational network limits.

Any natural formation interactions with the dynamic equilibria are based on the fact that all natural formations tend to limit the fundamental equilibrium that is unreachable. Any natural system of inorganic (nonliving), organic (living), and spiritual are approaching its identity. Elementary particles are linked into chemical elements, gas nebulae are transformed into galaxies, stars, and planetary systems. All spheres of living exist in the form of individual organisms, species, and populations with significant biotic potential. On the other hand, any natural formation seeks to find a balance with its environment, taking part in the formation of communicational network. A person cannot to communicate – it is their natural state. Parameters of the individual are determined by the flow of information, which is exchanged by all levels of organic (living). The essence of personality is being exchanged through a system of relations of dialogue, and cultures – through dialogues and relationships at appropriate levels. Communication equilibrium of the system makes it capable to interact with all spheres of existence: from the world of organic (nonliving) nature to the realm of the spiritual. Another important limit is achieved by balancing the system with a neighboring structural level of world organization, from viewpoint of which one can see its development as a whole. It determine by limit of the full lifetime of the system. All the specific identification and communication limits, as well as limit of lifetime, remain unattainable while looking from within the system, due to their opposing tendencies, providing a result of a consensus (an intermediate dynamic equilibrium or set of them).[2]

Thus, any phenomenon, process, thing, or structure may have a "cells interaction" with the coordinate system of nature, which themselves are elements of the system. All natural phenomena, processes, and structures can have a stable relationship with the coordinate system due to the nature of these equilibrium cells. On the one hand, they thereby participate in its formation, on the other hand – can be investigated in the same coordinate system. The coordinate system forms its basic scientific terms: the "related substances", "bound energy", "related information", emphasizing the passive part of nature: a set of some limits (attractors), which provides the process of self-organization for the rest of nature. This coordinate system is discovered directly and available to everyone.

A person's task is to take part in the formation of the natural coordinate system, and to ensure its sustainability, because only a person is able to develop the spiritual dimensions of this coordinate system. In the history of science, philosophy, and culture, trends of "development" and "striving for balance" were succeeded by each other. The early twentieth century was dominated by interest in the organization, dynamic equilibria; the end of the century and the present is prevailed by the interest in deterministic chaos, complexity, network equilibrium.

3. Consider the ontological levels of the world, which are comparable to "vacuum", "inertial systems", "thermodynamic equilibria", "spirituality", equilibrium parameters of which are well identified. "The limiting boundary surface" that we matched with vacuum (the basic state of quantum fields with minimal energy, zero momentum, angular momentum, electric charge and other quantum numbers) "closes" all the unfamiliar parts of the world by providing the equilibrium stable existence. The concept of "vacuum" is a fundamental in the sense that its properties determine the properties of all the major states. In some cases, such as spontaneous symmetry breaking, the vacuum state is not the only degenerate – there is a continuous spectrum of states differing from each other by the number of so-called "Goldstone bosons".[3] Bosons with zero mass and zero spin as the main parameters characterizing the vacuum cause the appearance of the quantum field, which modern science calls the most fundamental and universal form of matter, the foundation of all its concrete manifestations. All elementary particles are the quanta of certain physical fields, which continuously interact with each other (emit and destroy themselves). Thus, the vacuum can be regarded as the simplest system of reference (the limit of the fundamental equilibrium) for the world was created from elementary particles. In other words, the basic parameters characterizing the vacuum are bosons with zero mass and zero spin, i.e. above this frontier zero mass particles appear, causing the emergence of quantum field.

The next level deals with a steady substance and long-range physical interactions, and here, "the limiting boundary surface" separates the mechanisms that determine the existence of the most versatile of the physical interactions (gravitational, as well as the associated electromagnetic) from their phenomenological manifestations in many macroscopic processes. Parameters of the reference system for the remainder of this "boundary surface" of the world are the "inertia" and its measure – "mass". This allows forming a representation of the inertial systems as a basis for all other more complex frames of reference in that part of the world. The structure of the mass is extremely complex, which at first suggested in classical mechanics, and then similarly introduced in general relativity theory and quantum mechanics, thus ensuring its conformity with existing experimental facts and data.

Another "limiting boundary surface" separates complex macro objects, each consisting of a huge number of particles (of the order of Avogadro's number $N_A = 6.02 \times 10^{23}$), from the simplest parameters characterizing this new equilibrium. In this case temperature characterizes the thermal equilibrium, and entropy determines quality used by this energy. More sophisticated equilibria – thermodynamics are formed based on sets of thermal equilibria. The next stage of equilibrium concept development is related to the concepts of "dynamic chaos" on the basis of which the theory of dynamic equilibrium was originated in open non-equilibrium systems. The next stage equilibrium theory development leads to the formation of complex self-organizing systems, molecular circuits, pre-structured "before life" and "life".

The next "limiting boundary surface" is connected with spirituality, which is a complementary balance among all the subsystems of "I" and, above all, the four main ones: body, mind, subconscious, super-conscious. The spiritual man must have all these subsystems in harmony: his body and mind must be healthy, his subconscious – a well-organized and controlled by consciousness, interaction with super-consciousness (cultural codes, religion, ideology, ethnic group traditions) – humanistic. If these subsystems are in equilibrium with each other, then everything else in a person's mind and subconscious are closed by the "limiting boundary", so that the basic parameters of spirituality remain: the freedom of will and cultural-secular asceticism which is a bound state of intellectual, social, and individual manifestations of personality. Such asceticism is capable of forming the foundation of universal synthetic culture, so that all the unique and specific in different nations will have to supplement.

4. The coordinate system that connects all limiting dynamic equilibria of the world into a single entity; and it is in line with the concepts of the Absolute, the Universe, and God according to its ontological significance. During the New Age, the majority of these concepts have been thoroughly transformed, so that such a coordinate system may well occupy the leading place in this series in terms of mutually contemporary multilevel knowledge, by absorbing the existing universal ideas and the corresponding concepts.

The coordinate system is characterized by spatial location, time, structure, but no localization or no spatial limitations exist for it, that is present in every part of the Universe, at all levels of its structural organization. Time is continuously transformed from one form to another – the development process started during one type of time and then continued during the other. Since the coordinate system interacts with all the possible temporary structures and is pure existence that has no reason and is the origin of it. It interacts with various kinds of beings, enabling them, according to its "pure being", which contains the "grain" of any particular being. Locating sustainable relationship with the coordinate system begins with the ability to listen to silence in oneself, finding a fundamental mood to start the process of learning and investigation, touching the various types of "emptiness" – that is the way the initial cells of the coordinate system are formed, through which a person can establish a stable relationship with the system. The coordinate system is also approaching a person due to its high self-organization. The coordinate system is the ideal structure of all possible limit states of "fundamental emptiness." It cannot blend in with natural systems, but natural equilibrium processes sometimes interact, with it occasionally or at regular intervals due to the rhythms of the natural coordinate system.

Due to the coordinate system in question, a new epistemology is formed, which eliminates extreme opposition between scientific and non-scientific knowledge and thinking. All these mental and spiritual patterns that go beyond current knowledge, but can stably interact with the system of coordinates, can be considered under this new epistemology. Significant opportunities for dialogue and inter human communication are established. Since the most accurate and profound form of dialogue is not a communication between individuals, but a two-stage dialogue: a person – "coordinate system" and "coordinate system" – the other person. Value-oriented cognitive installation is focused on a stable relationship with the coordinate system, on the allocation of the passive part of nature, on the sets of some limits (attractors) with the help of which processes of self-organization are run at all nature's levels. This relationship is a prerequisite for the further development of man and mankind, and only in this case, humanity will not destroy the biosphere, atmosphere, but can develop harmoniously, optimally and practically indefinitely.

The coordinate system exists based on the interaction with the open systems, which have a tendency to self-organization, self-development and this openness allows them to be interconnected with the system. Interaction with the coordinate system of nature on the basis of the limiting dynamic equilibria involves the use of a pluralistic methodology, as well as some finished tools from special scientific disciplines (hydrodynamics, thermodynamics, quantum mechanics, statistical physics, evolution concepts, systems theory, synergetics, cybernetics, and ecology). In addition, the ultimate search of coordinate system and the maintenance of stable relationships with it independently form a "coordinate method" of philosophical research, which has an exceptionally broad versatility and whose possibilities are practically unlimited. This method has universal flexibility to interact with any natural phenomena and process. There are universal criteria and a unified methodology investigation for all these processes.

5. The system of interconnected dynamic equilibrium may be reached by developing sensory artistic representation using abstract and theoretical models, as well as creating various forms of synthesis in general representations of nature, society, man, consciousness, which can be regarded as the most complete picture of the life world, the genotype of social life, basic cultural and genetic code.

Cultural universals have considerable heuristic potential for forming invariants for further allowed development of mankind and, moreover, they are much more accessible to people who do not have special theoretical training, compared to the philosophical categories, universals, and general scientific concepts. This person does not need to learn how to treat a system of coordinates. The most important thing is to know and feel that such a coordinate system exists, and then it will find the individual, adjusts its rhythms, and will maintain a stable relationship with him, by opening up more and more to them.

For each individual it is most natural to interact with at least three levels of coordinate system: personal, relevant to their ethnic group culture and planetary one. A person becomes multilayered, enclosing equilibrium cells of all these levels. His state may be called person-planetary. A sustainable relationship between a mankind and these levels of the natural coordinate system is a prerequisite for their further development – a guarantee that mankind won't destroy the environment, but the mankind can develop harmoniously, optimally and practically indefinitely.

The coordinate system on the basis of limiting the dynamic equilibrium can effectively interact with different cultures in a globalizing world, to develop methodological approaches to education, to carry out the synthesis of the sciences and humanities, to solve complex contemporary problems. Culture in the XXI century will occupy a leading position among all other areas of the human spirit and interact with it will occur over the historically developing natural systems. Self-organization, openness and active interact with the environment are its main characteristics. These systems transform in "human-sized complexes" in which people actively promote the convergence of the sciences and humanities, suggesting complementarities influence of poetry and science, intuition and logic, the Western and Eastern types of thinking, rational and irrational methods of research, scientific and non-scientific approaches, cognitive and axiological criteria of knowledge.

In the global today's world, all cultures are exposed to two major trends. On the one hand they should identify themselves, that is, to clearly identify its boundaries, characteristics, and be transparent to the representatives of other cultures as much as possible to determine. The problem here is that many cultures are still not fully identified, are now in the middle of their development. On the other hand, they should promote the integration of mankind. It is no accident that the elimination of economic, political, national borders is now combined with the increasing cultural isolation (the paradox of cultural diversification). As part of the approach all cultures in the present conditions should be in a state of stable synchronic and diachronic fluctuations. In the first case this occurs between the nuclei of identity traditional cultures and emerging invariants of human world culture. In the second case this occurs among the best examples of the cultural world heritage and the relevant guidelines of the stage of the informational society. In the result of this process forming a specific dynamic equilibrium can able interaction between traditional cultures and information subsystems of the modern world (economic, financial, technological and so on). It is especially important because in the history of mankind the majority numbers of cultures were traditional and that among them there were two great mutations: the ancient and Christian culture. In all the above interactions can be distinguished limit the dynamic equilibrium and thus connect them with the natural coordinate system for all levels of emerging cultural supersystem. Every nation is able to make a unique contribution to the formation of the natural coordinate system and the more such contributions, the more stable it will the coordinate system. Such development has no alternative for mankind now.

Culture adequate to the needs of modern person should be networked, and it must include in this system traditional cultures as elements. Every modern individual should actively interact with multiple cultures, contributing to their closure in a single cycle (material, energy, information). The diversity provides for stability in nature; humanitarian and social spheres are no exception, and as the more cultures there will be on the planet the better for mankind. Of course, they all must be tolerant, humanistic-oriented and unique. All cultures must strive to identify their own worldly asceticism, which can become a key element in establishing a dialogue between cultures, forming a planetary reality, gradually transformed into planetary existence. Asceticism is a bound state of spirituality in the context of its ethnic and individual expressions that can be the foundation of synthetic human culture. The superstructure above this asceticism base joined all the unique and specific for different people.

Sustainable complementarily interaction between cultures associated with the formation of limiting their bases – the basic cells, a kind of cultural monads, and the totality of which forms a micro level of planetary culture. Such monads can be combined into modules (patterns of meso level), which is well correlated with phenomena of certain ages, artistic styles, trends. Also, the formation of cultural codes is convenient considered in the sphere of the concrete level. Fundamental cells will generate the code at the micro level, modules of the cell – at the meso level, the planetary envelopes and the network – at the macro level, based on the semiotic space – semi-sphere. All of the above cells (monads) keep its natural integrity, determined by the rhythm and coherence of spatial, temporal and spatial-temporal processes closed.

The education system connected with the natural coordinate system should be based on two basic principles: (1) self-identification of the individual and (2) the principle "to be near". Only under conditions for self-organization of the individual this first principle will be realized and personal formation can be achieved. The second principle is reduced to the activity of a pupil, while mentor add to this processes only the catalytic influence. They must send the pupil: correct, amplify, recommend the optimal literatures, and take part in analyzing the results, etc. Education should be continuous and consist of closed units (an average of 2–3 years each), within each of them the complete holistic world-outlook is formed. It is based on a systematic, structural knowledge, as well as on criteria to distinguish between the true, original information, the false information, incomplete and secondary knowledge that can destroy the natural system of coordinates. The main task of teachers within each of the above-mentioned steps (units) in the development of the educational process is to ensure interaction between students and the world coordinate system, which is possible at any age and for any volume of information. The process of this interaction is unique, created for each individual its own special method and is the most effective way of educating the individual.

6. We will never completely know the natural coordinate system, which always stays with us and at the same time is slipping away from us. From time to time it appears before us, bewitching our feelings and mind, directing the development of modern civilization. It's a kind of pure "gift", which nature has prepared for a person to carry out a special and unique function in the world around us. Only relying on the natural system of coordinates a synthesis of mutually complementary interactions will be provided in the fields of non-organic and organic nature, soul and spirit; and consensus among all the civilizations of modern humanity will be established, as well as among social, political, cultural processes that guide their development.

In the next few decades mankind must come in universal planetary sphere from system of interconnected networks of shells (social, scientific, technical, economic, financial, etc.), based on the limit culture structures. These networks will overlap, complement and develop each other by many ways, forming a planetary being, and coordinate system for themselves. The planetary being will be forming be self-organization and aimed to unit a natural and spiritual, material and ideal, real and virtual.[4]

Emerging planetary networks would be consistent with a new type of equilibrium – the balanced Wide-Web, formed on the basis of stochastic processes (metabolic processes of various types). Equilibrium-web requires many layers, depth, nexus of all civilization processes, because the network of interactions is more important than their sources, so that basic information resources, spirituality contained in the network of interactions forming a balance. Equilibrium-Wide-Web provides complementarily complexity and simplicity, the sciences and humanities, despite the fact that science works with a one number of limiting the basis of culture, and humanities – with other. The conceptual schemes of these series are different from each other. Equilibrium-Wide Web facilitates the identification of stable correlation between these series and the limiting cultural bases. The universal planetary cultural being and ensured formation its concept can be realized by these complementary relations.

At the present time it is necessary to accelerate the development of psychic and spiritual spheres of the individual through the ontological catalytic methods of education, involving a permanent combination of multiple views (from two to five) on the same concepts and phenomena, especially the combination of external and internal, synchronic and diachronic views. This makes it possible to combine different scales of research, the points of view, and the degree of abstraction. This provide with most fully identify any concept, to link it with the other in a continuous flow of knowledge. Only when every element of knowledge will be identified on the one hand, and on the other – is included in the system of knowledge of a higher structural level, we can talk about the greatest possible speed and quality of education. Increase or decrease this rate on optimal lead to significant irregularities and distortions in the formation, as a system of knowledge and sustainability of the person.

New mankind assimilation in the networks of culture flows consciously and voluntarily, but this is the only way for the mankind to maintain its real existence and is largely determined by self-organization of the world and all its ontological spheres (no-organic, organic, mental, and spiritual). "A man of networks" – is the nearest stage of human and mankind evolution.

North-East Federal University, Yakutsk, Russia
e-mail: nikkozh@gmail.com

NOTES

[1] Nikolay Kozhevnikov, *From equilibrium to equilibrium.* Moscow: Idea, 1997, 276p.
[2] Nikolay Kozhevnikov, "Phenomenological aspects of natural coordinate system." // Logos of Phenomenology and Phenomenology of Logos. Book Four: The Logos of Scientific Interrogation. Participating in Nature – Life – Sharing in Life, ed. A.-T. Tymieniecka – *Analecta Husserliana*, vol. XCI. Springer, 2006, pp. 45–55.
[3] Vacuum. // Physical Encyclopedia in 5 v., V.1. Moscow: Soviet Encyclopedia, 1988, p. 236.
[4] Vera Danilova, "The philosophical study of the concept noobiogeosphere." // Vestnik MSU, Series 7 "Philosophy" Moscow, 2004, No. 2, p. 53.

ATTILA GRANDPIERRE

THE BIOLOGICAL PRINCIPLE OF NATURAL SCIENCES AND THE LOGOS OF LIFE OF NATURAL PHILOSOPHY: A COMPARISON AND THE PERSPECTIVES OF UNIFYING THE SCIENCE AND PHILOSOPHY OF LIFE

ABSTRACT

Acknowledging that Nature is one unified whole, we expect that physics and biology are intimately related. Keeping in mind that physics became an exact science with which we are already familiar with, while, apparently, we do not have at present a similar knowledge about biology, we consider how can we make useful the clarity of physics to shed light to biology. The next question will be what are the most basic categories of physics and biology. If we do not want to cut laws of Nature into different parts, we obtain a constraint, and the remaining part of physics will be the input data to the equations of physics. In these terms, our question will be: if we keep biological laws intact, as indivisible units, what remains in case of biology? This approach, just because it is more fundamental, has significant consequences for philosophy, and obviously offers a new conceptual framework considering the relation between the ontopoietic principle of Anna-Teresa Tymieniecka and the biological principle. The quintessence of science, namely, the first essentially complete scientific world picture is presented in a detailed form.

INTRODUCTION

The aim of science is to understand, explain and predict the world of observable phenomena occuring in Nature; in its widest sense, to understand Man, Life and Nature in their full extension, depth and meaning, including the interrelations between Man, Life and the Universe.

In order to obtain well-founded, reliable knowledge, science requires a tool which gives a compact and transparent picture about the essence of our present knowledge, indicates which questions are interesting to consider and how to obtain well-founded knowledge. Such a tool is called as the "scientific world picture".[1] By the term "scientific world picture" we mean the *summarizing essence of all our scientific knowledge about the Universe in a compact, transparent, easy-to-use manner*, in a form which is able to yield ways of explanation and obtaining scientific knowledge. The scientific world picture is not only a map of the realm of Nature, enlisting what can be found there, but also a tool: (a) by which we can orientate ourselves about the present state of our knowledge, (b) which is able to tell us what are the important questions and (c) which indicates how they might be answered. Certainly, among the ingredients of the scientific world picture must be: (a) logic, (b) the theory and methodology of explanation, and (c) philosophy, answering such questions as "what is essence?" and "what is well-founded, reliable knowledge?" and "what is the scientific world picture"? Indeed, the scientific world picture must be in an important, explanatory sense "ultimate", because it cannot be based on something more simple or basic, since if it would, this latter one should serve as a better one. From our definition it is clear that such an inevitable tool of scientific research can be obtained only in an iterative process.[2] The scientific world picture can be regarded as the quintessence of science: the most perfect, simple, and elegant picture; the pure and concentrated essence of science.

The construction of a scientific world picture requires not only a deep understanding of Nature (in the widest sense), but also of science, and, especially, of scientific explanation. Nature is extraordinarily rich not only in the variety of phenomena, but also in its depth and meaning. Ultimately, Nature is one, it is in itself inseparable, we have to accept it as it is. Therefore, at its deepest level, which we call "core", Nature must be easily understandable, and so, the construction of a scientific world picture is possible, which is in itself an achievement. The recognition of the core of Nature makes it possible that through the scientific world picture we are able to see the picture of the core. The extraordinarily rich and deep nature of the Universe indicates that exploring its reality requires an extraordinary amount of attention, thoroughness, persistence, and devotion. We have always to keep in mind that the last word is not for us, picture builders, but for Nature, as it is.[3]

Of course, such a usage of the term "scientific world picture" requires that it has to represent an essentially complete, self-consistent and unified system of theoretically conceivable and empirically testable, scientific framework of the Universe.[4] From here on, by the term "fundamental" we mean the ultimate explanatory level in the system of explanations; by the term "general" we mean the widest possible scope of a given field of knowledge at a given level. The fact that Nature at its phenomenal level shows a breath-taking width and variety, while at its core a similarly breath-taking simplicity and conceptual compactness, indicates that Nature has an "inverted cone" explanatory structure, namely, all the innumerable and diverse phenomena can be explained by a minimum number of deep concepts.

If we will be able to find the quintessence of science, the essentially complete scientific world picture, then we may become able to envisage Nature in a unified and scientific manner, and so it will become possible to draw the outlines of a new, Universal Natural Science. Now we can introduce the term Universal Natural Philosophy contemplating the fundamental level of that new science.

We arrived to the stage where we have to consider what do we mean on the terms "essential" (and, later on, "complete"). The question "what is essential?" is a key question of the scientific world picture, because it is absolutely basic to build a summarizing picture about the world. The difficulty is that on its surface, Nature shows an unlimited variety. As we indicated above, simultaneously, at its core, Nature is one, undivided. It seems that it is the core what we must regard as the essence. In order to obtain a more concrete understanding of "essential", let us consider now how physics, the quintessential exact science makes this core of Nature explicit.

In physics, the "surface" of the realm of physics corresponds to observable phenomena, and "core" corresponds to physical laws. Starting with physical laws, the remaining part of Nature in physics is: data input that must be determined in advance; the input data give all necessary information about the physical system in its initial state. Input data can be obtained from observations of physical phenomena. Considering that phenomena occur occasionally, accidentally, while the physical laws are always the same; and, even more importantly, that one physical law can explain and predict an innumerable large number of phenomena, we can realize that the explanatory power (defined as the ratio of the number of explanandum to that of the explanant) of physical laws is practically infinite. The knowledge of one physical law is more valuable than the knowledge of an innumerable large number of occasional phenomena that are explained by the law. On that basis, we define "essential" from the angle of explanation, by the following meaning:

A definition of essential: One can regard as 'essential' a thing if and only if it has a (practically) infinite explanatory power in a scientific theory.

The key importance of the concept "essential" is illustrated by the fact that it directs our attention to those laws of physics which have the highest explanatory power.

As a first consequence of our result, we are led to a new question: which physical law has the highest explanatory power? Generally it is not acknowledged that all the fundamental physical laws may be derived from one single principle, the least action principle (e.g. Heron of Alexandria; Fermat; Maupertuis; Euler; Hamilton; Feynman 1942, 1994; Taylor 2010). The least action principle is the principle that determines the trajectory of a physical object between a given initial and final state. The least action principle turns out to be universally applicable in physics. All physical theories established since Newton can be derived from it. The action formulation is also elegantly concise. "The reader should understand that the entire physical world is described by one single action" (Zee 1986, 109). Therefore, we can introduce a specific meaning to the term "first principle".[5]

Definition of first principle: A fundamental law can be regarded as a "first principle" if and only if all of the fundamental laws of the given branch of science can be derived from it.

Due to our definition of "essential", we were able to recognize that in all physics, the most essential physical law is the first principle of physics. This recognition can make physics extremely transparent for scientists and philosophers, and makes it ideally suited as a pillar of the scientific world picture. Moreover, the insight given by physics, namely, that the world can be divided into three levels of reality, (a) the level of phenomena, (b) laws and (c) first principles, is ideally suited to the purpose to construct an essentially complete scientific world picture, because the number of first principles must be small. If all physics can be derived from the physical principle (a shorter expression for the first principle of physics), then all what remains in order to obtain the essentially complete scientific world picture is to find the first principles of fundamental natural sciences. Regarding that physics considers the realm of "inanimate" world, we consider that the second fundamental natural science is biology.[6]

At present, it seems that nobody knows the equivalent in biology of Newton's laws. In order to obtain a scientific world picture, we have to generalize our present picture about biology, and use the term "biology" in a new sense, including not only the presently popular form of it, but the Bauerian "theoretical biology", which gives the most general laws of living organisms. This use of the term "biology" will give it a status that is similar to that of physics. Theoretical physics worked out its fundamental laws and first principle, which is the least action principle (Taylor 2003; Moore 2004). The Bauerian theoretical biology already worked out its first principle, which is known as the Bauer-principle (Bauer 1967) which is shown to be equivalent to the greatest action principle (Grandpierre 2007). The universal law of biology, the Bauer principle tells that: "A system is living if and only if it invests work from the budget of its free energy initiated by itself against the equilibrium which should occur according to the physical and physico-chemical laws given the initial conditions of the system" (Bauer 1967, 51). We can re-formulate it in other words: living systems manifest continuously maximal mobilization of their free energy against inertness. The Bauerian theoretical biology concentrates on the fundamental law of biology and, because of that, it underlies all specific sub-branches of biology that are intensively investigated today.

The next question arises: Are there any other fundamental natural sciences, besides physics and biology? As I indicated in the Introduction, the deepest questions of existence are threefold, questioning the Universe, life, and self-consciousness. From that it follows that the third fundamental natural science should be the study of self-consciousness.[7] If we regard that psychology is the science of human psyche, and that the most characteristic property of human psyche is self-consciousness, we are led to the idea that the science of self-consciousness will be psychology. Of course, this interpretation present psychology as a science from a new angle, indicating a new direction for the future development of psychology, in which it can find its first principle also in a mathematical form. If the above three are the three fundamental questions, than these three must be the three fundamental sciences. This is an important point, because we wanted to outline the basis of an exact and essentially complete scientific world picture. If the ultimate first principles are those of physics, biology and psychology, then these first principles can be regarded as "ontological principles"—as such they have a special significance for philosophy which is the study of the most general aspects of reality.

This new scientific world picture, as a side-effect, unites the four different views of metaphysics.[8] Now we can conceive the idea of a new, universal natural philosophy studying the most fundamental aspects of the universal natural science.

AN IMPORTANT OBJECTION AGAINST RECOGNIZING THE FUNDAMENTAL SIGNIFICANCE OF THE PHYSICAL PRINCIPLE

Actually, the principle of least action currently attracts little attention among philosophers (Stöltzner 2003), despite the fact that it underlies everything in the realm of physics. I think this is because the role that the principle of least action has been played in physics and philosophy is still highly controversial. On the one hand, the principle reflects a so-called "apparent" economy or teleology, which most physicists presume to be alien to their branch of science. Yet, as I indicated, we must be aware of the fact that the last word belongs always to Nature. Actually, teleology is defined in the Encyclopedia Britannica as "explanation by reference to some purpose or end". Definitely, the least action principle is based on a relation between some initial and final state; therefore, reference to some end—i.e. to a subsequent physical state—is already explicit. Therefore, when we explain with the least action principle all physical phenomena, the explanation always refers to a final state, and so it is inevitably teleological, because that is what teleology is. It is another point that physical teleology is different from biological or human teleology, which admits purposeful behavior, too. In biological and human teleology there is an evidence of motivation that is obscured in physical teleology and replaced by an apparent mechanical teleology. Central to this controversy is the attempt to avoid any questions around the concept that Nature might use means to an end.

We illustrate the resistance against acknowledging the significance of the action principle by a quotation from James Woodward (2010): "For example, the mere fact that we can describe both the behavior of a system of gravitating masses and the operation of an electric circuit by means of Lagrange's equations does not mean that we have achieved a common explanation of the behavior of both or that we have 'unified' gravitation and electricity in any physically interesting sense." In contrast, we note that the physical principle is a relation between fundamental physical quantities. Physical laws express relations between observable physical quantities, while mathematical laws express a relation between mathematical quantities. Physical laws therefore can be tested by empirical observations, which is not the case for mathematical laws. If the observationally confirmed relation has a lawful character, it has an importance in a physically interesting sense. If the observationally confirmed physical relation expresses a law serving as the basis from which all the fundamental physical laws can be derived, it has a primary significance for physics as well as for the philosophy of science. One of the two basic requirements of a law of physics is that it has to be mathematically formulated. The other is that it refers to entities existing in Nature. Physical reality is based on two pillars: one is observational testability, and the other is its spatio-temporally detailed character that can be described by mathematically formulated physical laws.

As Carl Hempel (1966, 71–72) formulated: "Newton's theory includes specific assumptions, expressed in the law of gravitation and the laws of motion, which determine (a) what gravitational forces each of a set of physical bodies of given masses and positions will exert upon the others, and (b) what changes in their velocities and, consequently, in their locations will be brought about by these forces. It is this characteristic that gives the theory its power to explain previously observed uniformities and also to yield predictions and retrodictions." These two pillars appear in the practice of the physicists in the form of input data (a necessary minimum set of physical parameters of the initial state) for the equations of physics, and, on the other hand, in the form of the equations of physics. Moreover, Hempel adds: "A good theory will deepen as well as broaden that understanding. First, such a theory offers a systematically unified account of diverse phenomena." (ibid., p. 75). On that firm basis, we can draw the conjecture that the unification offered by the least action principle, since it is not only observationally testable, but is also fitting all observations, and is mathematically formulated in spatio-temporal details, therefore, in contrast to Woodward's opinion, has a primary importance for natural science and for the philosophy of natural science as well.

Yet the point raised by Woodward remains: we have to find the physical importance of the mathematical unification expressed by the Lagrange equations. First we point out that the unification by the Euler-Lagrange equations does not extend merely to gravitation and electricity, but also to mechanics, thermodynamics, and quantum physics, actually, to all the fundamental equations of physics. Second, the Euler-Lagrange equations represent only an intermediary step between the integral form of the least action principle and its applications. The real power of the action principle relies in its integral form. The Euler-Lagrange equations in general contain the Lagrange function; its application in gravitation, electricity or any branches of physics requires the specification of the interactions present in the given type of physical process that the physicist considers. Woodward is right in pointing out that the specific form of the Lagrange function has an important physical meaning, but lacks scientific basis when, implicitly, claiming that there is no physics beyond the special forms of physical interactions. The Euler-Lagrange equations in their general, unspecified form still express that all the fundamental laws of physics are equations of change that can be described by second order differential equations. We point out that, for instance, the integral form of the action principle represents an additional, physically important meaning, expressing the very economical aspect of the least action principle. Indeed, this integral aspect explains the "sum over all possible paths", which is so important at Feynman's path integral interpretation of the action principle (Feynman 1942; Feynman and Hibbs 1965). Actually, the "summing up" of quantum probability amplitudes is the result of the integral operation, represented by the integral form of the action principle. All types of interactions are based on that concrete physical "mechanism" indicated by Feynman: all quanta, independently from the type of interaction, acts through summing up all possible paths. This summing up seems to be mechanical, yet we point out that it requires explanation. It is a strange ability from a quantum, regarded as being absolutely inanimate, to behave mathematically, sum up anything, and solve mathematical equations in order to reach one point from another. How do they "perceive", how do they behave "as if" they "know" that they have to sum up anything, and how are they able to do that according to the least action principle?

By our opinion, these fundamental problems transcend beyond the superficial, mechanical framework of present-day physics. Anyhow, this concrete physical "mechanism", quantum exploration through the spontaneous emission of virtual particles to all possible paths, and their summing up, attaches a concrete physical meaning to the least action principle and to the unification it suggests, even implies.

THE ESSENTIALLY COMPLETE PICTURE OF THE STRUCTURE OF THE UNIVERSE

Regarding Nature from these deepest aspects of physics the universal natural philosophy considers that physical reality consists from three basic ingredients: (a) concrete "things", represented by the input data, and, (b) at a deeper level of reality, from physical laws, represented by the fundamental equations of physics, and, (c) at the fundamental level of physical reality, the least action principle. Therefore, the first significant achievement of the universal natural philosophy is that it succeeded to obtain the first essentially complete scientific world picture, which is the following.

The Universe has a primary fundamental hierarchy: a three-leveled structure of the Universe, apparently, not recognized until now. The three levels of reality are: (a) phenomena, (b) laws of Nature, and (c) first principles of Nature. The secondary fundamental structure of the Universe is its division into (a) physics, (b) biology and (c) psychology, which are all interrelated. This secondary fundamental structure is categorized by the character of the observable behavior, or, equivalently, by the first principles, or by the ultimate constitutive elements: (a) atom, (b) feeling, (c) thought, or (a) matter, (b) life, and (c) self-consciousness. Since there are no more first principles, the picture is essentially complete.

One last question is: is it possible to go beyond the first principles, and find a still deeper principle, the very first principle of the Universe? We think the correct answer is yes. The physical principle can be regarded as the special case of the biological principle in case when the freedom of selection of the endpoint shrinks to zero. Moreover, regarding that the relation between consciousness in general and self-consciousness is the relation between the general and the special case, self-consciousness is the special case of consciousness, and, therefore, the psychological principle is another special case of the biological principle. This means that the three principle is united in one, in the biological principle: we have a Trinity, in which the middle of the horizontally conceived triad is also the vertical element, the ultimately unifying principle, the principle of the One, which is, strangely, again the biological principle. Therefore, the picture is indeed essentially complete, no essential element is left out from it. We found two Triads: phenomena, laws, and first principles, versus physics, biology and psychology.

From this overall picture about the architecture of the Universe the present scientific world picture accepts only the physical realm. The main reason for it is that at present physics is the only exact natural science. We think that the first big question of the 21st century is how to make biology into a science similarly exact to physics. Our answer is outlined below.

BIOLOGY, THE SCIENCE OF THE 21ST CENTURY, IN A NEW LIGHT

"In the 21st century more and more biological data are accumulated. In the absence of a general theoretical biology, there is an increasing frustration between millions of biologists" (Brent and Bruck 2006, 416). Recently, following the groundbreaking work of Ervin Bauer (1967), who was the first to discover the biological principle and to work out the scientific basis of exact theoretical biology, we developed theoretical biology in the approach requiring it to be as close to theoretical physics as possible. We recognized that the minimal extension of physics into biology is possible by generalizing the least action principle, allowing the selection of the endpoint of its integral in accordance with the greatest action principle (Grandpierre 2007).[9] The difference between biological and physical behavior can be illustrated with the example of a fallen bird from the Pisa tower. If the bird is dead, its trajectory will be similar to that of all physical objects: a straight line vertical to the ground; the dead bird follows the law of free fall. Yet if the bird is living, its trajectory will be characteristically different. In the simplest case, when there are no any disturbing circumstances like a hawk around, the bird will follow a trajectory that allows it to regain its height above the ground within a suitably short time with the minimum effort.

This approach will ensure that the generalized physical principle becomes suitable to grasp the teleology so eminent in biology.[10] Indeed, teleology is the most characteristic aspect of biological functions and biological behavior. While in physics falling bodies as well as light travels on the shortest routes between their initial and end states, living organisms select the endpoint of their activities according to the greatest action principle. Action is a basic quantity having a dimension energy∗time, integrated for the given process between the final and initial states. Illustrating the greatest action principle we note that all living organisms tend to live as long as possible (maximizing the second term in the product energy∗time, and, in the meantime, to increase their vitality of quality of life (which, in a physical language, can be measured in terms of their free energy, therefore, maximizing the first term of the product energy∗time, and so, maximizing the product yielding the action in the period of their lifetime.[11] This example illustrates that living organisms, since behaving on the basis of the greatest action principle, cannot be governed by the least action principle. Indeed, since the greatest action principle of biology is an extension of the least action principle, it cannot be reduced into the physical principle; biology must be an autonomous science. Biological entities make use of the least action principle as a means to biological ends. Therefore, it is the primary task of science and philosophy to realize the importance of the Bauerian theoretical biology, and work out theoretical biology according to its actual weight in the new, essentially complete scientific world picture.

ABOUT THE RELATION BETWEEN BIOLOGY AND PHYSICS

Now if biology is not reducible to physics, then how can we conceive the fact that physical laws apply to all living organisms? How is it possible that the gross behavior of living organisms occurs accordingly to a different, biological principle, if physical laws apply to them? The paradox can be avoided if we allow that the initial conditions, which are the input data in physics, in case of biology have a further "degree of freedom": they can vary in time in a suitable manner to result in biological behavior when as input data are attached to the physical laws.

The situation is the following. Biological behavior can be described, equivalently, in two different languages. One is in the language of biology. It tells that biological behavior is governed by the biological principle. The other is in the language of physics. It tells that the observed biological behavior is the result of physical laws, admitting that the input data of the physical laws is variable in such way that it results in the observed biological behavior. The only question that remains in this second case: what causes the input data to vary in a way that is unpredictable on the basis of physical laws? We are led to the fundamental problem of control theory: to govern a cybernetic system's input in a suitable way to produce a given or prescribed output. Control theory considers problems like how to construct a rocket in order to make it able to follow an airplane governed by a human. In order to achieve that feat, control theory works with an additional free variable with values that correspond to the decisions of the agent. Certainly, if we allow that the input data are continuously injected into the equations of physics in a suitable manner to result in the prescribed biological behavior (for example, when you are thirsty and go for drink, you navigate yourself using many feedback processes), biology arises as the control theory of physics.

We found that we are living in a living Universe, which we distinguish from the physical universe with the capital letter. Yet, at the same time, it seems that life, as we know it, is rare or unique. Yet life should not be protein-based, since plasma life forms are also possible (Grandpierre 2008a, b). Indeed, if we look after life forms with the help of the exact criteria of life given by Bauer (above), then it is possible to see that even apparently inanimate matter can carry hidden, transient life forms on extremely long or short time scales. Indeed, absolutely sterile inanimateness seems to be a mere abstraction from the actual reality present in Nature. The Universe can be full with an extreme variety of cosmic life forms (Grandpierre 2008a). If so, life can be literally more widespread than exactly inanimate matter.

In this way, surprisingly, one can recognize that the three first principles we found plays a similar role to the ancient Chaldean first principles of (material) existence, life (or power) and Act (Majercik 2001); the primordial first principle Ilu (the One or the Good), unites three first principles, his three first manifestations: Anu (time, the universe, or matter), Hea (reason and life) and Bel (the creator, the governor of the organized universe; Lenormant 1999, 114). Moreover, the first principles of matter, life and self-consciousness were also recognized in ancient China (e.g., the jing, the material principle, chi, the life principle, and shen, the principle of spirit;

see e.g. Beinfield and Korngold 1991). In ancient Hindu philosophy, a similar trinity is known under the term "three gunas" (the sattva, the quality of spirit; the rajas, the quality of life, and tamas, the quality of matter; Bhagavad Gita, Chap. 7, verses 12–14).

BIOLOGICAL PRINCIPLE, LOGIC AND LOGOS OF LIFE

My point is that the universal natural philosophy promises clearer understanding of the nature of logic, logos, and the "logos of life", proposed by Tymieniecka. Logic is frequently equated with Aristotelian logic: the laws of logic are applied to the premise in order to obtain the logical conclusion. We point out that this approach shows a remarkable similarity to the approach of physics, in which the equations of physics are applied to input data. Machines work in a similar manner. We insert a coin, and the result comes out at the output; push a button, and the Mars bar appears. Machines are working mechanically, step-by step, linearly in an immutable order. On that basis, we can classify Aristotelian logic as mechanical. Now if biology is the control theory of physics, generalizing the input data, and injecting further input into the equations of physics during the process, than the following interesting idea surfaces: is it possible to generalize mechanical logic in the same sense which makes biology the control theory of physics? We think that the answer is: yes, and the generalized form of mechanical logic is nothing else but the logos of the ancient Greeks.

In order to proceed, we have to prepare the stage, at first we have to consider the following questions: What is the difference between mathematical and physical laws? "What is it that breathes fire into the equations and makes a universe for them to describe?" (Hawking 1988, 174) We consider here that the essential difference between mathematical and physical laws is that mathematical laws represent lawful relations between abstract, mathematical properties, while physical laws represent lawful relations between observable, physical properties. The relation between the equations of physics and the physical laws is that the former exist in our mind, while the latter in Nature. In other words, the difference between physical equations and laws is that of map and territory.

All empirical sciences are built on the concept of "fact". Facts, in contrast of non-facts, are manifestations of some existent entities. Therefore, it is necessary to discern correctly "facts" from things that are not facts. There are some universal criteria for that, like the criteria of consistency. When we consider whether a thing is a fact or not, we know a priori that a fact cannot contradict to the existence of other facts. Another criterion is systematic and universal confirmation or validation by observations as well as by theoretical knowledge. In order to illustrate the importance of theoretical knowledge in evaluating what counts as "fact" and what not, we note that e.g. the life principle is not yet accepted in science. The reason to reject it is not its immaterial nature, since all laws of Nature are immaterial. Yet, as Hempel (1966, 72) pointed out, the assumptions made by a scientific theory about underlying processes must be definite enough to permit the derivation of specific implications concerning the phenomena that the theory is to explain. The doctrine about the life principle (Hempel, apparently, does not know Bauer's work; he refers to the ancient idea of "entelechy") fails on this account. It does not indicate under what circumstances the life principle will go into action and, specifically, in what way it will direct biological processes. This inadequacy of the life principle doctrine does not stem from the circumstance that the life principle is conceived as nonmaterial agency which cannot be seen or felt. This becomes clear when we contrast it with the explanation of the regularities of planetary and lunar motions by means of the Newtonian theory. Both accounts invoke nonmaterial agencies: one of them vital "forces", the other, gravitational ones. But Newton's theory includes specific assumptions, expressed in the law of gravitation and the laws of motion, which determine (a) what gravitational forces each of a set of physical bodies of given masses and positions will exert upon the others, and (b) what changes in their velocities and, consequently, in their locations will be brought about by these forces. It is this characteristic that gives the theory its power to explain previously observed uniformities and also to yield predictions and retrodictions. Thus, the theory was used by Halley to predict that a comet he had observed in 1682 would return in 1759, and to identify it retrodictively (Hempel 1966, 72). On that basis, we can deduce that gravity has a factual existence, its existence is a fact.

I point out that if theoretical biology can be formulated also in a mathematical form, and if it will be confirmed by all available empirical evidences, and capable of predicting yet unexplained phenomena, then, if applying the same kind of considerations as accepted in the case of theoretical physics, theoretical biology has to become an established science. This means that although we all experience the evidently observable facts that the behavior of living organisms is fundamentally different from that of physical objects, at present science does not accept the life principle just because it seems for most scientists and philosophers, including Hempel, that we do not know it in such an exact and empirically testable mathematical form as we know the laws of physics. I point out that the role of our–frequently incomplete—theoretical knowledge is many times decisive in our judgments about what we count as "fact" and what not. It is clear that Hempel did not know the work of Ervin Bauer, because for the Bauer-principle of life all the criteria he presented fulfils. It is clear that such a life principle should be accepted in science since it is not only known in a mathematical and testable form, but is consistent with all observations. In that case, the existence of the life principle must be regarded as a fact.

LOGIC IS THE BASIS AND PARTNER OF LAWS OF NATURE

From this point onward I want to regard logic in a wider sense, including not just only human logic. I mean that human logic is only an aspect of "natural logic" that belongs to the core of Nature. Natural logic acts on natural processes. Similarly to our human logic, which determines the right inferences, natural logic determines what will occur in Nature. Now because we defined Nature as the self-consistent system of relations with observable phenomena, therefore natural logic must contain the rules by which the future events can be realized and built up into the self-consistent body of Nature. Among others, natural logic has two basic functions: it generates the possibilities and it selects from these possibilities the ones that are consistent with the whole body of Nature and the given situation in a way that its realization can be regarded as optimal on the basis of the first principles. Therefore, natural logic is in the following intimate relation with physical laws: it generates the possibilities of the world process, and selects from them the ones that can be realized by the physical laws, and so the function of physical laws is to realize them, i.e. attach the suitable physical properties to these possibilities selected by natural logic. The consequence of that is that physical laws cannot function separately from natural logic. Natural logic is the basis and a partner of physical laws. That part of natural logic, which generates the physical possibilities, will be termed as physical logic.

It becomes clear that it is natural logic that prepares the ground for establishing the relations (like physical laws) between such specific entities as the physical properties. Or, to put it differently, natural logic belongs to the physical laws. Regarding that human logic is suitable to reveal the conditions of truth, and put severe constraints on what can be realized and what not, assuming a parallelism between human and natural logic we can conceive natural logic as a basis and partner of the laws of Nature working out the conditions of realization of natural processes. We can conceive "physical logic" as working out the preconditions of realization of physical processes. In other words, "physical logic" (i.e. the logical aspect of the inseparable logic-physical law organic unit) can be regarded as the very basis of physical reality.

MECHANICAL LOGIC AND BIOLOGICAL LOGIC

In general, one can distinguish three versions of logic that correspond to the three fundamental natural sciences: the physical, the biological and the psychological. Since biological logic acts in Nature, it can act within our organisms, as we are members of the biological species Homo Sapiens, a part of Nature; therefore, biological logic can be present within us and shape our internal mental processes, so it can work in the process of our thinking. In this way, it can modify, if it is necessary, continuously the input conditions of mechanical logic, in co-operation with the biological principle. Moreover, the co-operation of natural logic and the laws of Nature that is responsible for the generation of Homo Sapiens, including self-consciousness, that is, psychological logic and the psychological principle, is responsible for the generation of human logic as a phenomenon of Nature, as a phenomenon of self-consciousness. Therefore, physical, biological and psychological logic acting in Nature can be regarded as the physical-biological-psychological basis of our human logic. Our result is that human logic is driven not only by the autonomous part of self-consciousness, but also by a natural "force": by natural logic and the laws of Nature.

We note that the riddle of creativity presents a paradox at the level of mechanical logic, since mechanical logic is programmable into a software of a computer, it represents only the surface of our knowledge. Actually, since self-consciousness is ultimately a natural phenomenon, there is a parallelism between natural and human logic. Therefore, in many cases it is not necessary to distinguish them when speaking about "logic", at least in cases when what we say can refer to both context, the natural and the human as well. From now onwards, when we do not indicate about which logic are we speaking, the sentence can refer to both cases, either to the natural or to the human logic, or both.

Since mechanical logic works mechanically, it does not have a room for creativity. Although mechanical logic, like software programs, represents algorithmic complexity (Grandpierre 2008a), and so it is suitable to solve physical problems, it is not deep enough in order to account about creativity. We can realize that creativity must correspond to a deeper level of reality. Since the principle of creativity must be also consistent with the laws of logic, therefore this "creativity principle" represents the logic of reality in a fuller sense than the physical laws and mechanical logic. Therefore, it is useful to distinguish this more general creativity principle of logic from the usual term denoted by "logic" (which refers usually to mechanical logic).

We think that the most suitable term for this deeper creative logical principle is "logos". Since we can regard that such creative principles like logos exist at a deeper level of reality than laws, we can regard that logos is the creative source of logic. Now since logos can be regarded as universally valid, it can be conceived as the basis and partner of the laws of Nature; therefore, we propose to consider it as the common basis and partner of the physical, biological and psychological laws.

Actually, the self-renewing logic that can recharge its input in the process is not mechanical; it can be conceived that self-renewing logic stands in a similar relation to mechanical logic as biology with physics. The creativity principle is what governs the renewal of logic within the continuously changing inner and outer conditions. The deepest level of logic can be conceived as being the creative logic.

We can consider that logos, in a narrow sense, can be identified with creative logic, or, in a wider sense, we can select the option to regard logos as logic in its dynamic, vital, organic fullness, the organic unit of creative, self-renewing and mechanical logic. We will refer to the former with the term "creative logos", and to the latter simply as "logos". Therefore, we propose to regard logos as extending from the creative, principal level of reality, through the level of laws of Nature, until the phenomenal level. At the level of laws logos has three versions: physical (or mechanical, formal), biological and psychological (or self-conscious) logos. At the phenomenal level logos is not creative and is not problem-solving, but simply perceptive, self-consistently and consistently with all the deeper levels of logos (we can refer to this kind of phenomenal logos with the term perceptive logos).

ANIMATING PRINCIPLE

The origin of the animating principle goes back to prehistoric animism, frequently regarded as the first religion or wisdom of mankind (Kirk, Raven and Schofield 1983, 154). Heraclitus (ca. 535–475 B.C.) considered that "the Logos is a component of all existing thing, yet has a single collective being: it is a component of order or structure or arrangement, not the whole of an object's structure or shape but that part of it which connects it with everything else. Since there is one common rule or law which underlies the behavior (ginestai) of all things, then men are subject to this law and, if they want to live effectively, must follow it" (Kirk 1975, 58). This ancient idea fits well to our proposal about the existence of natural logic. It became a familiar saying, frequently attributed to Einstein: "The most incomprehensible thing about the world is that it is comprehensible." The solution of this problem is not complicated: the world is comprehensible because we are a part of Nature, and so the universal laws of Nature are present also in our organism.

Recently, Anna-Teresa Tymieniecka (2010/11) developed a remarkable system of idea about the "ontopoietic principle", which is also called as the "logos of life" (Tymieniecka 2009). The first naming seems to indicate an ontological principle characterized by its creativity (poiesis). The latter term indicates the twofold character of the "logos of life", being reasonful and playing the role of the life principle. She claims that the root of the logos is in its creative imaginative metamorphosis (Tymieniecka 2010/11, p. 12). This fits our view to regard logos as including its deepest level ingredient, the creative principle, yet including something more as well, namely, in our picture it includes physical, biological, psychological, mechanical and perceptual logic. She considers that the living agent's experience advances along the steps of the logos following its constructive devices from one step to the next, timing their deployment according to its constructive completion, that these processes reach the point of tying the knot in a synthesizing objectifying act of the logos. (ibid., 18). Another remarkable and detailed agreement between our results corresponds to the question what is the relation of human logic, natural logic and the ontological principles. Tymieniecka points out that "The cognitive/conscious constitution of objectivity is convertible with the natural functional root of existential generation. In fact, these movements are inseparable, even if in abstraction they are distinct." (ibid., 19–20) We find here again a surprisingly detailed agreement with our picture. Tymieniecka speaks about the natural functional root of existential generation, which in our terms is natural logic, or natural logos. She found that the cognitive/conscious constitution of objectivity is convertible with this natural entity. This is interpreted in our framework as the psychological (self-conscious) aspect of the natural logic acting in Nature is convertible with the joint working of the natural logic, co-operating with the first principles of Nature, with the ontological principles of physics, biology and psychology. This means that Tymieniecka found that the logos and laws of Nature are acting in co-operation. As we found, logos is the basis and partner of the laws of Nature.

Tymieniecka (2010/11, 23–24) writes: "the logos of life in its intrinsic metamorphosis during the evolutionary course of the individualizing genesis of beingness unfolds numerous modalities that reach realms beyond those geared to survival and which culminate in the full-fledged unfolding of the human creative virtualities." This translates in our picture into the indication that the first principle of biology acts on the same manner as the least action principle of physics, by virtual particles that are suitable to map instantaneously the whole of the Universe (because they exist not in the usual 3+1 dimensional space-time, but in the infinite dimensional Hilbert space, see Grandpierre 2007), securing a kind of instantaneous "primary perception" (Grandpierre 1997). Tymieniecka adds that "Having reached beyond the existential/evolutionary parameters of vitally significant (survival-oriented) horizons to the spheres of communal/societal life, the creative logos now throws up spiritual and, lastly, sacral horizons of experience that actually surpass the now narrow confines of the existential horizon." All these findings of Tymieniecka nicely fits with our indications telling that the biological principle is the "greatest action principle" (in terms of physical properties) and the "greatest happiness principle" (Grandpierre 2010/11), in terms of biological properties. From our formulation of the greatest happiness principle (Grandpierre 2007, 2011) it is clear that the greatest happiness principle has an integral character, summing up happiness for our lifetime, therefore it has two basic ingredients, one is lifetime, the other is life's quality or happiness. This latter factor is the one that point out beyond survival, towards communal/societal life, throwing up spiritual and sacral horizons.

In summarizing our comparison of the biological principle and Tymieniecka's logos of life, we found that both have a twofold nature, conceived as consisting from two basic constituents, (a) logos, having a metaphysical status, preparing the conditions for the activity of the first principles of Nature, being the basis for the actions of laws of Nature, and (b) the first principle of life or the "natural law" aspect of the "logos of life", having an ontological status and belonging to the natural sciences. Both our results and of Tymieniecka's indicate that these two factors, logos and the ontological principle, are in actual reality inseparable, they are partners of each other, co-operate in their activity. In other words, we can say that the biological principle has a basic logical or logoic character, or that the "logos of life" can be identified with the biological principle.

Acknowledgements It is a special pleasure for me to express my thanks to my dear friend, Rene van Peer, journalist, who contributed to this chapter with genuine proposals, including the "core of Nature" and "the inverted cone" structure of Nature in the three-days-long consultation during which this chapter, due to him, gained its final form and its better English.

Konkoly Observatory of the Hungarian Academy of Sciences, Budapest H-1515, Hungary
e-mail: grandp@iif.hu

NOTES

[1] I prefer to use the term scientific "world picture" instead of "world view" because I want to arrive to a picture that we can agree on, even when using different views. I regard worldview as the world picture plus the factors arising from our personal angle.

[2] In a process that repeats itself in a loop-like manner until it distils to the most concentrated and clear form.
[3] The picture is always less detailed than Nature itself.
[4] At some point, astronomy must come into the picture. Since the basis of the world picture is the Universe, it must give a scientific picture about the world in its entirety, therefore you cannot omit astronomy.
[5] The fundamental laws in physics, namely, that of classical mechanics, electromagnetism, thermodynamics, theory of gravitation, and quantum physics, including quantum field theories and string theory. In classical mechanics, the Euler-Lagrange equations, in electromagnetism, the Maxwell equations, the second law of thermodynamics, the Schrödinger-equation of quantum mechanics etc.
[6] At present, biology, the science of life, is widely conceived in a restricted manner.
[7] The question of self-consciousness can only be dealt with after the question of biology, which we are discussing in this chapter, is solved.
[8] According to Encyclopedia Britannica, these four views present metaphysics as: (1) an inquiry into what exists, or what really exists; (2) the science of reality, as opposed to appearance; (3) the study of the world as a whole; (4) a theory of first principles.
[9] The integral refers to a sum total between the initial and final states. In the following example the sum total is of the quantity of "action", which arises if you add up all the energy invested in each of the time intervals of the flight, multiplied with the length of each corresponding individual time interval.
[10] Biological teleology is a teleology of consciousness, so it can be different from human teleology which can be a self-conscious teleology, too. We do not have to underestimate consciousness, which in many cases can be much more efficient than the self-consciously controlled and narrowed self-consciousness.
[11] The first thing we as humans would automatically opt for is to prolong our lives; but we do also take the quality of that life into consideration. Quantity (length) is then also a function of quality (happiness, energy, vitality).

REFERENCES

Bauer, E. 1935/1967. *Theoretical biology* (1935: in Russian; 1967: in Hungarian) Akadémiai Kiadó, Budapest, 51.
Beinfield, H., and E. Korngold. 1991. *Between Heaven and Earth: A guide to Chinese medicine*. Ballantine Books.
Brent, R., and J. Bruck. 2006, Can computers help to explain biology? Nature 440:416–417.
Feynman, R.P. 1942. The principle of least action in quantum mechanics. Ph. D. Thesis. Princeton University. Source: Dissertation Abstracts International, Vol. 12–03, p. 0320.
Feynman, R.P. 1994. *The character of physical law*, 97–100, Chapter 4. New York: Random House.
Feynman, R.P., and A.R. Hibbs. 1965. *Quantum mechanics and path integrals*. New York: McGraw-Hill.
Grandpierre, A. 1997. The physics of collective consciousness. *World Futures* 48:23–56.
Grandpierre, A. 2007. Biological extension of the action principle: Endpoint determination beyond the quantum level and the ultimate physical roots of consciousness. *Neuroquantology* 5(4):346–362, http://arxiv.org/abs/0802.0601.
Grandpierre, A. 2008a. Cosmic life forms. In *From Fossils to astrobiology. Records of life on Earth and the search for extraterrestrial biosignatures*, eds. J. Seckbach and M. Walsh, 369–385. Springer. http://www.springerlink.com/content/qh5r5664348n0032/.
Grandpierre, A. 2008b. Fundamental complexity measures of life. In *Divine action and natural selection: Questions of science and faith in biological evolution*, eds. J. Seckbach and R. Gordon, 566–615. Singapore: World Scientific. http://www.worldscibooks.com/lifesci/6998.html, http://www.konkoly.hu/staff/grandpierre/Complex.htm.
Grandpierre, A. 2011. On the first principle of biology and the foundation of the universal science. In *Astronomy and civilization in the new enlightenment, Budapest, August 10–13, 2009*, eds. A.-T. Tymieniecka and A. Grandpierre. Analecta Husserliana, Vol. 107, pp. 19–36.

Hawking, S. 1988. *A brief history of time*, 174. New York: Bantam.
Hempel, C.G. 1966. *Philosophy of natural science*, 70–72. Englewood Cliffs, NJ: Prentice-Hall Inc.
Kirk, G.S. 1975. *Heraclitus: The cosmic fragments*, 1971. London: Syndies of Cambridge University Press.
Kirk, G.S., Raven, J.E., and Schofield, M. 1983. *The presocratic philosophers: a critical history with a selection of texts*. London: Cambridge University Press.
Lenormant, F. 1877/1999. *Chaldean magic. Its origin and development*, 114. York Beach: Samuel Weiser, Inc.
Majercik, R. 2001. Chaldean Triads in Neoplatonic Exegesis: Some reconsiderations. *The Classical Quarterly*, New Series 51:265–296.
Moore, T.A. 2004. Getting the most action out of least action: A proposal. *American Journal of Physics* 72:522–527.
Stöltzner, M. 2003. The principle of least action as the logical Empiricist's Shibboleth. *Studies in History and Philosophy of Modern Physics* 34:285–318.
Taylor, E.F. 2003. A call to action. Guest Editorial. *American Journal of Physics* 71:423–425.
Taylor, E.F. 2010. *Principle of least action*. http://www.eftaylor.com/leastaction.html.
Tymieniecka, A.-T. 2009. The Fullness of Logos in the Key of Life, Anal. Huss. 100.
Tymieniecka, A.-T. 2010/11. The new enlightenment: Cosmo-transcendental positioning of the living being in the Universe. In *Astronomy and civilization in the new enlightenment, Budapest, August 10–13, 2009*, eds. A.-T. Tymieniecka and A. Grandpierre. Analecta Husserliana, Vol. 107, pp. 19–36. Springer.
Woodward, J. 2010. Scientific explanation. In *The Stanford Encyclopedia of Philosophy*, ed. N. Zalta. http://plato.stanford.edu/archives/spr2010/entries/scientific-explanation/.
Zee, A. 1986. *Fearful symmetry. The search for beauty in modern physics*, 107–109, 143. New York: Macmillan Publ. Co.

NAME INDEX

Aarø, A. F., 337
Abram, D., 333–335
Acik, T., 51–57
Adams, G. P., 167n4
Addams, J., 160, 168n43
Adorno, T. W., 178–179, 517, 519n11, 597n70
Afham, W., 119
Agamemnon, 487, 490–491
Ajax, 487, 489, 491–493, 493n5
Aldhous, P., 101
Ales Bello, A., 36n48, 423
Algar, H., 436n43
Alighieri, D., 628n15
Al-Saji, A., 48n15
Ambrose Cole, R., 671
Aminrazavi, M., 436n24
Anaximander, 4, 410
Anaximenes, 4
Anderson, T. C., 63, 73n4–n5
Apel, K. O., 34n9
Arasteh, A. R., 435n16, 436n44
Arberry, A. J., 436n33, n37
Arbib, M. A., 435n2
Arendt, H., 480, 482n48
Aristotle, 21n1, 24, 37n66, 46, 79, 91, 93, 100n2, 104, 111, 115, 181–186, 208–209, 215n5, 243–245, 285, 296, 417, 420n25, 442, 462n7, 475, 481n2, 497, 503–504, 506n30, n32, n36, 511, 523–524, 570, 574–575, 608, 615–616, 619, 622, 627n1, n5–n6, n10
Arnold, M., 56, 56–57n8
Assmann, A, 51–52
Assmann, J., 51, 56n3, 232
Augustine, 44, 95, 393n13, 442–445, 456, 523–524, 681
Austin, 296–297
Auxier, R. E., 34n9
Averintsev, S., 681
Ayada, F., 650

Bachelard, G., 675
Bagley, W., 150
Bagnall, 356
Bakhtin, M., 677
Barbaras, R., 287
Barbieri, M., 339n6
Barnes, A., 153
Barnes, H. E., 352n12
Barnes, J., 215n2
Barthes, R., 173–174, 471, 475, 481n10, n16
Bartnik, C. S., 641, 644n28
Barzun, J., 420n8
Bauer, E., 714, 718–719, 721
Bauman, Z., 71, 74n25, 356, 560, 576n1, n10
Beck, C. H., 56n3, 133
Beck, U., 73n13
Bednarz, N., 365
Beer, S., 95, 101n17
Beinfield, H., 720
Beliy, A., 175
Benediktson, D. T., 277n3
Benjamin, R., 671n3
Benjamin, W., 44, 49n21, 506n21, 519n3
Benoist, J., 131–132, 135n41, 136n47–n50
Berdyaev, N. A., 631–642, 643n3, n6, n9, n11–n19, n21–n22, n24–n25, n27, 644n30
Bergson, H., 27, 152, 290, 399, 471, 504, 570
Berkeley, 101n14, 403, 435n2, 600–601
Bernar, 356, 440
Bernstein, M. S., 161–162
Bertalanffy, L. von., 96, 101n19
Bertie, G., 155
Best, 356
Bichelmeyer, B. B., 358–359
Biesta, G. J. J., 156–157, 167n1, 168n16, n29, n36
Biestat, 167n1
Binkley, R., 577n28
Blamey, K., 627n14, 628n17
Blanchot, M., 467–480, 485n6, n8–n9, n13, n18, 482n29, n32–n33, n36, n38–n43

NAME INDEX

Blonski, P., 169n45
Blonsky, P. P., 150, 160–162
Blumenberg, H., 498, 505n11, 628n8
Blumenthal, H. J., 244–247, 249n3
Bocchi, G., 356, 359–361, 363–364
Bochenski, J., 419, 421n37
Bohr, N., 648–649, 700
Bolton, N., 291
Bolzano, B., 131, 136n42, n44, n47–n48
Bonnett, M., 280
Bool, F. H., 101n16
Borges, J. L., 600–601, 610
Bosch, H., 696, 698–699
Brainard, M., 198
Braque, G., 700n4
Brenner, A, 91–102
Brent, R., 718
Brentano, F., 48n5, 136n48, 587, 590–591, 594n10
Bretony, K. J., 155
Brezzi, F., 36n48
Brickman, W. W., 151, 167n9, n11, 168n37, n40
Bronk, A., 576n10
Brough, J. B., 48n15, 49n18
Brown, C. S., 334
Bruce, V., 544
Bruck, J., 718
Bruegel, P., 318
Bruner, J., 364–366
Bruzina, R., 597n69
Buceniece, E., 39–49
Buch, Z., 593n5
Buckley, P. R., 197–198, 203n10, 596n52
Bulgakov, S., 631
Butrica, J. L., 277n3
Büyükdüvenci, S., 168n39

Cairns, D., 36n28, 90n1, 109n5, 200, 215n18, 240n45, 506n20, 593n8
Campbell, B., 152, 281
Caputo, J. D., 42, 48n9
Carloni, D., 35n15
Carr, D., 136n52, 238n1, 526, 536n32, 553–555, 557, 597n66
Carr, W., 168n19
Caruana, W., 666, 669–671, 671n6
Carus, P., 35n21
Cassirer, E., 28, 35n15, n29, n32, 36n36–n37, 437n72, 522, 536n2

Cech, T. R., 652
Ceruti, M., 356–357, 359–361, 363–364
Ceserani, 356
Chambliss, J. J., 627n6
Chatwin, B., 665, 668, 671n1
Chen Xiao, 652
Chernyshevsky, 282
Chittick, W. C., 436n23, n27, n32, n38, n44
Chiurazzi, 356
Chodkiewicz, C., 436n27
Chodkiewicz, M., 436n27, n33
Chrestensen, J., 436n33
Cicero, M. T., 416, 420n20
Claesges, U., 202n4
Clapared, E., 156
Claparede, F., 156
Clark, T., 461n1, 478, 482n39, n42
Clarke, M. L., 56n5
Claviez, T., 627n6
Clinbell, H., 560
Cole, M., 365–367
Coleridge, S. T., 56, 95, 101n16
Collange, J. F., 576n5
Collingwood, R. G., 525
Confucius, 54, 645–650, 653, 655–658
Congar, I., 440
Copernicus, 154
Corbin, H., 432, 437n48, n61
Costa, F., 36n38
Costa, V., 35n27
Coucerio-Bueno, J. C., 290
Cozma, C., 284–285, 289, 293, 413–421, 577n26
Crapulli, G., 34n6
Creed Meredith, J., 627n4
Cristin, R., 36n38
Cronus, 609
Crowell, S. G., 482n47, 593n8, 597n68
Curtius, E. R., 56n4, n7

d'Olivet, T., 53
da Vinci, L., 282, 695–696, 700n2
Dalledalle, G., 155–156
Damiano, E., 357, 360–365, 368
Daniels, D. N., 666
Daniels, M., 73n3
Danilova, V., 709n4
Darling, J., 168n21
David, L., 336
Davie, D., 273–274, 276n2, 278n24–n27

NAME INDEX

Davis, K., 285
de Chardin, T., 570
De Garmo, C., 152
de Lubac, H., 440
de Montaigne, M., 499
de Tienne, A., 332
Debesse, M., 155
Decroly, 156
Deferrari, R. J., 462n10–n11
Deleuze, G., 41, 43, 48n2, n15, 536n4
Delledalle, G., 168n25
Dembińska-Siury, D., 247
Democritus, 498–499, 608
Dent, J. M., 22n6
Derrida, J., 22n7, 207, 211, 215n1, 251–253,
    256–259, 262n1–n2, 263n6, n10, n12–n13,
    n15–n17, 475, 482n39, n42, 508, 512, 518,
    519n2, n4, n6, n8, n12, 558n4, 560
Descartes, R., 7, 23–25, 27–28, 31, 34n3–n4, n7,
    n9–n10, 35n15, 81, 92, 100n4–n5, 242, 296,
    469, 479, 481n4, 503, 507, 552, 600
Deutscher, G., 172–174
Dewey, J., 145n2–n3, 147–169, 280–283, 360,
    541
Diderot, 52, 345
Dilthey, W., 27, 318, 558n1
Diogenes, 421n36, n38, 502–503, 506n26, n29
Djui, I., 167n2
Djui, P., 169n46
Dodd, J., 193, 202n1, 218, 225n2, 597n70, 611n6
Dodde, N. L., 156, 168n28
Doede, R., 175–176
Doney, W., 34n9
Dong Zhongshu, 645, 649, 651, 655
Donohoe, J., 593
Drewermann, E., 311, 314n4
Dreyfus, H., 235, 298–299
Drottens, R., 151
Duddington, N., 643n11
Duffy, 356
Durkheim, É., 230, 232, 239n12
Dybel, P., 595n26
Dykhuizen, G., 168n17, n32

Eagleton, 356
Eccles, J. C., 230, 239n13–n15
Edwards, J. C., 352n41, 356
Edwards, L. C., 543
Edwards, M. W., 270–273, 276, 277n7–n9

Einstein, 724
Elden, S., 675, 683n1
Eliade, M., 680, 683n10
Eliot, T. S., 271, 277n5, 588
Elveton, R. O., 90n2
Embree, L., 435n1, 596n52
Empiricus, S., 505n2
Eremita, V., 118, 120
Escher, M. C., 95, 101n16
Euler, 713, 716, 726n5
Ewald, P. W., 652

Fermat, 713
Ferrara, A., 621, 624, 628n16, n23
Ferrarello, S., 217–225
Ferrier, 156
Fevre, L., 166
Feynman, R. P., 713, 716
Fichte, 242, 580, 582, 594n10, 701
Filograsso, N., 360–364
Filosifija, D., 167n2
Findlay, J. J., 153
Findlay, J. N., 110n13
Fink, E., 21n1, 22n7, 28, 36n42, 193, 199–202,
    589, 593n6, n8, 597n69, 598n87
Fleischer, M., 239n17
Foerster, H. von., 364–365
Føllesdal, D., 235–236, 240n53
Forman, R. K. C., 356, 435n5
Fornari, A., 373–394
Foster, R., 179
Foucault, M., 207, 323, 326n3, 475, 553–554,
    558n4, 600–603
Frank, S. L., 631
Frege, 212, 215n17
Freud, 263n4, 307, 311–313
Fricke, H., 56n4
Frings, M., 35n26
Fu Xi, 646

Gadamer, H.-G., 22n7, 54, 172–173, 175–178,
    467, 475–477, 482, 482n28, n35, n37
Gale, E. J., 356–357
Gallagher, S., 437n73
Ganesan, S. V., 435n13
Gardner, J., 522, 526–528, 530, 534, 536n34–n35,
    537n55, n70
Garrison, J., 143, 145n2–n3, 168n39, 365
Gatti, M. L., 241

Gattico, E., 362–363
Gawron, L., 576n1
Geertz, C., 230
Gelven, M., 301
Gendlin, E., 429, 432, 436n42
Gergen, K. J., 357
Gerhardt, C. J., 35n30, n33
Ghigi, N., 36–37n50, 225n10
Ghosh, M., 103–110
Ghoush, M., 700n6
Giaconi, C., 355–368
Gibson, D. G., 101n20
Glenn Gray, J., 90n5
Gniazdowski, A., 595n26
Goddess Athena, 414, 487, 491–492
Goethe, B., 53–55, 97, 303
Gogochuri, H. G., 643n6
Goldberger, S., 429
Goleman, D., 437n63
Gong Ye Chang, 650
Goodman, 356
Goold, G. P., 277n4
Göttingen, 436n43, 582, 594n10
Gram, L. M., 269–278
Grampa, G., 628n17
Grandpierre, A., 714, 718–719, 723, 725
Grassom, B., 507–519
Green, B., 695–696, 700n5
Griffin, P., 358, 365
Gril, D., 436n27
Groth, M., 437n76
Gründer, K., 56n4
Guardini, R., 377
Guba, E. G., 356, 358
Gubser, M., 594n10, n17
Gumbert, H. U., 56n6
Gunn, A. E., 437n76

Habermas, J., 73n19, 152, 299, 303, 558n6, 597n70
Hagedorn, L., 593n4
Hahn, L. E., 34n9, 167n14
Halfwassen, J., 242, 248
Halley, 721
Hamilton, 506n20, 713
Hamlet, 473, 485
Hannay, A., 115
Hans Urs von Balthasar, 440, 456–457, 460
Hardt, M., 74n23, 241

Hardy, T., 273, 511
Hare, R. M., 565–566, 572, 574–575, 577n28, n34
Harper, W., 90n3, n5, 149, 215n17, 352n20, 353n42, 420n8, 437n72, 481n5, 519n9, 527, 529, 595n30, 596n43
Harrigan, J. S., 437n65, n69
Harrington, A., 331
Harry Passow, A., 167n13
Hart, J. G., 201, 580–581, 586, 593n8, n10, 594n13, n16, 596n42, n50, n52
Hartmann, N., 702
Hartner, A., 168n19
Hartnett, 153
Hastie, W., 627n4
Havelaar, M., 71
Hawking, S., 720
Hegel, F., 152, 165, 173, 177, 210, 232, 241–242, 282, 517, 554–555, 558n3, 656, 675
Heidegger, M., 22n7, 34n9, 44, 76–90, 134n20, 144–145, 172, 177–179, 209–215, 241–242, 290, 295–305, 341–342, 351n2, 352n20, 399, 408, 457, 468–472, 475, 478–480, 511–512, 519, 521, 523–524, 558n3, 588–589, 593n4, n8, 596n43, 567n68, 600, 603–604, 608, 680
Heidel, W. A., 144
Held, K., 261–262, 264n20–n21, n27–n31, n33, 598n89
Hempel, C. G., 715–716, 720–721
Henckmann, W., 132, 133n1, 136n51
Henry, M., 43, 49n16, 438n82
Heorot, 526, 528, 530, 532–533
Heraclitus, 3–6, 12n2, 45–46, 103–104, 109n4, 208–211, 402, 405, 410, 410n1, 433, 496, 498–499, 675, 679, 724
Herman, D., 352n36, 521–522, 536n6–n7
Heron, J., 358, 713
Heyer, E., 53
Hibbs, 716
Hilary, 442–443
Hintikka, J., 34n9
Hobbes, T., 68, 556
Hoffmeyer, J., 335, 337, 339n6
Homer, 54–55, 218, 418, 485–492
Honneth, A., 73n15
Hoogestraat, J., 275–276, 278n37–n41
Hooper, D., 648
Hopkins, B. C., 134n23

NAME INDEX

Hopkins, J., 164
Horner, R., 451–452
Howlett, C. F., 169n51
Hryniewicz, W., 632–633, 641, 643n2, n4, 644n29
Hughes, R. I. G., 648
Hume, 26, 127, 134n15, n22, 552, 595n20, 600–601
Husserl, 39–49, 207–215, 251–265, 659–663
Hutcheson, 503

Ibn, M., 436n27, n29, n34, 437n54
Ibrahim, E., 436
Ijabs, I., 49n32
Ingarden, R., 31, 475, 521, 565–567, 570, 576n12, n14, 577n19, 595n23
Ionesco, E., 603
Iribarne, J. V., 43, 48n14
Iyer, L., 435n9, 481n6

Jackson, P. W., 167n6, 169n59
James, W., 93, 100n8, 164, 501n5, 606
Jameson, F., 72, 74n31, 356
Jamme, C., 598n89
Janson, H. W., 519n10
Jaspers, K., 141, 145n4
Jastrow, J., 348, 353n42
Jauss, H. R., 475
Jaynes, J., 652
Jensen, T. Ø., 74n27, n32
Joardar, K., 207–215
Job, 54
Johannes Servan, 551–558
Johnson, H. F., 671n3
Johnson, S. M., 36n47
Johnson-Davies, D., 436n28
Joisten, K., 623–624, 628n19–n20
Jonas, H., 23, 25–26, 28, 34n1, n12, 35n16, n22, 36n35
Jonassen, 356
Jordan, D., 543
Josipovic, Z., 435n18
Jowett, B., 506n21, 519n3, n7
Judge William, 119–120
Julina, N., 167n2

Kafka, F., 473–474
Kahn, C. H., 12n2, n74
Kaiser, J., 101n20
Kalashnikov, A., 161

Kandinsky, W., 665
Kane, G., 648
Kant, I., 22n6, 26, 34, 35n21, 40–41, 48n1, 94, 98–99, 101n24, 103, 105, 123, 126–127, 129–130, 182, 184–185, 210, 261, 322, 328–332, 337, 338n2, 344–345, 352n14–n16, 396, 503, 522–525, 536n4, 552, 558n3, 570, 574–575, 584, 591, 595n35, 600, 614, 627n4
Karadas, F., 521–538
Kaufmann, F., 596n61–n62
Kaufmann, W., 90n6
Kay, E., 156
Kearney, R., 525, 536n20
Kellner, 356
Kelly, M. R., 43, 49n17
Kelly, S. D., 352n36
Kendo, N., 169n56
Kenner, H., 272–276, 278n17–n18, n20
Kenner, K., 277n4
Kermode, F., 637n10
Kerner, G. C., 73n4, n10
Kerr, F., 435n2
Kershensteiner, G., 152
Kersten, F., 240n50, 254, 263n5
Khlebnikov, V., 175
Kierkegaard, S., 111–112, 114–121
Kilpatrick, W. H., 150, 152, 154, 165, 167n14
Kimmel, L., 420n14, 699, 700n8
King Alfred, 510
King Da Yu, 646
King Hrothgar, 526
Kingsley, P., 431, 436n20, 437n56–n58
Kingsmill Abbot, T., 627n4
Kirk, G. S., 724
Kisiel, T., 482n47
Klee, P., 665
Kngwarreye, E. K., 671
Kobayashi, V. N., 165, 169n50, n53, n55, n57
Koestler, A., 497, 505n8
Kohák, E., 592n1, n2, 593n6, 594n9, 595n22, n36, 604, 610n2, 611n6
Köhler, J., 683n7
Kojiri, S., 165
Komarovski, B. B., 169n46
Korngold, E., 720
Kozhevnikov, M., 435n18
Kozhevnikov, N. N., 709n1–n2
Kraft, J., 125, 134n9

Krasińska, E., 576n10
Kruks, 555, 558n6
Krupskaya, N., 160–161
Kuhn, T., 281, 553
Kull, K., 335, 339n6
Kung ye Chang, 655
Kurenkova, R., 141–146, 282, 289
Kurenkova, R. A., 141–146, 282, 289
Kuster, F., 203n10
Kuwakin, W. A., 643n9

Laclau, E., 69–70, 73n21
Laertius, D., 421n36, n38, 502, 506n26, n29
Laforgue, J., 269–276, 277n5, 278n31
Lai, G., 314n2
Landgrebe, L., 90n2, n9, 595n22, 596n61
Langeveld, M. J., 157
Langsdorf, W. B., 505n14
Larochelle, M., 365
László, J., 628n24
Lawlor, L., 256, 263n6
Lefebvre, H., 675–676, 683n1, n3
Legge, J., 647
Lehto, O., 339n6
Leibniz, G. W., 27–28, 31, 34, 35n30, n33, 600, 691
Lenormant, F., 719
Leont'ev, A. N., 364
Leopold, A., 560
Leroux, G., 245
Lessing, 282
Levinas, E., 39, 85, 439–441, 445–454, 456–459, 462, 462n16, 470, 481n6, 508, 519n2, 558n3, 570
Lewis-Williams, D., 672n21
Lian zhi, 654
Liang neng, 654
Liddell, 104
Ligthart, J., 156
Lincoln, Y. S., 356, 358
Lingis, A., 519n2, 678, 683n8
Lo Verso, G., 357
Locke, J., 68
Lohmar, D., 134n22, n24, 135n27, n29–n30, n34
Lom, P., 607, 611n7–n8, n10
Lonergan, B., 440
Lorenc, K., 560
Louchakova, O., 435n16–n18, 437n46, n49, n64, n67, n70

Louchakova-Schwartz, O., 423–438
Lovibond, S., 617, 627n11
Lovinger, J., 147
Luhmann, N., 318
Luisi, M., 332
Luisi, P. L., 93, 101n12
Lunacharski, A. V., 169n47
Lunacharsky, A., 160–162
Lurija, 364, 366
Lynch-Fraser, D., 539, 542
Lyotard, J.-F., 356, 554, 558n4

MacIntyre, A., 595n20
Mackiewicz, W., 576n2
Macquarrie, J., 90n3, 352n20
Madonna, 695
Madsen, O. J., 73n14, 74n26, n29
Majercik, R., 719
Majolino, C., 136n50
Małecka, A., 495–506
Mall, R. A., 200, 202n6
Mallarmé, S., 473–474, 476, 478
Mamarika, M., 666
Mandelshtam, I., 175
Manning, T., 436n33
Marbach, E., 41, 48n4, n6, n11, 49n22
Marion, J.-L., 439–441, 445–446, 449–454, 456, 459–461, 462n18
Marlowe, C., 97
Martin, U. N., 666
Martinelli, D., 339n6
Martin Wieland, C., 502
Masaryk, T., 596n61
Maturana, H., 91–99, 101n9–n10, n13, n17–n18
Maturana, U., 31, 36n37, 37n63
Maupertuis, 713
Mauss, M., 239n12
McGuinness, B. F., 351n1, 352n10, 353n51
McIntyre, 235
McKenzi, W. R., 149, 167n3
McMurry, C., 152
McMurry, F., 152
Mead, G. H., 149, 290
Mecacci, 356
Meiklejohn, J. M. D., 22n6, 35n21, 627n4
Meinhardt, H., 241
Meinhart, C., 101n21
Melaney, W. D., 467–482
Melle, U., 202n5, 225n11, 593n8

NAME INDEX

Mencius, 646, 648, 654, 656–657
Menelaus, 487, 490–491
Mengus, C., 576n5
Merleau-Ponty, M., 290, 329–330, 335, 399, 429, 432, 436n40, 471, 478, 481n11, 597n72–n73, 598n79, 659, 661–663
Merlo Ponti, M., 47
Mettepenningen, J., 461n1
Meyendorff, J., 639, 643n23
Meyering, T. C., 435n2
Miedema, S., 156–157, 167n1, 168n16, n29, n36
Mifflin, H., 652
Millon, J., 598n78
Milosh, J., 528, 532, 536n35, 537n70
Mink, L., 521, 524, 526–527, 536n10
Miró, J., 665
Mohanty, J. N., 105, 109n6
Mohr, R. D., 350, 353n49
Molchanov, V., 44, 49n20
Moll, L., 365
Monk, D., 273–275, 278n21–n23, n36
Montague, W. P., 167n4
Montessori, M., 156–158
Moore, G. E., 405
Moore, T. A., 714
Morf, A., 364
Morphew, V. N., 281
Morreal, J., 495, 497, 503, 505n6, 506n31, n34–n35
Morris, J., 436n27
Motes, M. A., 435n18
Mouffe, C., 69–70, 73n21
Mounier, E., 325, 326n6, 570
Mullis, K. B., 648
Murphy, R. T., 134n15
Murphy, T. C., 435n2
Murr, J. S., 530, 537n55
Mutschler, F. H., 277n3

Nabors, 543
Næss, A., 333
Nancy, J.-L., 472, 481n17
Napangarti, P. L., 667
Napangarti, S. B., 668
Nasr, S. H., 436n24
Natanson, M., 192, 203n9
Negri, A., 74n23
Nelson, P. J., 668
Nenon, Th., 134n23, 595n24

Newman, D., 365
Newton, I., 713–715, 721
Neymayer, R. A., 357
NI, L.-K., 135n34
Nicholas, A., 101n14
Nida-Rümelin, M., 101n15
Nietzsche, F., 27, 45–46, 49n26, n32, 81–82, 85, 90n6, 209, 300, 302–303, 323, 326n2, 390, 474, 502, 601, 608, 675, 677–678, 680, 682, 683n7, n14, 686
Nijhoff, M., 36n38, 48n4, 49n22, 109n5, 134n15, 135n29, 215n18, 238n1, 239n17, 240n50, 263n9, n11, 264n33
Norton, R. C., 53, 56n5
Nöth, W., 339n6
Nucho, F., 631, 643n1
Nussbaum, M. C., 556, 627n5

O'Hear, A., 155, 168n24
O'Kane, J., 436n36
Odo, T., 165
Odysseus, 485–493, 493n4
Oelkers, J., 151–152, 167n1, n8
Öhlschläger, C., 628n19
Oktar, S., 341–353
Olbromski, C. J., 251–264
Oliver, K., 73n9
Ong, W. J., 56n2
Orage, A., 277n1
Orth, E. W., 202n2
Østerberg, D., 73n12
Ou Tsui-Chen, 151
Overgaard, S., 201
Øyen, S. A., 61–74
Öztürk, A., 539–547

Paige, D. D., 277n1
Panda, R. K., 110n18
Paragon, N. Y., 647
Park, In-C., 202n4
Parker, F., 150
Parkhurst, H., 154, 156
Parmenides, 4, 207–208, 210–211, 301, 326n5, 405, 410, 431, 433, 462n5
Passow, A. H., 167n13, 169n52, n54
Pattison, G., 115–116
Pearce, D., 672n21
Pears, D. F., 352n10, 353n51
Peirce, C. S., 327, 331–332, 334, 337, 339n5

Penrose, R., 576n9
Peterson, P., 156
Pettegrove, J. P., 35n32
Philippi-Siewetz van Reesema, C., 157
Phillips, D. C., 356
Philoctetes, 487–489, 492, 493n1
Piaget, J., 281, 360–366
Pierce, C. S., 168n25
Pilkington, E., 101n20
Pineda, J. A., 437n73
Pinkevich, A., 160–161
Pistrak, M., 162
Plato, 15–22, 39–40, 46, 75–80, 83, 86, 88–89, 91, 93, 95, 102n1, 104, 111–114, 118–121, 142–143, 186, 209–211, 241, 244–245, 296, 341, 353n47, n52, 416, 419, 420n19, n24, 441, 449, 457, 475, 481n2, 497, 499–503, 506n15–n16, n20, 507, 509, 511–512, 519n4, 575, 586, 596n51, 598n90, 603, 611n5, 615, 627n6, 680, 686, 701
Plekhanov, E., 141–145
Plessner, H., 498, 505n13
Plotinus, 241–248, 249n1–n5, 449
Płotka, M., 241–249
Płotka, W., 191–203
Pöggeler, O., 598n89
Pojaghi, B., 364
Polanyi, M., 330
Polt, R., 215n16
Popper, K. R., 230, 239n13–n15
Portmann, A., 98, 101n23
Pos, H. J., 596n61
Poulson, C. N., 666
Poulson, M. N., 666
Poulson, P. N., 666
Pound, E., 269–276, 277n6, 278n29–n31
Pozncr, V. M., 163
Prendergast, C., 54
Promieńska, H., 576n6, n8
Przywara, E., 440
Pufall, 356

Queen Elizabeth II, 666

Radtke, B., 436n36
Rahner, K., 440
Ramsey, W., 274–276, 277n5, 278n31, n32
Rappe, S., 242–243, 246–247
Raven, J. E., 724

Reavey, G., 643n12, n17
Regan, T., 101n14, 570
Rendtorff, J. D., 73n4
Rezek, P., 593n4
Rhees, R., 351n4, 352n22, 410n6
Rhyn, H., 167n1, 167n8
Riceour, P., 44, 536n9
Richardson, D. B., 634, 640, 643n7
Richmond, S., 436n36
Ricoeur, P., 106, 109n10, 373–374, 376–379, 383, 386–388, 393, 393n1–n6, n9, n11–n13, 394n16, n17, n20, n24, 521, 523–528, 536n12, n15–n17, n20, 619, 622–623, 627n14, 628n17, n18
Rijichiro, H., 165
Rikiso, N., 164
Rilke, R. M., 474, 479
Ritter, J., 56n4
Robinson, E., 90n3, 352n20, 519n5
Roeg, N., 668
Rogacheva, E., 141–145, 147–169
Rojcewicz, R., 37n51, 238n6, 596n42
Rokstad, K., 15–22
Rolston, H., 570
Rolston, T., 560
Roochnik, D., 109n2, n4, n9
Rorty, A. O., 627n5
Rorty, R., 341–344, 351n1, 352n9, n11, n18, 560
Rose, N., 73n16, 74n24
Rosen, S. M., 429, 436n41
Ross, B., 34n3
Rossi, A., 34n9
Rousseau, 160, 282, 627n6
Rovatti, 356
Rudd, N., 270–273, 276, 277n5, n10–n12
Rumi, J., 665
Russell, B., 107, 110n14, 208, 344, 348, 352n10, n40, 396, 401
Russell, R. J., 435n2, 545
Ryan, A., 167n12
Ryba, T., 439–462

Saeed, K. M., 285
Saint Athanasius, 639
Saint Augustin, 683n11
Sallis, J., 350–351, 353n46, n50
Sanders Peirce, C., 327, 331
Sandkühler, H. J., 595n26

Sartre, J.-P., 61–73, 73n1, n7, 74n30, 209, 214, 341–345, 351n3, n5–n7, 352n12–n13, n17, n21, 374
Scheler, M., 27, 35n26, 123–126, 129, 132, 133n4, 134n20, 136n51
Scheller, M., 570
Schelling, W. J., 32, 37n71, 242
Schlick, M., 123–136, 341
Schmidt, D., 177
Schmidt, M., 101n21
Schneider, H., 168n26
Schonfield, M., 627n6, 724
Schürmann, R., 481n2
Schuwer, A., 37n51, 238n6, 596n42
Schwartz, M., 436n25
Schweitzer, A., 570
Scotland, J., 153–154, 168n20, n22
Scott, 104
Searle, J., 295–297
Sebeok, T. A., 335, 339n6
Seducer, J., 118, 120
Seebohm, Th., 129, 134n16, 135n35
Sehdev, M., 307–314
Selvi, K., 279–294
Sempio, O. L., 364
SenSharma, D. B., 435n6
Sepp, H. R., 593n4
Servan, J., v
Sexton, T. L., 358
Sezgin, E., 395–411
Shah-Kazemi, R., 436n23, n26, n35, n39
Shao Yan, 648
Shapiro, A. V., 163
Shatsky, S. T., 149, 151, 160–162
Shatzky, S., 147
Shaurr, C. L., 646
Shayer, M., 365
Shelley, M., 97
Shelton, J., 134
Shields, C., 243
Shleger, L., 160
Shotter, J., 357
Shulgin, V. N., 163
Shutz, A., 145
Sidorsky, D., 151, 167n7
Simmel, G., 73n18
Sims, B. N., 666
Sims, P. J., 668

Singer, P., 560, 565–566, 570, 572–573, 576n11, 577n27, n29, n34
Skinner, Q., 148, 157
Skolimowski, H., 560, 576n3
Smid, R. N., 36n50, 264n23
Smith, A., 235, 244–245, 249n5
Smith, C., 627n7
Smith, J., 530, 537n55
Smuts, J. C., 100n6
Socrates, 18–20, 21n2, 40, 46, 89, 111, 114, 116, 118–119, 142, 415, 497–499, 501, 503, 505n10, 519n7
Sokolowski, R., 135n29, 596n52
Solomon, R. C., 54, 215n17, 352n36
Sophocles, 55, 486–489, 491–492, 493n1, n5
Sorkin, D., 56–57n8
Spaemann, R., 326
Spencer, L. J., 668
St. Anne, 696
Stafecka, M., 171–180
Staudigal, M., 49n16
Steffe, L. P., 357
Stein, E., 595n20, n22
Steinbock, A. J., 136n52, 239n17, 551–552, 555–557, 562n2, n3, 579, 593n5, n8, 595n27, 597n66–n67
Steiner, R., 156
Stöltzner, M., 134n5, n7, 715
Storari, G. P., 362–363
Stowe, W. W., 167n15
Strauss, 22
Ströker, E., 136n46, 193, 196, 203n9, 240n42, n45, n50
Styczyński, M., 638
Sullivan, J. P., 271–276, 277n4, n13–n15, 278n16, n19
Sun Qing Zi, 652
Svadkovsky, G. F., 162
Switzer, R., 75–90
Szakolczai, A., 611n8
Szmyd, J., 559–577
Sztorc, Cf. W., 505n9

Tagore, R., 107–108, 110n15–n17
Tai Tsung Li Shimin, 647
Talmy, L., 536n5
Tanner, L. N., 150, 167n6
Tarkovsky, A., 665
Tatarkiewicz, V., 49n28

Tatarkiewitz, V., 46
Taylor, C., 95, 101n16, 104, 109n3, 627n9
Taylor, E. F., 713–714
Taylor, M. C., 121
Taylor, P. W., 570
Terrosi, 356
Thales, 4, 499–500
Thompson, E., 153
Thompson, J. B., 536n15, 628n17
Thompson, K., 596n52
Thomson, I., 241
Timms, F. N., 666
Tito, J. M., 198
Tjakamarra, A., 670
Tjampitjinpa, K. M., 669–670
Tjampitjinpa, R., 666
Tjampitjinpa, W., 667
Tjapaltjarri, B. S., 668–669
Tjapaltjarri, C. P., 669–670
Tjapaltjarri, M. N., 669–670
Tjapaltjarri, T. L., 667, 669–670
Tjapangarti, T. P., 667
Tjapangarti, W., 668
Tjungurrayi, C. T., 666
Tjungurrayi, L., 666
Tjungurrayi, Y., 667
Tjupurrula, T. T., 670
Toadvine, T., 327, 333–335, 435n1, n12
Tønnessen, M., 327–339
Torjussen, L. P., 328, 339n8
Torrey Harris, W., 152
Totaro, F., 317–326
Trask, W. R., 56n4, n7
Trevi, M., 314n1
Trewavas, A., 101n14, n27
Trotsky, L., 149
Trutty-Coohill, P., 700n2
Tucker, A., 596n54
Tugan-Baranovsky, M. I., 631
Tugendhat, E., 124, 133n3
Tuomi, I., 280, 290
Turan, H., 485–493
Turanli, A., 295–305
Tymieniecka, A.-T., 28–31, 33, 35n28, n31, 36–37, 39, 45–48, 51, 100–101, 105, 109, 111–113, 136, 145, 147, 217, 285, 291–292, 313, 317, 327–328, 337, 339, 399, 409–411, 413–421, 423, 433–435, 437–438, 440–441, 446, 452–456, 460–461, 463, 496–497, 504–505, 560, 565–566, 568–572, 575–577, 634–635, 643, 676–677, 680–681, 683, 685, 687–693, 695–700, 709, 720, 724–726

Uebel, T., 134n5, n7
Uexküll, J. von., 100, 100n6, 101n29, 328–337, 338n2–n3
Ullmann, T., 611n4
Usher, 356

Valmiki, 356
Van De Pitte, M. M., 134n4
Van Gelder, 157
Varela, F. J., 31, 36n37, 37n63, 92–99, 100n7, 101n9–n10, n13, n17–n18, 360
Varisco, B. M., 360
Varisco, M. B., 365
Vattimo, 356
Vaughan, M. K., 160, 168
Verducci, D., 23–37
Vestre, B., 73n2
Vevere, Velga, 111–122
Viik, Tõnu, 227–240
Voltaire, 56n6, 282, 345
von Foerster, H., 364–365
von Glasersfeld, E., 356, 360–361
von Goethe, W., 97
von Humboldt, W., 55–56
von Kerckhoven, G., 597n65
von Uexküll, J., 100, 100n6, n29, 327–328, 333, 335–336
Vospitanii I obrazovanii, O., 169n47
Vyasa, 54
Vygotskij, L. S., 364–367
Vygotsky, 281

Waismann, F., 135n38, 341, 351n1
Waldenfeld, B., 42
Wallis, R. T., 462n8
Walton, R. J., 43, 48n14
Warner, A., 437n67
Watanabe, J., 480n1
Weber, M., 230
Wei Ling Kung, 655
Wei Zhang, 123–136
Weinberg, S., 648
Wellek, R., 56n4
Welton, D., 90n2, 133n3, 135n29, 587, 593n8, 595n27, 596n56, 597n66

NAME INDEX

Wertch, J. V., 365
Westbrook, R., 152
White, H., 521, 525, 536n22, n29–n30
Wiener, P. P., 56n4, 134n5
Wilhelm, G., 241
Williams, R., 51, 56n1, 576n9
Wilson, D. J., 167n15
Winckelmann, J. J., 53–55
Wittgenstein, L., 130–132, 135n38, n40, 296–298, 300–301, 303–304, 341–351, 351n1–n2, n4, 352n8, n10, n23–n28, n31–n34, n39, 353n43–n44, n48, n51, 395, 398–402, 405–407, 409, 410n2, n7
Witzany, G., 101n28
Wolcher, L. E., 347, 352n35
Wolf, F. A., 55
Wolff, C., 25–26, 28, 35n17, n33
Wolfsdorf, Cf. D., 505
Wood, J., 501–502, 506n22, n24

Woodward, J., 715–716
Wordsworth, W., 95, 101

Xiong Zi, 652–653

Yan Yuan, 653
Yazdi, M. H., 435n19, 436n22
Yulina, N., 148

Zahavi, D., 348, 352n37, 435n3, n15, 593n8
Zapffe, P. W., 327, 333–334
Zee, A., 713
Zelenko, A., 156, 160, 162
Zhu Xi, 647, 653, 655
Zi Kung, 650
Zi Si, 646
Zimmerman, M. E., 560
Zolli, C., 56
Zuckert, C. H., 22n7

PLEASE POST!

The World Institute for Advanced Phenomenological Research and Learning

1 Ivy Pointe Way
Hanover, New Hampshire 03755, United States
Telephone: (802) 295-3487; Fax: (802) 295-5963
Website: http://www.phenomenology.org

Anna-Teresa Tymieniecka, President; Thomas Ryba, Vice-President

The 60th International Congress of Phenomenology

Place: Hosted by the University of Bergen, Norway
Dates: August 10–13, 2010

P R O G R A M

Topic: **LOGOS AND LIFE PHENOMENOLOGY/ONTOPOIESIS REVIVING ANTIQUITY**

Place: The Congress begins with an opening reception on the afternoon of Tuesday, August 10th 16:00 – 18:00 to be held in the Reception Hall at Jusbygget (The Law Faculty Building) address: Magnus Lagabøtesplass 1.

Dates: The scholarly sessions will take place on August 11, 12 and 13, 2010. All of the scholarly sessions will be held at Jusbygget (The Law Faculty Building) address: Magnus Lagabøtesplass 1.

Scientific Committee:
ARGENTINA: Anibal Pedro Luis Fornari
ITALY: Daniela Verducci
JAPAN: Tadashi Ogawa, Kiyoko Ogawa
NORWAY: Konrad Rokstad, Vigdis Songe-Moller
ROMANIA: Carmen Cozma
TURKEY: Erkut Sezgin, Halil Turan
UNITED STATES: Kathleen Haney, Thomas Ryba, Robert Sweeney, Patricia Trutty-Coohill, Anna-Teresa Tymieniecka, President

Place: Hosted by University of Bergen, Norway, Department of Philosophy

Local Organization Committee: Ane Faugstad Aaro, Anne Granberg, Egil H. Olsvik, Johannes Servan

Chaired by: Konrad Rokstad

Program Presided by: Anna-Teresa Tymieniecka, World Phenomenology Institute

PROGRAM

Wednesday, August 11, 2010
9:00 – 9:30, Jusbygget (The Law Faculty Building), address: Magnus Lagabøtesplass 1, Auditorium 3

Registration

Wednesday, August 11
10:00 – 13:30, Jusbygget, Auditorium 3

> *INAUGURAL LECTURE*
> Chaired by: Vigdis Songe-Moller, University of Bergen, Norway

FROM HERACLITUS TO THE ONTOPOIETIC LOGOS AND THE ALL
Anna-Teresa Tymieniecka, World Phenomenology Institute, United States

> **PLENARY SESSION I**
> Chaired by: Patricia Trutty-Coohill, Sienna College, United States

WAS PLATO A PLATONIST?
Konrad Rokstad, University of Bergen, Norway

THE LIFE OF BEING, FOUND AGAIN. WITH THE PHENOMENOLOGY OF LIFE OF ANNA-TERESA TYMIENIECKA.
Daniela Verducci, Universita degli Studi di Macerata, Italy

THREE CRITIQUES OF REASON PROJECTS WITH REFERENCE TO ANTIQUITY: I. KANT AND THE PLATONIC IDEAS, E. HUSSERL AND THE MNEMOSINEAN ENTICEMENT, A-T. TYMIENIECKA AND THE DYONISIAN LOGOS
Ella Buceniece, University of Latvia

THE 'MODERN' CONCEPT OF THE INDIVIDUAL AND THE REVIVAL OF ANTIQUITY
Oliver W. Holmes, Wesleyan University, United States

13:30 – 14:30   Lunch

WORKING SESSIONS

Wednesday, August 11
14:30 – 18:45, Jusbygget, Seminarrom 1 (Room 404)

> ### *SESSION I*: LOGOS AND LIFE
> Chaired by: Thomas Ryba, Purdue University, United States

THE EXISTENTIALISTIC SUBJECT OF TODAY
Simen Oyen, University of Bergen, Norway

RE-TURING TO THE REAL: PHENOMENOLOGICAL APPROPRIATIONS OF PLATO'S 'IDEAS' AND THE ALLEGORY OF THE CAVE
Robert Switzer, The American University in Cairo, Egypt

LIVING LIFE AND MAKING LIFE
Andreas Brenner, Universitat Basel, Switzerland

MAN'S WORLD AND LOGOS AS FEELING
Manjulika Ghosh, University of North Bengal, India

THE FEAST OF REASON (PLATO) OR THE REASON OF LIFE (S. KIERKEGARRD AND A-T. TYMIENIECKA)?
Velga Vevere, University of Latvia

GIBT ES EIN MATERIALES APRIORI?" MIT SCHLICKS KRITIK AN DER PHANOMENOLOGIE UBER DAS VERHALTNIS ZWISCHEN SPRACHE UND VERNUNFT NACHZUDENKEN ANFANGEN
Wei Zhang, Sun Yat-sen University, China

Wednesday, August 11
14:30 – 18:45, Jusbygget, Seminarrom 3 (Room 412)

SESSION II: LOGOS AND EDUCATION
Chaired by: Witold Plotka, University of Gdansk, Poland

THE IDEA OF PAIDEA IN THE CONTEXT OF ONTOPOIESIS OF LIFE
Rimma Kurenkova, Evgeny Plekhanov, Elena Rogacheva, Vladimir Pedagogical Institute, Russia

ESSENTIAL CHARACTERISTICS
Semiha Akinci, Anadolu University, Turkey

HUMAN SPIRITUALITY IN THE FACE OF CONTEMPORARY DILEMMAS OF THE AXIOLOGY OF LIFE
Zofia Fraczek, University of Rzeszow, Poland

THE ATTEMPT TO UNDERSTAND OPTIMISTS AND PESSIMISTS
Ayse Sibel Turkum, Andolu University, Turkey

INTERNATIONAL DIMENTION OF JOHN DEWEY'S PEDAGOGY: LESSONS FOR TOMORROW
Elena Rogacheva, Vladimir Pedagogical Institute, Russia

THINKING CONDITIONED BY LANGUAGE AND TRADITION
Mara Stafecka, United States

HOW TO CONDUCT LIFE (ARETE AND PHRONESIS)
J. C. Couceiro-Bueno, Univ. de la Coruna, Campus Elvina s/n, Spain

Wednesday, August 11
14:30 – 18:45, Jusbygget, Seminarrom 4 (Room 413)

SESSION III: **HUSSERL IN THE CONTEXT OF TRADITION**
Chaired by: Clara Mandolini, Universita degli studi di Macerata, Italy

OBSCURITIES ON THE IDEALISTIC SIDE"? - PLATONIC MISCONCEPTIONS IN HUSSERL
Egil H. Olsvik, University of Bergen, Norway

THE REASON OF THE CRISIS. HUSSERL'S RE-EXAMINATION OF THE CONCEPT OF RATIONALITY
Witold Plotka, University of Gdansk, Poland

LOGOS AS SIGNIFIER: HUSSERL IN THE CONTEXT OF TRADITION
Koushik Joardar, University of North Bengal, India

THE AXIOLOGY OF ONTOPOIESIS AND ITS RATIONALITY
Susi Ferrarello, University of Rome, La Sapienza, Italy

ANTIQUITY AS A MEANING-AUTOMATON: A CULTURAL PHENOMENOLOGY OF HISTORICAL CONSCIOUSNESS
Tonu Viik, Tallinn University, Estonia

THE RECOVERY OF THE SELF. PLOTINUS ON LOGOS AND SELF-COGNITION
Magdalena Plotka, Department of the History of Ancient and Medieval Philosophy, Warsaw, Poland

THE EVOLUTION OF THE CATEGORY OF THE <<NOW>> AND THE CONSCIOUSNESS OF TIME IN HUSSERLIAN PHENOMENOLOGY
Cezary J. Olbromski, University Marii Curie-Sklodowskiej, Poland

Wednesday, August 11
14:30 – 18:45, Jusbygget, Seminarrom 5 (Room 414)

SESSION IV: COGNITION, CREATIVITY, EMBODIMENT
Chaired by: Egil H. Olsvik, University of Bergen, Norway

HUMAN MANIFESTATION OF THE LOGOS OF LIFE: CREATIVITY, SPEECH, THINKING
Zaiga Ikere, Daugavpils University, Latvia

POUND, PROPERTIUS AND LOGOPOEIA
Lars Morten Gram, University of Bergen, Norway

ANTIQUITIES OF THE BODY: A PHENOMENOLOGICAL ANALYSIS OF THE CULTURAL DISCOURSE ON EMBODIMENT BEHIND MODERNITY
Mary Jeanne Larrabee, DePaul University, United States

PHENOMENOLOGY: CREATION AND CONSTRUCTION OF KNOWLEDGE
Klymet Selvi, Anadolu University, Turkey

PERSPICUOIUS REPRESENTATION
Aydan Turanli, Istanbul Technical University, Turkey

ORIGIN AND FEATURES OF PHYSICAL CREATIONS IN AN ONTOPOIETIC PERSPECTIVE
Mina Sehdev, Macerata, Italy

Wednesday, August 11
Location to be Announced

SESSION V: NATURE, WORLD, ORDER
Chaired by: Ane Faugstad Aaro, University of Bergen, Norway

NATURE AND ARTIFICE IN MANIFESTING/CONSTITUTING THE BEING
Francesco Totaro, Universita degli studi di Macerata, Italy

SEMIOTICS OF BEING AND UEXKULLIAN PHENOMENOLOGY
Morten Tonnessen, University of Tartu, Estonia

THE BEGINNING OF THE CONTINUOUS MOVEMENT IN DESCARTES' FIRST WORKS
Daria Carloni, Italy

MOVEMENT AND RHYTHM AS PART OF THE FOUNDATIONAL DYNAMIC PRINCIPLE OF LIFE (FROM PRE-SOCRATIC PHILOSOPHY TO PHENOMENOLOGY OF LIFE)
Maija Kule, University of Latvia

LINES OF REVISITING AND REINTEGRATION ABOUT THE ANCIENT SENSE OF CONTINUITY BETWEEN MAN AND NATURE BY THE CONTEMPORARY CONSTRUCTIVISM.
Catia Giaconi, Universita degli Studi di Macerata, Italy

Thursday, August 12, 2010
9:00 – 13:30, Jusbygget (The Law Faculty Building), address: Magnus Lagabøtesplass 1, Auditorium 3

> ### *PLENARY SESSION II*
> Chaired by: Halil Turan, Orta Dogu Teknik Universitesi, Turkey

THE CRITICAL CONSENT OF LOGOS TO LIFE – PAUL RICOEUR REVIVING ANTIQUITY –
Anibal Fornari, Universidad Catolica de Santa Fe, Argentina

LOGOS AND LIFE, PHENOMENOLOGY/ONTOPOIESIS REVIVING ANTIQUITY AND PHILOSOPHY IN ITS ORIGINATING SENSE
Erkut Sezgin, Istanbul Kultur University, Turkey

'SOPHIA' AS 'TELOS' IN THE 'ONTOPOIETIC PERSPECTIVE'
Carmen Cozma, University "Al.I.Cuza", Romania

GIVENNESS AFTER LEVINAS, MARION, AND TYMIENIECKA; LEVINAS: CONSTITUTION, GIVENNESS AND TRANSCENDENCE
Thomas Ryba, Purdue University, United States

THE WORLD IN THE HEART: ANTIQUITY AND TYMIENIECKA'S ONTOPOIETIC LOGOS
Olga Louchakova, Institute of Transpersonal Psychology, United States

Thursday, August 12
14:30   SIGHTSEEING

Friday, August 13, 2010
9:00 – 13:30, Jusbygget (The Law Faculty Building), address: Magnus Lagabøtesplass 1, Auditorium 3

PLENARY SESSION III: UNIVERSAL ORDER, NATURE, REASON
Chaired by: Konrad Rokstad, University of Bergen, Norway

THE PLACE: WHERE WE SEE THE WORLD AS A LIMITED WHOLE,
Sibel Oktar, Ozyegin University, Turkey

POWERFUL FORCE BETWEEN VIRTUALITY AND ENACTMENT
Clara Mandolini, Universita degli studi di Macerata, Italy

VISUALIZING ANNA-TERESA TYMIENIECKA'S APPROACH WITH STRING THEORY
Patricia Trutty-Coohill, Sienna College, United States

THE BIOLOGICAL PRINCIPLE OF NATURAL SCIENCES AND THE LOGOS OF LIFE OF NATURAL PHILOSOPHY: A COMPARISON AND THE PERSPECTIVES OF UNIFYING THE SCIENCE AND PHILOSOPHY OF LIFE
Attila Grandpierre, Konkoly Observatory, Hungary

UNIVERSAL PRINCIPLES OF THE WORLD AND THE COORDINATE SYSTEM ON THE BASIS OF LIMIT DYNAMICAL EQUILIBRIUM
Nikolay Kozhevnikov, Yakut State University, Russia

Friday, August 13
14:30 – 18:45, Jusbygget, Seminarrom 1 (Room 404)

SESSION VI: LOGOS AND ILLUMINATION
Chaired by: Simen Oyen, University of Bergen, Norway

WHAT WAS A CLASSIC UNTIL THE BEGINNING OF THE 20th CENTURY?
Tansu Acik, University of Ankara, Turkey

PLATO'S CONCEPT OF TIME IN THE PHENOMENOLOGICAL INTERPRETATION OF EDMUND HUSSERL AND ANNA-TERESA TYMIENIECKA
Maria Bielawka, Poland

EDITH STEIN AND THE SEMANTICS OF *LOGOS:* THE TRANSPOSITION IN ONTOPOIETICAL TERMS OF ARISTOTLE'S LOGOIC CONCEPTION WITHIN *FINITE AND ETERNAL BEING*
Maria-Chiara Teloni, Macerata University, Italy

THE EFFECT OF ILLUMINATION ON THE WAY BACK FROM ARISTOTLE TO PLATO
Salahaddin Khalilov, Azerbaijan Universiteti

THE RETURN TO PLATO'S THEORY OF IDEAS
Konul Bunyadzade, East West Research Center, Azerbaijan

LOGOS AND HUMOR IN ANCIENT GREEK PHILOSOPHY. A PHENOMENOLOGICAL APPROACH
Anna Malecka, AGH - University of Science and Technology in Krakow, Poland

Friday, August 13
14:30 – 18:45, Jusbygget, Seminarrom 3 (Room 412)

SESSION VII: INTERSUBJECTIVITY, FREEDOM, JUSTICE
Chaired by: Anne Granberg, University of Bergen, Norway

MAKING HISTORY OUR OWN – APPROPRIATION AND TRANSGRESSION OF THE INTENTIONAL HISTORY OF HUMAN RIGHTS
Johannes Servan, University of Bergen, Norway

SWINGING NOTIONS AND VITALITY OF UNIVERSAL MORAL VALUES IN THE POST-MODERN WORLD: CREATIVITY OF THE LOGOS OF LIFE
Jan Szmyd, University of Krakow, Poland

GOODNESS OF GOOD AS LIFE. A REREADING OF LEVINAS'S USE OF NEOPLATONISM
Sergio Labate, Universita degli Studi di Macerata, Italy

THE WORLDLY ETHICS OF HUSSERL AND PATOCKA
Michael Gubser, James Madison University, United States

THE QUESTION OF THE SUBJECT: JAN PATOCKA'S PHENOMENOLOGICAL CONTRIBUTION
Saulius Geniusas, James Madison University, United States

ONTOPOIESIS AND NARRATIVE ETHICS: AN ACUTUALIZATION OF ARISTOTLE'S ACCOUNT OF IMAGINATION
Silvia Pierosara, University of Macerata, Italy

Friday, August 13
14:30 – 18:45, Jusbygget, Seminarrom 4 (Room 413)

SESSION VIII: CREATIVITY AND THE ONTOPOIETIC LOGOS
Chaired by: Johannes Servan, University of Bergen, Norway

BLANCHOT'S INAUGURAL POETICS: VISIBILITY AND THE INFINITE CONVERSATION
William D. Melaney, The American University in Cairo, Egypt

LOVE OF LIFE, TRAGEDY AND SOME CHARACTERS IN GREEK MYTHOLOGY
Halil Turan, Orta Dogu Teknik Universitesi, Turkey

THE IDEAL AND THE REAL: BRIDGING THE GAP
Brian Grassom, Gray's School of Art, United Kingdom

HISTORICITY, NARRATIVE AND THE CONSTRUCTION OF MONSTROSITY IN JOHN GARDNER'S "GRENDEL"
Firat Karadas, Mustafa Kemal Universitesi, Turkey

THE POWER OF DANCE/MOVEMENT AS A MEANS OF EXPRESSION
Ali Ozturk, Andolu University, Turkey

Friday, August 13
14:30 – 18:45, Jusbygget, Seminarrom 5 (Room 414)

> *SESSION IX*: **SEEKING THE LOGOS IN DIFFERENT CULTURES**
> Chaired by: Zaiga Ikere, Daugavpils University, Latvia

RABINDRANTAH TAGORE AND HIS RELIGION OF POET
Grzegorz Okraszewski, Uniwersytet Jagiellonski, Poland

TATTVAM ASI" - THE RELATION BETWEEN SELF AND THE UNIVERSE IN ADVAITA VEDANTA SYSTEM (ESPECIALLY IN SANKARA'S PHILOSOPHY)
Kinga Kleczek-Semerjak, Uniwersytet Jagiellonski, Poland

THE CONCEPT OF THE 'COMPLETED MIND' (CHENG XIN) IN PHILOSOPHY OF ZHUANGZI
Ada Augustyniak, Uniwersytet Jagiellonski, Poland

SELF SUSTAINED BY THE OTHER AS AN AUTHENTIC MODE OF EXISTENCE: THE THOUGHT OF SHIRAN
Robert Szuksztul, Uniwersytet Jagiellonski, Poland

"THEOSIS" AND LIFE IN NICOLAI BERDYAEV'S PHILOSOPHY
Katarzyna Stark, Poland

A SELF-CREATING AS THE PROCESS OF CREATING THE UNIVERSE - SOME REMARKS ON VEDIC COSMOGENIC HYMNS AND UPANISHADIC CONCEPTIONS OF ATMAN
Malgorzata Ruchel, Uniwersytet Jagiellonski, Poland

THE PROJECT OF LIFE IN BUDDHISM AND EARLY CHRISTIANITY
Remigiusz Krol, Uniwersytet Jagiellonski, Poland

THE WORLD(S) IN THE WORD(S): ZOROASTER'S ORAL POETRY AS AN ICONIC MEDIUM OF REVELATORY REALIZATION
Martin Schwartz, University of California, Berkeley, United States

Friday, August 13
19:00, Jusbygget, Auditorium 3

> *CLOSING SESSION*
> Presided by: Thomas Ryba, Purdue University, United States
> Carmen Cozma, University "Al.I.Cuza", Romania
> Erkut Sezgin, Istanbul Kultur University, Turkey